# Advances in Industrial Control

Springer
*London*
*Berlin*
*Heidelberg*
*New York*
*Barcelona*
*Budapest*
*Hong Kong*
*Milan*
*Paris*
*Santa Clara*
*Singapore*
*Tokyo*

P.G. Howlett and P.J. Pudney

# Energy-Efficient Train Control

With 76 Figures

Springer

Philip G. Howlett
Peter J. Pudney

Scheduling and Control Group
School of Mathematics
University of South Australia
The Levels Campus
Pooraka
South Australia 5095
Australia

British Library Cataloguing in Publication Data
Howlett, Philip G.
    Energy Efficient Train Control. - (Advances in Industrial Control)
    I. Title  II. Pudney, Peter J.  III. Series
    385.1
    ISBN-13:978-1-4471-3086-4

Library of Congress Cataloging-in-Publication Data
Howlett, Philip, 1944-
    Energy efficient train control / Philip Howlett and Peter Pudney.
        p.    cm.  - - (Advances in industrial control)
    Includes bibliographical references (p.      -    ) and index.
    ISBN-13:978-1-4471-3086-4      e-ISBN-13:978-1-4471-3084-0
    DOI: 10.1007/978-1-4471-3084-0


    1. Railroads - - Automatic train control.   I. Pudney, Peter, 1961-. II. Title. III. Series.
    TF638.H69   1995                                                        95-31200
    625.1 - - dc20



Typesetting: Camera ready by authors

69/3830-543210 Printed on acid-free paper

# SERIES EDITORS' FOREWORD

The series *Advances in Industrial Control* aims to report and encourage technology transfer in control engineering. The rapid development of control technology impacts all areas of the control discipline. New theory, new controllers, actuators, sensors, new industrial processes, computer methods, new applications, new philosophies,......, new challenges. Much of this development work resides in industrial reports, feasibility study papers and the reports of advanced collaborative projects. The series offers an opportunity for researchers to present an extended exposition of such new work in all aspects of industrial control for wider and rapid dissemination.

The number of texts reporting an in-depth study of the control of transportation systems is not so many. In this contribution to the *Advances in Industrial Control* Series, Phil Howlett and Peter Pudney report on well over a decade of research into the energy efficient aspects of rail transportation. The work described is characterised by the development of a sound theoretical framework for the optimal control problems to be solved.

The presentation of modelling, and optimisation problems forms the main thrust of the text after the introductory chapters. These first chapters describe the background to the train efficiency research work and introduce the reader to the theoretical concepts required later in the text. Also of considerable interest to practitioners is the description of the in-cab technology, the METROMISER, which resulted from some of the research work.

Not surprisingly, the pioneering theoretical and technological research reported in this volume has now won several awards and we are delighted that a full presentation of this long programme of research is available for wider dissemination as an AIC volume.

M.J. Grimble and M.A. Johnson
*Industrial Control Centre*
*Glasgow, Scotland, U.K.*

# PREFACE

Rail is potentially a very efficient form of transport, but must deliver attractive alternatives to road and air transport if it is to be competitive. The challenge for rail is to meet the needs of potential users. For heavy-haul freight, railways will need to be reliable and cost-effective. For intercity passengers, rail must provide a fast and comfortable alternative to air travel, or a relaxed and luxurious holiday service. Suburban and mass transit systems will need to overcome the problems of peak congestion and match the perceived convenience of the car. In theory, rail can provide energy-efficient transport on a dedicated corridor free of external interference. Can this theory be turned into effective practice?

For railway operators there are many factors to consider. The customers want prompt and reliable service that must also be cost-effective. The railways must make a profit. On congested lines, scheduling problems are seen as dominant. For the long haul across continents, arrival on time and reduction of fuel consumption are more important. There are of course other concerns. Safe operation is a critical factor and the adoption of new signalling technology is high on the agenda. In this book, however, we restrict our attention to the problem of energy-efficient control.

### *Beginnings*

In 1980 Ian Milroy returned from Loughborough University in the UK to the School of Electrical Engineering at the South Australian Institute of Technology, now the University of South Australia. He had just completed his PhD thesis, *Aspects of Automatic Train Control*, which included some novel work on driving strategies [1, 2]. In particular he showed that for short journeys an energy-efficient driving strategy has three control phases: maximum acceleration, coast, and maximum brake.

The work on driving strategies was continued by a post-graduate Electrical Engineering student, Kim Tyler, supervised by Ian Milroy and by David Lee from the School of Mathematics and Computer Studies. They discovered a fourth optimal control mode, speed-hold, applicable on longer journeys [3] and used an analogue computer to simulate efficient speed profiles for journeys of varying distance and duration.

### The Transport Control Group

In 1982 Ian Milroy formed the Transport Control Group at the South Australian Institute of Technology to work on a project funded by the South Australian Department of Transport. The aim of the project was to determine whether the suggested driving strategies were effective in practice and, if so, to develop a system for achieving fuel savings on suburban trains in Adelaide.

The first part of the project involved calculating efficient speed profiles for various sections of track on the Adelaide rail network. This was done by Basil Benjamin from the School of Mathematics and Computer Studies, Jerry Kautsky from the Flinders University of South Australia and Nancy Nichols from the University of Reading. The project manager, Andy Long, then boarded a train with a stopwatch and a pre-computed speed profile and advised the driver when to change control. On each trip the train completed the section within a few seconds of the desired time, and the time spent accelerating was much less than normal.

### Metromiser

The next stage of the project was to design and build a system that could compute an efficient driving strategy in real time and display appropriate driving advice to the driver. Basil Benjamin and a post-graduate Computer Studies student, Peter Pudney, developed algorithms and computer software for the system. Computer hardware was designed and built by Tony Gelonese from the School of Electrical Engineering. The resulting system, now known as *Metromiser*, monitored the state of a journey and, using stored timetable and route information, advised the driver when to coast and brake so that the train arrived at each stop on time and consumed as little energy as possible.

*Metromiser* was first evaluated in Adelaide in February 1985. The unit was fitted to a diesel-hydraulic railcar running in normal service on an 80-minute round trip. Twenty trips were evaluated—ten trips with driving advice and ten trips without advice. *Metromiser* achieved a fuel saving of 15% and significant improvements in timekeeping [4].

Since then *Metromiser* has been evaluated on suburban trains in Adelaide, Melbourne, Brisbane and Toronto. For each trial a Metromiser unit was installed on a test train and operated in normal service. Fuel consumption and timekeeping data were collected over a three week period. All trials showed significantly improved timekeeping and substantial fuel savings.

### Theoretical Developments

During the period 1982–85 the theoretical basis for the work was also extended. After early discussions with Ian Milroy it was decided by the Group that Phil Howlett, from the School of Mathematics and Computer Studies, should seek a more rigorous mathematical justification for the strategies suggested by Milroy. Following discussions with Jim Michael at Adelaide University and Bruce Craven at Melbourne University in 1984, Howlett did indeed give a rigorous justification of the work and at the same time produced the first theoretical confirmation that speed-holding should be used if the journey time is relatively large, and that an optimal driving strategy used an *accelerate–hold–coast–brake* control sequence. He also found a special relationship between the holding speed and the speed at which braking should begin [5, 6]. This relationship was found independently and at much the same time by Asnis et al in the Soviet Union [7].

### Long-Haul Trains

Although much had been achieved with the *Metromiser* project, the strategy and algorithms used were only effective when stops were less than about 10 km apart. The group began planning a second project to develop a fuel conservation system for long-haul freight trains. In 1986 the Transport Control Group was awarded a National Energy Research, Development and Demonstration Council (NERDDC) contract to develop such a system.

Long-haul freight trains presented several new problems:

- the performance of a train is initially unknown, and changes with the direction of the wind and when wagons are loaded or unloaded
- the trains have a low power to mass ratio, and so gradients have a greater effect
- there can be many different speed limits between stops

A different train model was now required. The original model had assumed continuous control of acceleration. This is not the case for the diesel-electric locomotives used by *Australian National*, where each locomotive has eight discrete control notches, and each notch determines a constant rate of fuel supply to the diesel generator. An appropriate model was developed by Benjamin et al [8]. Benjamin also directed work to develop parameter estimation techniques for the new model.

The Australian long-haul rail network has long sections of single-line track and only occasional crossing loops. This means that design of appropriate crossing schedules is critical in improving the performance of a long-haul system. Graham Mills and Sonya Perkins joined the group to work on the scheduling aspects of the project. This work is described in an article in the *Asia-Pacific Journey of Operational Research* [9].

Further NERDDC grants were awarded to the Group in 1987–88 and in 1989–90. Development of the long-haul system was subsequently completed and commercial production begun. In 1989 the Transport Control Group received a National Energy Innovation Award for Outstanding Achievement in Research and Development into, and the Effective Management of, Transport Energy Use.

In 1989 Howlett and Benjamin began a systematic consideration of their new model for the train control problem. They assumed a fixed number of discrete control settings with each setting corresponding to a constant rate of fuel supply. Total fuel consumption was used as the cost functional. The first serious theoretical work with this model began with the arrival of Cheng Jiaxing, a visiting scholar from Anhui University in the Peoples Republic of China. By December 1989 preliminary results were presented by Cheng and Howlett at the Conference of Australian Institutes of Transport Research at the Flinders University of South Australia. Even with the new model, the research confirmed the fundamental optimality of the *power–hold–coast–brake* strategy for level track.

### *Scheduling and Control of Trains on a Network*

By early 1990 it was clear that the heuristic algorithms used to compute efficient long-haul speed profiles could be improved by incorporating the most recent theoretical work. Further investigations by Howlett and Cheng had suggested that the effects of gradient and speed limits could be incorporated into the new model. At the same time Graham Mills believed that much more could be done to improve long-haul scheduling.

In 1991 the South Australian Institute of Technology was incorporated into the new University of South Australia. In the same year the Group, now renamed the Scheduling and Control Group, and again lead by the redoubtable Ian Milroy, received a Generic and Industrial Research and Development (GIRD) grant. The project proposed two major areas of investigation. The first, under the direction of Graham Mills, would develop algorithms for scheduling trains on a long-haul network. The second, directed by Phil Howlett, would develop energy-efficient control strategies and devise numerical algorithms for calculating these strategies.

The original commercial collaborator ceased operation in 1991, soon after the project began, and it was necessary to find another partner to provide technical support. Fortunately the Group had done some work on automatic braking systems for another railway systems supply company, who were keen to develop stronger ties with the University, and they became the new partner for the GIRD project early in 1992. The research program and specifications for the long-haul system were completed in 1993, but commercial development was hampered by licence negotiations with the original partner. Nevertheless, the Group received another National Energy Innovation Award and a survey paper by Howlett, Milroy and Pudney [10] was selected as a finalist for the applications prize at the Twelfth World Congress of the International Federation of Automatic Control (IFAC) in Sydney, Australia, 1993.


### Solar Cars

More recently the Scheduling and Control Group, now led by Phil Howlett, has worked with the Aurora Vehicles Association to determine energy-efficient driving strategies for a solar car. Aurora Vehicles are the leading Australian developers of solar car technology. After some preliminary work in cooperation with the Commonwealth Scientific and Industrial Research Organisation (CSIRO) in 1993, the Group obtained a three-year Australian Research Council Collaborative grant with Aurora to develop energy management strategies for solar and electric cars. The Aurora *Q1* solar car finished the 1993 World Solar Challenge in fifth position overall, and was the leading Australian entry. The first five of the 52 entries averaged more than 70 km/h for the 3000 km journey, breaking the record 67 km/h set by the GM *Sunraycer* in 1987. In November 1994 the *Q1* was driven 4000 km across Australia from Perth to Sydney in eight days, beating the previous best time of 20 days set by Hans Tholstrup and Larry Perkins in 1982.

# ACKNOWLEDGEMENTS

We would like to thank present and past members and associates of the Scheduling and Control Group for their contributions to the solution of the train control problem. They are: Ian Milroy, Kim Tyler, David Lee, Basil Benjamin, Bob Northcote, Doug Seeley, Andy Long, Tony Gelonese, Jerry Kautsky, Nancy Nichols, Bruce Craven, Jim Michael, Andrew Skinner, Peter McWhirr, Bob Payne, Sally Rice, Graham Mills, Sonya Perkins, Bernie Chandler, Cheng Jiaxing and Tania Tarnopolskaya. In particular Ian Milroy, Cheng Jiaxing and Basil Benjamin have co-authored some of the original papers and assisted with proof-reading. We would also like to thank our colleagues in Trans Adelaide, Australian National, The Canadian Institute of Guided Ground Transport, Teknis Systems, Queensland Rail, National Rail Corporation, and Westinghouse Brake and Signal Australia.

Parts of this book are based on previously published material. Chapters 7, 11 and 13 are rewritten from papers [6, 11, 12] published in the *Journal of the Australian Mathematical Society, Series B*, and some of the material in Chapter 12 appeared in a paper [13] published in *Computational Techniques and Applications, CTAC-91*; this material appears with the kind permission of the Australian Mathematical Society, Australian National University, Canberra ACT 0200. Chapter 8 is rewritten from a paper [14] published in *Automatica*, and Chapters 1, 2, 3 and 12 contain material from a paper [10] published in *Control Engineering Practice*; this material appears with the kind permission of Elsevier Science Ltd, The Boulevard, Langford Lane, Kidlington OX5 1GB UK. Chapter 10 contains material from a paper [15] published in *Transactions on Automatic Control*, and appears with permission of IEEE Publishing Services, 445 Hoes Lane, Piscataway NJ 08855-1331.

# CONTENTS

# Section B:  Analysis of the Mechanical Energy Model

# Section C:  Analysis of the Fuel Consumption Model

Section A

# Introduction

## CHAPTER 1

# THE TRAIN CONTROL PROBLEM

In 1977-78 Milroy [1] considered the problem of driving a train from one station to the next along a level track within a given allowable time in such a way that energy consumption is minimised. He used the *energy flows* in the traction and braking systems of the train to derive state variable equations with time as the independent variable and position and speed as the dependent state variables. He used an heuristic application of the *Pontryagin Principle* to conclude that the optimal driving strategy consisted of a *maximum acceleration–coast–brake* control sequence. Subsequent studies confirmed the optimality of this control sequence for short journeys, and showed that a speed-hold phase should be included on longer journeys.

These principles were used by the Scheduling and Control Group at the University of South Australia to design and build *Metromiser*, an on-board system providing driving advice to reduce fuel consumption on suburban trains. In-service trials of *Metromiser* demonstrated substantial fuel savings and significant improvements in timekeeping.

## 1.1 The Original Formulation

Energy is usually supplied to a train as electrical energy or by the combustion of fuel. This energy is converted by the traction system to mechanical energy, which is used to drive the train. Energy is lost to frictional resistance and braking. The remaining energy is stored as kinetic or potential energy. The energy flows are illustrated in Figure 1-1.

Milroy assumed that the control variable was the applied acceleration and that the level of control could be varied continuously. In his formulation $T$ is the time allowed for the journey, $X$ is the distance between two stations, $u(t)$ is the acceleration applied to the train, $v(t)$ is the speed of the train, and $-r[v(t)]$ is the resistive acceleration due to friction. This acceleration is given by the *Davis formula* [16]

$$r(v) = a + bv + cv^2$$

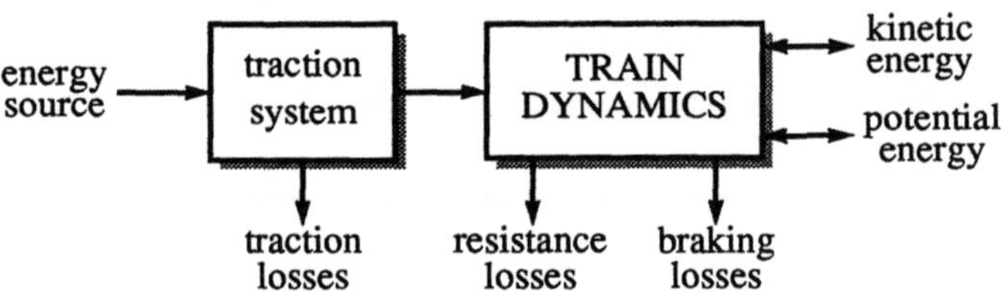


Figure 1-1: Energy flows for a train

where $a$, $b$ and $c$ are known positive real numbers. The graph $y = r(v)$ is strictly increasing and convex in the region $v \geq 0$. The net acceleration of the train is given by the equation

$$v'(t) = u(t) - r[v(t)],$$

where the notation $v'$ denotes the derivative of the function $v$. For convenience, we will also use the notation

$$v'(t) = \frac{dv}{dt}.$$

It is assumed that the *applied acceleration* $u(t)$ is limited by the constraint

$$|u(t)| \leq 1$$

and that only positive acceleration consumes energy. The cost of the journey is measured by the total amount of mechanical energy consumed by the train. Thus the cost functional is given by

$$J(u, v) = \int_0^T u_+(t)v(t)dt$$

and contains the term

$$u_+(t) = \frac{u(t) + |u(t)|}{2}$$

which is the positive part of $u(t)$. Milroy formulated the train control problem as follows.

### *Train Control Problem*

*Minimise the energy consumption*

$$J(u, v) = \int_0^T u_+(t)v(t)dt \tag{1.1}$$

*subject to the differential equation*

$$v'(t) = u(t) - r[v(t)] \tag{1.2}$$

*with boundary conditions*

$$v(0) = v(T) = 0$$

*and subject to the equality constraint*

$$\int_0^T v(t)dt = X$$

*and the inequality constraint*

$$|u(t)| \le 1. \tag{1.3}$$

□

By applying the Pontryagin Principle he obtained a speed profile which he suggested was optimal [2]. This conjecture was supported by subsequent practical tests [4]. The Pontryagin Principle will be discussed in the next chapter.

## 1.2 Solution of the Original Problem

Milroy used an heuristic application of the Pontryagin Principle to obtain conditions for a minimum-cost strategy. He used the additional equation

$$x'(t) = v(t), \tag{1.4}$$

where $x(t)$ is the position of the train, to relate the state variables $x$ and $v$, and used (1.1), (1.2) and (1.4) to form a *Hamiltonian* function

$$H(x, v, u) = -u_+ + z_1 v + z_2[u - r(v)]$$

with co-state variables $z_1$ and $z_2$ satisfying the adjoint differential equations

$$z_1'(t) = 0 \tag{1.5}$$

and

$$z_2'(t) = u_+ - z_1(t) + z_2(t)r'\,[v(t)]\,. \tag{1.6}$$

From (1.5) it follows that $z_1(t) = A$ where $A$ is a real constant. Thus (1.6) can be rewritten as

$$z_2'(t) = u_+ - A + z_2(t)r'\,[v(t)]\,. \tag{1.7}$$

For an *optimal strategy* it is necessary to choose $u$ so that the Hamiltonian is maximised. If $H$ is written in the form

$$H(u) = \begin{cases} (z_2 - v)\,u + \ldots & 0 < u \le 1 \\ z_2 u + \ldots & -1 \le u < 0 \end{cases}$$

then it can be seen that the control $u$ required to maximise $H(u)$ depends on the value of $z_2$. There are five cases to consider:

1.   $z_2 > v \;\Rightarrow\; u = 1$
2.   $z_2 = v \;\Rightarrow\; u \in [0, 1]$
3.   $0 < z_2 < v \;\Rightarrow\; u = 0$
4.   $z_2 = 0 \;\Rightarrow\; u \in [-1, 0]$
5.   $z_2 < 0 \;\Rightarrow\; u = -1$

The five cases are illustrated in Figure 1-2.

### *Maximum Acceleration, Coast, and Maximum Brake*

Cases 1, 3 and 5 correspond respectively to maximum acceleration, coast, and maximum brake. These cases all occur in a typical optimal control strategy.

### *Speed-Hold*

The singular control mode corresponding to Case 2 requires $z_2(t) = v$ to be maintained on a non-trivial interval, and hence $z_2'(t) = v'(t)$ on this interval. From (1.7) it follows that

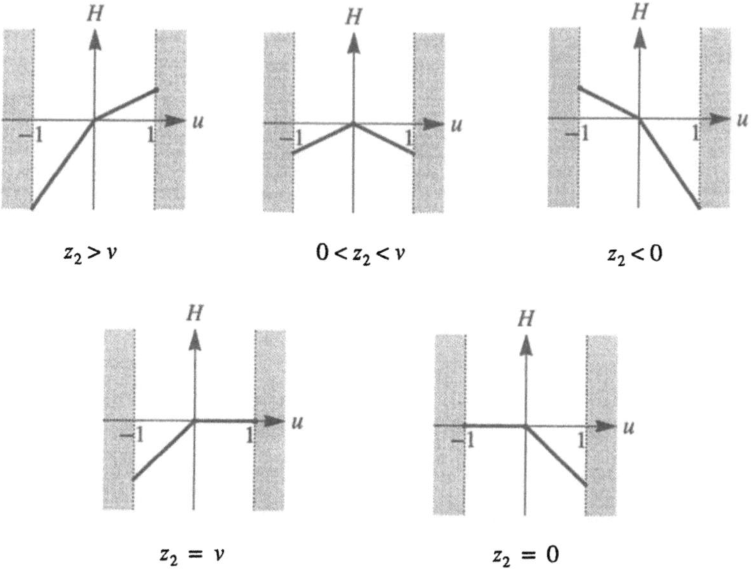



Figure 1-2: Maximising $H(u)$ subject to $|u| \leq 1$

$$\varphi' [v(t)] = A$$

where $\varphi(v) = vr(v)$. Since $\varphi(v)$ is convex the speed is constant on this interval and is given by $v(t) = V$ where $V$ is the unique solution to the equation

$$\varphi'(V) = A.$$

Thus Case 2 corresponds to speed-holding. Since $v(t) \geq 0$ and since

$$\varphi' [v(t)] \geq r(0) = a$$

this condition can only occur if $A \geq a > 0$.

## *Partial Brake*

The singular control mode in Case 4 corresponds to partial braking, and requires $z_2(t) = 0$ to be maintained on a non-trivial interval. Thus

$$z_2'(t) = 0$$

on this interval and from (1.7) it follows that $A = 0$. This condition implies that (1.7) has the general form

$$z_2'(t) = u_+ + z_2(t)r'\,[v(t)] \tag{1.8}$$

on each control interval. Because $v(0) = 0$, an optimal strategy can only begin with $u = 1$. Since there is no solution $V > 0$ to the equation $\varphi'(V) = 0$ there is no speed-holding phase. Thus from $u = 1$ with $z_2 > v > 0$ we can only switch to $u = 0$ with $0 < z_2 < v$. Suppose a switch occurs at $t = t_1$. If we set $\eta = z_2/v$ then $\eta(t_1) = 1$ and for $t > t_1$ it follows from (1.2) and (1.8) that

$$\eta'(t) = \frac{\eta(t)\varphi'\,[v(t)]}{v(t)} > 0.$$

This gives $\eta(t) > 1$ for $t > t_1$ and contradicts the condition $0 < \eta < 1$ required when $u = 0$. Thus we cannot switch from $u = 1$. The speed $v(t)$ increases for all $t$ and the journey can never be completed because the condition $v(T) = 0$ cannot be met. Hence $A \neq 0$ and Case 4 cannot occur.


## 1.3 Initial Results

From his original analysis, Milroy suggested that an energy-efficient speed profile should consist of three phases: maximum acceleration, coast, and maximum brake. Initial experiments with suburban passenger trains confirmed that this strategy was indeed energy-efficient. The results of the initial research were first presented by Milroy at a railway engineering conference [2] in Sydney in 1981. At this early stage, however, it was not understood that speed-holding should be used in some circumstances. The speed profile for a typical *accelerate–coast–brake* strategy is shown in Figure 1-3.

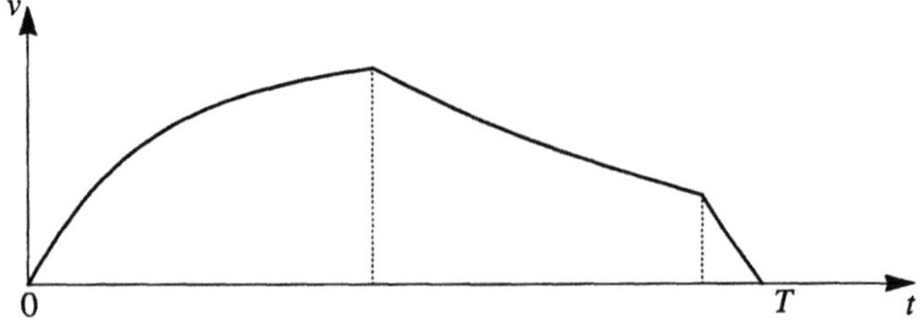

Figure 1-3: An *accelerate–coast–brake* speed profile

## 1.4 Further Early Results

Although Milroy had recognised the possible existence of singular control modes in an optimal strategy, he was not able to quantify these modes. Subsequent simulation trials by Tyler and Lee suggested that speed-holding should be included after the initial acceleration phase in an optimal strategy if the journey time was sufficiently large. They also realised that the speed-hold phase was one of two possible singular control modes that could be obtained from the original Pontryagin analysis. The simulation results were reported at a control engineering conference [3] in Newcastle, in 1982. The speed profile for a typical *accelerate–hold–coast–brake* strategy is shown in Figure 1-4.

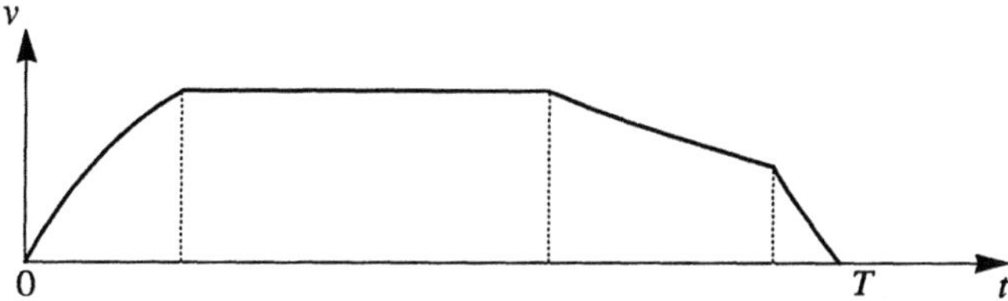

Figure 1-4: An *accelerate–hold–coast–brake* speed profile

## 1.5 Metromiser

In 1982 Ian Milroy formed the Transport Control Group at the South Australian Institute of Technology to work on a project funded by the South Australian Department of Transport. The Group showed that their proposed driving strategies were effective in practice, and subsequently developed and demonstrated a system for achieving fuel savings on suburban trains. The system is now known as *Metromiser*.

A timetabled suburban rail journey can be considered as a sequence of separate segments between consecutive stops. The timetable specifies an arrival time at each stop. At each point it is necessary to calculate a driving strategy that gets the train to the next stop on time and as efficiently as possible.

A typical suburban journey segment is between one and five kilometres in length. The time allowed between stops is about 5% more than the minimum possible time. For such segments an *accelerate–coast–brake* strategy is the most efficient, and the problem becomes one of determining when to switch from accelerate to coast and from coast to brake. On-board trials have confirmed that it is difficult for even the most experienced drivers to judge the point where coasting should begin. The correct coasting point depends on both the time remaining to complete the segment and on the speed of the train. Furthermore, small changes in the time at which coasting begins can result in large changes in the stopping time.

To illustrate this last point, consider the following situation. Suppose the train speed is 25 ms$^{-1}$ when coasting begins and 5 ms$^{-1}$ when the brakes are applied. If coasting begins five seconds too soon then the coasting point is moved back 125 m. The train now reaches the braking speed about 125 m short of the desired braking point. The train speed is now less than 5 ms$^{-1}$, and the train will take another 25 seconds to reach the braking point. The nett effect is an increase of 20 seconds in the journey time. Thus the train arrives at the stop 20 seconds late.

On the Dry Creek to Gawler line in South Australia there are 16 stops in a total journey time of 40 minutes. A 20 second delay on each segment means that the train reaches the final station more than 5 minutes late.

*Metromiser* performs two main tasks. The first task is to monitor the state of the journey. The second is to calculate and display appropriate driving advice.

### Journey State

Journey state is defined by

- the current time and the time remaining until the train is due at the next stop;
- the location of the train and the distance to the next stop; and
- the speed of the train.

Distance and speed are measured using an odometer attached to a wheel. Station locations and arrival times are stored in the *Metromiser* unit.

### Driving Advice

Driving advice is based on two key decisions:

- whether braking is required to avoid overshooting the next stop; and
- whether the train is able to coast and brake to arrive at the next stop on time.

### Brake Decision

The brake decision depends on the distance remaining to the next stop, the speed of the train, the braking performance of the train, and the gradient of the track. Braking performance is specified by a negative acceleration rate $-K$ for level track and an acceleration $a_g$ due to the gradient of the track. The distance required to stop the train is

$$s = \frac{v^2}{2(K + a_g)} \tag{1.9}$$

where $v$ is the speed of the train. Let the distance remaining to the next stop be denoted by $d$. If $s \leq d$ then braking is not required. If $s > d$ then the brakes must be applied.


### *Coast Decision*

The coast decision is more complex. Like the brake decision, it depends on location, speed and the coasting performance of the train. However, it also depends on the time remaining to complete the section. If the train is early the coast phase can be extended. If the train is late, the coast phase must be reduced in duration or even omitted.

The coasting performance of the train is specified by a negative acceleration $-a_c$ that is influenced by track gradient and frictional resistance, and hence depends on the location and speed of the train. An important feature of *Metromiser* is that it automatically learns coasting performance for every speed at each point on the track.

The coast decision is made in the following way. Let the state of the train be $(t, d, v)$, where $t$ is the time remaining until the train is due at the next stop, $d$ is the distance to the next stop, and $v$ is the speed of the train. Let $v_b$ be the speed at which the brakes will be applied. *Metromiser* calculates a suitable negative coast acceleration $-a_c$ so that the distance travelled during the coast phase is given by

$$s_c = \frac{v^2 - v_b^2}{2a_c}. \tag{1.10}$$

Let $(t_b, d_b, v_b)$ be the state of the train when the brakes are applied. From (1.9) the braking distance is given by

$$d_b = \frac{v_b^2}{2(K + a_g)}.$$

Using (1.10) and

$$d_b = d - s_c$$

to eliminate $v_b$ gives

$$d_b = \frac{2a_c d - v^2}{2\left[a_c - (K + a_g)\right]}.$$

If $d_b < 0$ the train stops while coasting, before the next stop. If $d_b \geq 0$ the braking speed is

$$v_b = \sqrt{2(K + a_g)\,d_b}$$

and the duration of the brake phase is

$$\tau_b = \frac{v_b}{K + a_g}$$

The total time spent coasting and braking will be

$$\tau = \tau_b + \frac{v_b - v}{a_c}.$$

If $t \geq \tau$ the train can coast, and will arrive at the next stop before the scheduled arrival time. If $t < \tau$ it is not possible to coast and arrive on time.

### *Driving Advice Display*

*Metromiser* advises one of three driving modes: *drive, coast* or *brake*. The appropriate driving mode is indicated by a corresponding light on a small display in front of the driver. An audible warning is given each time the advice changes. The Metromiser display panel is shown in Figure 1-5.

Figure 1-5: The *Metromiser* display panel

The usual advice sequence for a journey segment is *drive–coast–brake*. The drive light comes on at the beginning of the segment, and indicates to the driver that it is too early to coast. The train should be driven as quickly as permitted. The advice changes to coast when the coast decision indicates that it is possible to reach the next stop on time by coasting and braking.

Sometimes it may be necessary to slow the train unexpectedly after coasting has begun. If the change in speed is severe enough to mean that coasting is no longer appropriate, the advice will change back to drive.

In practice, the coast phase is never omitted. If the train is running so late that it cannot reach the next stop on time, the duration of the coast phase is reduced to an agreed minimum. This gives the driver a consistent control sequence and time to prepare for braking.

### *Optimal Timetables*

*Metromiser* calculates driving advice that, where possible, will get the train to each stop within a few seconds of the scheduled arrival time. To achieve good energy savings, *Metromiser* requires good timetables. A journey can be considered as a sequence of station-to-station segments. For each segment, the cost of the optimal control profile will decrease as the time available increases. The *cost–time curve* for a typical segment is shown in Figure 1-6.

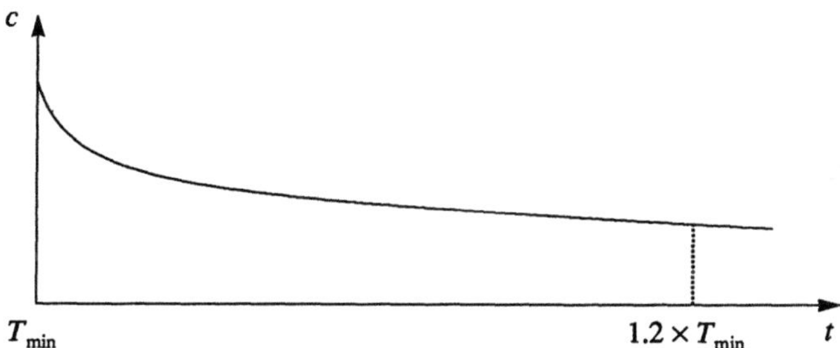

Figure 1-6: Typical cost–time curve

To optimise the timetable it is necessary to find the segment times that give the desired total running time with minimum total cost. If $c = c_i(t)$ is the cost-time curve for the $i$th segment and if $T_i$ is the optimal time for this segment then

$$T_1 + T_2 + \ldots + T_n = T$$

is the total running time and

$$c_1'(T_1) = c_2'(T_2) = \ldots = c_n'(T_n)$$

where $c_i'(t)$ is the slope of the cost-time curve $c_i$ at time $t$. The *Metromiser* system includes a software package that solves these equations and enables timetable planners to generate optimal timetables for various journey times.


## *Trial Results*

In February 1985 a prototype *Metromiser* was evaluated in Adelaide, South Australia. The unit was fitted to a two-car diesel-hydraulic railcar, which was operated in normal service. Testing occurred on a 30 km route with 17 stations. Data was collected for 10 runs with advice and 10 runs without advice for each direction on the route. Most of the drivers had not used *Metromiser* before the tests. The drivers were able to follow the advice without any difficulty.

Since the initial evaluation, *Metromiser* has been further tested in Adelaide, Melbourne, Brisbane and Toronto. The results of these tests are shown in the tables below. The tables show the mean earliness of arrival at the final station, the standard deviation of arrival times at the final station, and mean fuel or energy consumption for journeys with and without driving advice. For realistic timetables, energy savings are in the range 15–30%.

**Table 1-1: Metromiser Evaluation: Adelaide, 1985**

| Journey | earliness (seconds) | arrival s.d. (seconds) | fuel (litres) | |
|---|---|---|---|---|
| Dry Creek—Gawler | | | | |
| without advice | 24 | 85 | 64.5 | |
| with advice | 0 | 27 | 54.2 | 16% saving |
| Gawler—Dry Creek | | | | |
| without advice | 28 | 82 | 59.3 | |
| with advice | 10 | 2 | 51.5 | 13% saving |

**Table 1-2: Metromiser Evaluation: Adelaide, 1987**

| Journey | earliness (seconds) | arrival s.d. (seconds) | fuel (litres) | |
|---|---|---|---|---|
| Gawler Central—Adelaide | | | | |
| without advice | — | — | 86.5 | |
| with advice | — | — | 69.0 | 20% saving |

**Table 1-3: Metromiser Evaluation: Toronto, 1988**

| Journey | earliness (seconds) | arrival s.d. (seconds) | energy (kWh) | |
|---|---|---|---|---|
| Toronto—Oakville | | | | |
| without advice | — | — | 102 | |
| with advice | — | — | 86 | 16% saving |
| Oakville—Toronto | | | | |
| without advice | — | — | 85 | |
| with advice | — | — | 72 | 15% saving |
| Toronto—Pickering | | | | |
| without advice | — | — | 84 | |
| with advice | — | — | 60 | 29% saving |
| Pickering—Toronto | | | | |
| without advice | — | — | 78 | |
| with advice | — | — | 60 | 23% saving |

**Table 1-4: Metromiser Evaluation: Melbourne, 1989**

| Journey | earliness (seconds) | arrival s.d. (seconds) | energy (kWh) | |
|---|---|---|---|---|
| Flinders Street—Sandringham | | | | |
| without advice | 82 | 58 | 937 | |
| with advice | 1 | 76 | 657 | 30% saving |
| Sandringham—Flinders Street | | | | |
| without advice | −55 | 62 | 869 | |
| with advice | −20 | 21 | 821 | 5% saving |
| Flinders Street—Glen Waverley | | | | |
| without advice | 191 | 96 | 1034 | |
| with advice | 44 | 61 | 754 | 27% saving |
| Glen Waverley—Flinders Street | | | | |
| without advice | 79 | 72 | 760 | |
| with advice | 29 | 26 | 541 | 29% saving |

**Table 1-5: Metromiser Evaluation: Brisbane, 1990**

| Journey | earliness (seconds) | arrival s.d. (seconds) | energy (kWh) | |
|---|---|---|---|---|
| Roma Street—Ipswich | | | | |
| without advice | −190 | 359 | 239 | |
| with advice | −11 | 28 | 202 | 15% saving |
| Ipswich—Roma Street | | | | |
| without advice | −246 | 230 | 224 | |
| with advice | −173 | 82 | 239 | 7% saving |
| Roma Street—Caboolture | | | | |
| without advice | −240 | 192 | 264 | |
| with advice | −108 | 146 | 236 | 11% saving |
| Caboolture—Roma Street | | | | |
| without advice | −280 | 93 | 256 | |
| with advice | −109 | 198 | 240 | 6% saving |

# 1.6 Long-Haul Operations

In Australia, long-haul rail networks are predominantly single-line track. Infrequent crossing loops allow trains to cross or overtake. Typically, crossing loops are only slightly longer than a train. For two trains to cross, one must stop on the loop while the other passes. Energy is wasted when one of the trains arrives at a crossing loop earlier than the other. This is a very common occurrence on large networks. Considerable savings are possible if the trains are coordinated to arrive at the loop at approximately the same time, and if each is driven in an energy-efficient manner. Further savings are possible by considering all trains on the network, and calculating the location and time for each cross to minimise the total cost of energy and lateness.

The Scheduling and Control Group has designed a system for saving energy in the operation of long-haul trains. This system includes a strategic scheduler, a dynamic rescheduler and a driving advice system.

*Strategic Scheduling*

The strategic scheduler is used for developing efficient working timetables and for network planning. To create a timetable, the timetable planner specifies details of the network, the required train services and a cost function. Using current network and train performance data, the scheduler produces an efficient working timetable. By considering a variety of the network or train performance data, long-term strategies for network development can be evaluated.


### Dynamic Rescheduling

The dynamic rescheduler is used in a control centre to generate and maintain an energy-efficient crossing schedule for all trains on the rail network. It schedules future train movements to minimise the overall cost of train lateness and energy consumption. The scheduler calculates the best location and time for each cross, and determines which trains should take the sidings.

Traditionally, crossing decisions have been made by train controllers. Train controllers, however, are only able to make local decisions and are not able to assess the effects of an individual decision on the entire schedule. The dynamic rescheduler manipulates the schedule to reduce the total cost over all journeys.


### Driving Advice

Schedule data is passed from the rescheduler to a driving advice unit on each train. Each unit calculates an energy-efficient driving strategy and monitors the actual performance of the train. This information is sent back to the rescheduler, and is used to maintain or update the schedule. This allows the entire system to respond to the progress and performance of all trains on the network. On each train, the driving advice unit calculates and maintains an energy-efficient driving strategy to meet the schedule provided by the dynamic rescheduler.

The driving strategy for a long-haul journey differs from the strategy for a suburban journey, mainly because the distance between key points is much greater. An individual segment can extend over several hundred kilometres of track, and may contain many changes of gradient and many changes of speed limit. The simple *drive–coast–brake* strategy no longer applies. However, efficient strategies still require only maximum power, speed-hold, coast and brake. If the performance characteristics of the train are known, efficient speed profiles can be calculated.

For long-haul journeys, the performance characteristics of a train are generally not known before a journey commences—in fact, they will change during a journey as wagons are loaded or unloaded, or with changes in wind direction. Thus the future performance of the train must be estimated during the journey from observations of current and past performance. The advice unit uses an extended Kalman filter to estimate the power-to-mass ratio of the train and the three coefficients of frictional resistance.

As a journey progresses, it is inevitable that the speed of the train will depart from any pre-calculated profile. The advice unit therefore continually recalculates an efficient speed profile using the current train model, location, and speed.

Figure 1-7 shows a proposed design for the driver display unit. The lower portion of the display shows the train moving along the track. The dashed portion of the track indicates the current braking distance. Imminent speed restrictions are also shown. The upper portion of the display shows the current speed limit, the next target, and the current driving advice.

Figure 1-7: Proposed display unit

## CHAPTER 2

# MODELLING THE TRAIN CONTROL PROBLEM

The train control problem was originally formulated with applied acceleration as the control, and with journey cost measured by the mechanical energy required to drive the train. The *mechanical energy model* is physically sound, but does not properly describe the real control mechanism and does not represent the real financial cost of a journey. The *fuel consumption model* was designed to model the control mechanism of a typical diesel-electric locomotive, and uses the total fuel consumption to measure the cost of a journey.

Both models were developed for a point-mass train, but can also be applied to trains with distributed mass by defining a suitably modified track height profile.

## 2.1 The Mechanical Energy Model

A point mass train moving along a smooth track under the influence of an applied force satisfies the differential equations

$$x'(t) = v(t)$$
$$mv'(t) = F(t) - R[v(t)] + G[x(t)]$$

where $t$ is the time, $x(t)$ is the position of the train, $v(t)$ is the speed of the train, $m$ is the mass of the train, $F(t)$ is the component of the applied force along the track, $-R[v(t)]$ is the component of resistance along the track, and $G[x(t)]$ is the component of gravitational force along the track. If we divide through by the mass of the train the above equation can be written in the form

$$v'(t) = u(t) - r[v(t)] + g[x(t)] \tag{2.1}$$

where $u(t)$ is the acceleration along the track corresponding to the *applied force* $F(t)$, $-r[v(t)]$ is the *resistive acceleration* along the track, and $g[x(t)]$ is the *gradient acceleration* along the track. In practice,

$$g[x(t)] = -g\sin\theta[x(t)]$$

where $g$ is the acceleration due to gravity and $\theta\,[x(t)]$ is the angle of slope of the track. For convenience, we will refer to $u(t)$ as the *applied acceleration*. On physical grounds, it is reasonable to assume that the applied acceleration is bounded. Thus we assume that $|u(t)| \le U$ for all $t$. If the train starts from rest at time $t = 0$ and finishes at rest at time $t = T$ then $v(0) = v(T) = 0$. If we assume that $x(0) = 0$ then the distance travelled by the train at time $t = \tau \ge 0$ is given by

$$x(\tau) = \int_0^\tau v(t)dt. \tag{2.2}$$

If the total distance to be travelled is $X$ then we require $x(T) = X$. We will normally assume that $v(t) \ge 0$ for all $t \in [0, T]$.

Let

$$F_+(t) = \begin{cases} F(t) & F(t) > 0 \\ 0 & F(t) \le 0 \end{cases}$$

denote the positive part of the applied force and let

$$F_-(t) = \begin{cases} 0 & F(t) > 0 \\ -F(t) & F(t) < 0 \end{cases}$$

denote the negative part of the applied force. When $F(t) > 0$, energy is supplied to the train. When $F(t) < 0$ energy is dissipated, mainly as heat, by the brakes. The cost of the journey is the total energy supplied to the train, and is calculated from the work done on the train by $F_+(t)$, the positive part of the applied force. The cost of the journey is therefore

$$J = \int_0^T F_+(t)v(t)dt. \tag{2.3}$$

The advantage of this model is that it is consistent with known physical laws and it does not depend on the detailed structure of the train. An *energy balance equation* can be obtained by integrating the equation of motion.

The *height profile* $y(\xi)$ of the track at $\xi = x(\tau)$ can be defined using the formula

$$y(\xi) = y(0) + \int_0^\xi \sin\theta(x)dx.$$

If the equation of motion is written in the form

$$mv(t)v'(t) \;=\; mv(t)\,\{\,u(t) - r\,[v(t)] - g\sin\theta\,[x(t)]\,\}$$

then integrating from $t = 0$ to $t = \tau$ and using $dx(t) = v(t)\,dt$ gives

$$\tfrac{1}{2}mv(\tau)^2 - \tfrac{1}{2}mv(0)^2$$

$$= m\int_0^\tau \{\,u(t) - r\,[v(t)] - g\sin\theta\,[x(t)]\,\}\,dx(t)$$

$$= m\int_0^\tau \{\,u(t) - r\,[v(t)]\,\}\,dx(t) - mg\,\{\,y(\xi) - y(0)\,\}$$

and hence

$$\{\tfrac{1}{2}mv(\tau)^2 - \tfrac{1}{2}mv(0)^2\} + \{\,mgy(\xi) - mgy(0)\,\}$$

$$= m\int_0^\tau \{\,u(t) - r\,[v(t)]\,\}\,dx(t).$$

where the left hand side represents the changes in kinetic and potential energy, and the right hand side represents the net work done on the train by the applied force and resistance. That is

*change in kinetic energy + change in potential energy*
*= net work done on the train.*

Equations (2.1)–(2.3) with $\theta(x) = 0$ for all $x$ describe Milroy's original formulation of the train control problem on level track. While the equations in this formulation satisfy the normal Newtonian laws of motion, they do not model the usual control mechanism for a train. In practice, it is power rather than applied acceleration that is controlled, and it is reasonable to assume that it is the maximum applied power that is bounded. Because the control mechanism in this model does not correspond to the real control mechanism, it is difficult to construct realistic examples. This difficulty is apparent from the speed profiles in Examples 7.1–7.3, and is discussed briefly in Section 7.5.

## 2.2  The Fuel Consumption Model

### 2.2.1  Traction Characteristics for a Diesel-Electric Locomotive

By observing the traction characteristics of diesel-electric locomotives, Benjamin et al [8] noted that only certain discrete control settings are possible, that each control setting determines a constant rate of fuel supply, and that tractive power is directly proportional to the rate of fuel supply. Figure 2-1 shows a graph of applied force against speed for the eight traction control settings on the General Motors JT26C-2SS locomotive. For each notch, except at low speeds, the tractive effort is inversely proportional to the speed and hence the power is almost constant.

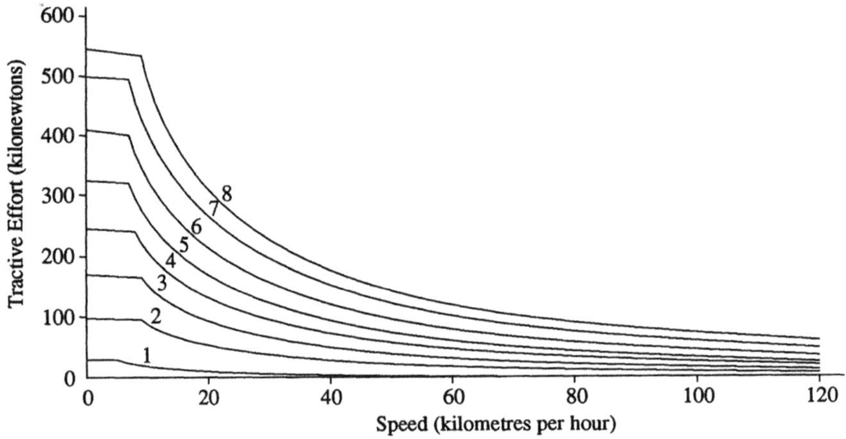


Figure 2-1: Tractive effort curves for a General Motors JT26C-2SS locomotive

Figure 2-2 shows a graph of power against fuel supply rate in the various notches for the General Motors JT26C-2SS locomotive. These results were obtained by measurements on a real train. The assumption of a linear relationship between power and fuel supply rate is reasonable. In practice, it is often true that this curve is convex, with greater efficiency at the higher rates of fuel supply.

### 2.2.2  Braking Characteristics for a Diesel-Electric Locomotive

Braking is more difficult to model, as the driver can select any combination of mechanical and dynamic (electrical) braking. In particular, mechanical braking is

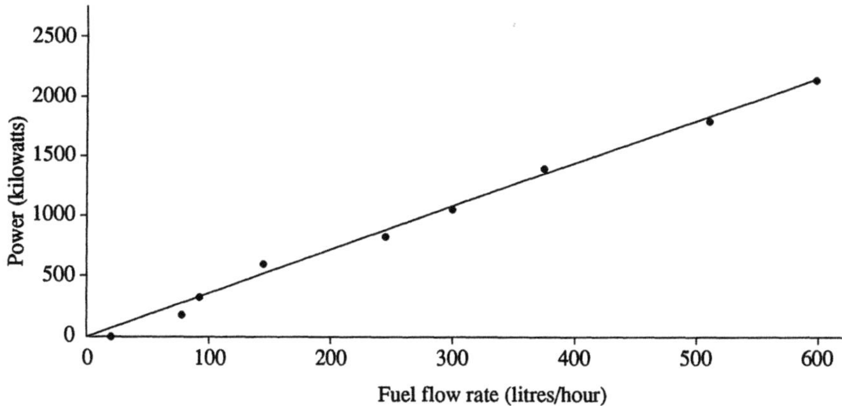

Figure 2-2: Power against fuel flow rate for a General Motors JT26C-2SS locomotive

used to supplement dynamic braking at low speed. It is reasonable, however, to assume that the driver can apply a constant negative brake force. In our models the rate of fuel supply is assumed to be zero during braking, and so the precise nature of the braking does not effect the overall fuel consumption. In practice, a low notch setting is often used to operate the electrical brakes. Figure 2-3 shows the graph of dynamic braking effort against speed for the General Motors JT26C-2SS locomotive.

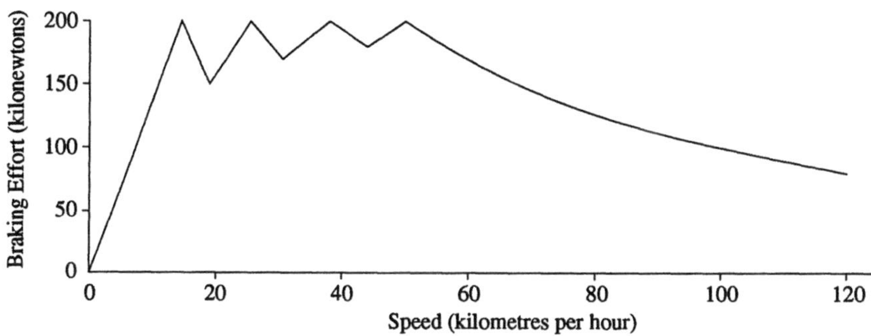

Figure 2-3: Dynamic braking effort curves for a General Motors JT26C-2SS locomotive

### 2.2.3 Modelling the Control Mechanism

For trains with discrete control notches, such as the diesel-electric locomotives described above, the position of the control mechanism can be represented by an integer control variable $j$. The control setting $j$ determines a constant rate of fuel supply $f_j$. Each non-negative value of the control variable determines a traction control and each negative value determines a brake control. It is assumed that $f_j > 0$ when $j > 0$ and that $f_j = 0$ when $j \leq 0$. When $j \geq 0$ the power developed by the locomotive is directly proportional to the rate of fuel supply. When $j < 0$ there is a constant negative force applied to the train. If $K_j$ denotes the brake acceleration then $K_j = 0$ for $j \geq 0$ and $K_j < 0$ for $j < 0$.

### 2.2.4  The Equations of Motion

At this stage it is prudent to re-consider the energy flow patterns in the traction and brake systems. The energy flows are illustrated in Figure 1-1.

The equations of motion can be formulated with time as the independent variable and with position and speed as the dependent state variables. Although this formulation may seem more natural, it is often convenient to use an equivalent formulation with position as the independent variable and with time and speed as the dependent state variables. In the latter formulation it is necessary to control the evolution of time and speed as functions of position. The equations of motion should support and be supported by the energy flow patterns described in Section 1.1.

The equations of motion for a point-mass train are

$$x'(t) = v(t) \tag{2.4}$$

and

$$v'(t) = \frac{Hf_j}{v(t)} + K_j - r[v(t)] + g[x(t)] \tag{2.5}$$

where $j = j(t)$ is the control setting, $f_j$ is the fuel supply rate, $H$ is a constant and $K_j \leq 0$ is the brake acceleration. It is assumed that $r(0) > 0$, $r(v)$ is strictly increasing, and that the graph $y = vr(v)$ is strictly convex. The cost of the journey is the total fuel consumption given by

$$J = \int_0^T f(t)\,dt$$

where $f(t) = f_{j(t)}$ is the fuel supply rate and $T$ is the total time taken. It is often convenient to write the equations of motion in the form

$$t'(x) = \frac{1}{v(x)} \tag{2.6}$$

and

$$v(x)v'(x) = \frac{Hf_j}{v(x)} + K_j - r\,[v(x)] + g(x) \tag{2.7}$$

where $j = j(x)$ is the control setting. Fuel consumption is given by

$$J = \int_0^X \frac{f(x)}{v(x)}\,dx$$

where $f(x) = f_{j(x)}$ is the fuel supply rate and $X$ is the total distance travelled.

Although (2.4)–(2.7) are directly applicable to a point-mass train, it is important to remark that similar equations, with a modified average gradient acceleration, can be used to describe a train with distributed mass. This is discussed in the next section.

## 2.3 Trains with Distributed Mass

Consider a train with distributed mass and with equation of motion given in general form by

$$v(x)v'(x) = u\,[x, v(x)] - r\,[v(x)] + \frac{1}{M}\int_0^S \rho(s)g(x-s)\,ds \tag{2.8}$$

where $u\,[x, v(x)] = u_j(v)$ is the controlled acceleration of the train for control setting $j = j(x)$ and speed $v = v(x)$, $M$ is the mass of the train, $\rho(s)$ is the mass per unit length at distance $s$ from the front of the train, and $S$ is the length of the train.

If the *modified gradient acceleration* for the given train is defined by the formula

$$\bar{g}(x) = \frac{1}{M}\int_0^S \rho(s)g\,(x-s)\,ds$$

then (2.8) can be rewritten in the form

$$v(x)v'(x) = u\,[x, v(x)] - r\,[v(x)] + \bar{g}(x) \tag{2.9}$$

This allows the train to be treated as a point mass, with the actual gradient acceleration replaced by the modified gradient acceleration. We remarked earlier that in practice the gradient acceleration for a point mass train is given by $g(x) = -g \sin\theta(x)$ and that the height profile of the track is given by

$$y(\xi) = y(0) + \int_0^\xi \sin\theta(x)\,dx.$$

By writing $\bar{g}(x) = -g \sin\bar{\theta}(x)$ a *modified height profile* $\bar{y}(\xi)$ can be defined using the formula

$$\bar{y}(\xi) = y(0) + \int_0^\xi \sin\bar{\theta}(x)\,dx$$

where $y(0)$ is the initial height. Integration of (2.9) gives the energy balance equation

$$\left\{\frac{1}{2}Mv(\xi)^2 - \frac{1}{2}Mv(0)^2\right\} + \left\{Mg\bar{y}(\xi) - Mgy(0)\right\}$$

$$= M\int_0^\xi \left\{u[x, v(x)] - r[v(x)]\right\}\,dx$$

which can be interpreted as

*change in kinetic energy + change in potential energy*
*= net work done on train.*

This equation describes the energy balance illustrated in Figure 1-1.

# CHAPTER 3
# PRACTICAL DRIVING STRATEGIES

There are two aspects of train control that can be addressed before we consider the optimisation problems. First, we can show that the discrete control mechanism of the fuel consumption model is not a practical limitation, since any speed profile can be followed as accurately as we please using alternate *coast–power* pairs. Second, by considering the energy balance equations we can show that speed-holding, where possible, is the most efficient driving mode.

## 3.1 Approximation of Measurable Control

In this section we will show that any interval of non-negative measurable fuel supply rate $f(t)$ can be approximated to any desired accuracy by a sequence of *coast–power* control pairs. Let $j \in C = \{-1, 0, 1, ..., m\}$ denote the level of control, where the control mode $j = -1$ corresponds to braking, and the control modes with $j \geq 0$ correspond to different traction settings. The coast control mode $j = 0$ corresponds to a zero fuel supply rate, and the power control mode $j = m$ corresponds to the maximum fuel supply rate.

We will show that the *coast–power* strategy can be used to produce a speed profile that is as close as we please to the speed profile produced by the given measurable fuel supply rate, and that the cost of the *coast–power* strategy is also as close as we please to the cost of the given control strategy.

Consider solutions $(x, v)$ to the equations of motion starting at time $t = \tau$. Let $(x_f[\tau, y, w](t), v_f[\tau, y, w](t))$ denote the solution to the system of differential equations

$$\frac{dx}{dt} = v$$

and

$$\frac{dv}{dt} = \frac{Hf(t)}{v} - r(v) + g(x)$$

with $x(\tau) = y$ and $v(\tau) = w$. Thus the required solution passes through the point $(y, w)$ at time $t = \tau$. The solutions for the coast control mode with $f(t) \equiv 0$ and the power control mode with $f(t) \equiv 1$ are denoted by $(x_0[\tau, y, w](t), v_0[\tau, y, w](t))$ and $(x_1[\tau, y, w](t), v_1[\tau, y, w](t))$ respectively.

Consider a time interval $[\tau_a, \tau_b]$ and a control strategy

$$S_f = S([f(t); [\tau_a, \tau_b]])$$

and suppose that the initial position is $y_a$ and the initial speed is $w_a$. The equations of motion can be solved to find

$$(x_f(t), v_f(t)) = (x_f[\tau_a, y_a, w_a](t), v_f[\tau_a, y_a, w_a](t))$$

for all $t \in [\tau_a, \tau_b]$.

Now consider an associated subdivision $\{[\tau_k, \tau_{k+1}]\}_{k=0,1,\ldots,n}$ of the interval $[\tau_a, \tau_b]$ with $\tau_0 = \tau_a$ and $\tau_{n+1} = \tau_b$, and write

$$(x_f(\tau_k), v_f(\tau_k)) = (y_k, w_k).$$

Define upper and lower bounds $\bar{v}_k(t)$ and $\underline{v}_k(t)$ for $v_f(t)$ on the interval $[\tau_k, \tau_{k+1}]$ as follows. Let

$$\bar{v}_k(t) = \min\{v_1[\tau_k, y_k, w_k](t), v_0[\tau_{k+1}, y_{k+1}, w_{k+1}](t)\}$$

and

$$\underline{v}_k(t) = \max\{v_0[\tau_k, y_k, w_k](t), v_1[\tau_{k+1}, y_{k+1}, w_{k+1}](t)\}$$

respectively. It can be seen that there are points $s_k$ and $t_k$ in the interval $[\tau_k, \tau_{k+1}]$ such that

$$\bar{v}_k(t) = \begin{cases} v_1[\tau_k, y_k, w_k](t) & t \in [\tau_k, s_k] \\ v_0[\tau_{k+1}, y_{k+1}, w_{k+1}](t) & t \in [s_k, \tau_{k+1}] \end{cases}$$

and

$$\underline{v}_k(t) = \begin{cases} v_0[\tau_k, y_k, w_k](t) & t \in [\tau_k, t_k] \\ v_1[\tau_{k+1}, y_{k+1}, w_{k+1}](t) & t \in [t_k, \tau_{k+1}]. \end{cases}$$

For each point $q \in [\tau_k, s_k]$ there is a uniquely defined point $r \in [t_k, \tau_{k+1}]$ such that

$$v_0\left[ q, x_1 [\tau_k, y_k, w_k](q), v_1 [\tau_k, y_k, w_k](q) \right](r)$$
$$= v_1 [\tau_{k+1}, y_{k+1}, w_{k+1}](r)$$

From Figure 3-1 it can be seen that it is possible to choose $q = q_k$ and the corresponding $r = r_k$ so that

$$\int_{\tau_k}^{q_k} v_1 [\tau_k, z_k, w_k](t)dt$$

$$+ \int_{q_k}^{r_k} v_0 [q_k, x_1 [\tau_k, z_k, w_k](q_k), v_1 [\tau_k, z_k, w_k](q_k)](t)dt$$

$$+ \int_{r_k}^{\tau_{k+1}} v_1 [\tau_{k+1}, z_{k+1}, w_{k+1}](t)dt \;=\; \int_{\tau_k}^{\tau_{k+1}} v_f [\tau_a, z_a, w_a](t)dt$$

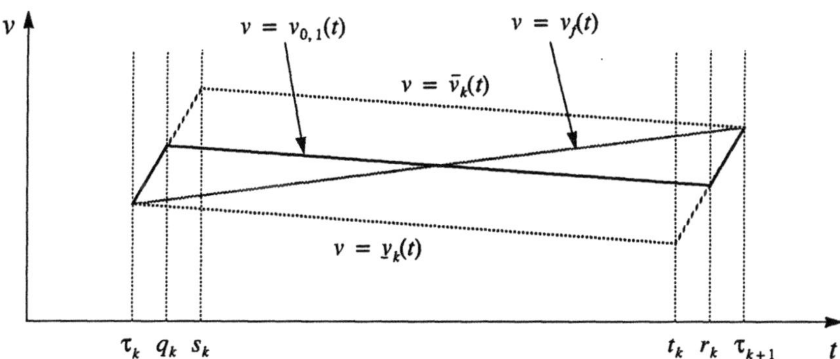


Figure 3-1: Stylised speed profiles $\bar{v}$, $\underline{v}$, $v_f$ and $v_{0,1}$ on $[\tau_k, \tau_{k+1}]$

If an alternative control strategy

$$S_{0,1} = S(\{ [1;(\tau_k, q_k)], [0;(q_k, r_k)], [1;(r_k, \tau_{k+1})] \}_{k=0,1,...,n})$$

is defined then the corresponding position and speed on the interval $[\tau_k, \tau_{k+1}]$ are given by

$$
x_{0,1}(t) = \begin{cases}
x_1[\tau_k, y_k, w_k](t) & t \in [\tau_k, q_k] \\
x_0[q_k, x_1[\tau_k, y_k, w_k](q_k), v_1[\tau_k, z_k, w_k](q_k)](t) & t \in [q_k, r_k] \\
x_1[\tau_{k+1}, y_{k+1}, w_{k+1}](t) & t \in [r_k, \tau_{k+1}]
\end{cases}
$$

and

$$
v_{0,1}(t) = \begin{cases}
v_1[\tau_k, y_k, w_k](t) & t \in [\tau_k, q_k] \\
v_0[q_k, x_1[\tau_k, y_k, w_k](q_k), v_1[\tau_k, z_k, w_k](q_k)](t) & t \in [q_k, r_k] \\
v_1[\tau_{k+1}, y_{k+1}, w_{k+1}](t) & t \in [r_k, \tau_{k+1}]
\end{cases}
$$

and clearly $x_{0,1}(\tau_k) = x_f(\tau_k)$ and $v_{0,1}(\tau_k) = v_f(\tau_k)$ for all $k = 0, 1, \ldots, n+1$.

To complete this section it will be shown that the fuel consumption for the alternative strategy is essentially the same as the fuel consumption for the original strategy. On each time interval $[\tau_k, \tau_{k+1}]$ the respective increments of fuel consumption are given by

$$
\Delta_k J_f = \int_{\tau_k}^{\tau_{k+1}} f(t)\,dt
$$

and

$$
\Delta_k J_{0,1} = [q_k - \tau_k] + [\tau_{k+1} - r_k].
$$

By integrating the appropriate equations of motion, corresponding energy balance equations are obtained. By using the function

$$
G(x) = -\int g(x)\,dx = -\int g(x)v\,dt
$$

it can be shown for the original strategy that

$$
\left[\frac{1}{2}v_f^2 + G(x_f)\right]\Bigg|_{\tau_k}^{\tau_{k+1}} = \int_{\tau_k}^{\tau_{k+1}} [Hf(t) - v_f r(v_f)]\,dt \tag{3.1}
$$

and for the alternative strategy that

$$\left[\frac{1}{2}v_{0,\,1}^2 + G(x_{0,\,1})\right]\Bigg|_{\tau_k}^{\tau_{k+1}}$$

$$= H\{[q_k - \tau_k] + [\tau_{k+1} - r_k]\} - \int_{\tau_k}^{\tau_{k+1}} v_{0,\,1}r(v_{0,\,1})dt. \tag{3.2}$$

Since $(x_f(t),\, v_f(t))$ and $(x_{0,\,1}(t),\, v_{0,\,1}(t))$ are equal at $t = \tau_k$ and $t = \tau_{k+1}$ it follows from (3.1) and (3.2) that

$$\{[q_k - \tau_k] + [\tau_{k+1} - r_k]\} - \int_{\tau_k}^{\tau_{k+1}} f(t)dt$$

$$= \frac{1}{H}\int_{\tau_k}^{\tau_{k+1}} [v_{0,\,1}r(v_{0,\,1}) - v_f r(v_f)]\,dt.$$

By choosing a sufficiently fine subdivision it is therefore possible to ensure that $v_{0,\,1}(t)$ is uniformly as close as we please to $v_f(t)$ and hence that

$$\left| [q_k - \tau_k] + [\tau_{k+1} - r_k] - \int_{\tau_k}^{\tau_{k+1}} f(t)dt \right| < \frac{\varepsilon}{\tau_b - \tau_a}\Delta\tau_k.$$

Therefore

$$|\Delta_k J_{0,\,1} - \Delta_k J_f| < \frac{\varepsilon}{\tau_b - \tau_a}\Delta\tau_k$$

and hence the fuel consumption for the alternative strategy can be made arbitrarily close to the fuel consumption for the original strategy.

## 3.2 Speed-Holding

To maintain a given constant speed on level track a corresponding constant rate of fuel supply is required. With discrete control this will generally not be possible. The driver may, however, approximate a constant speed strategy using periods of control with fuel supply rate alternately below and above the desired rate. For example, the driver could use the *coast–power* strategy described in the previous section.

If the tractive acceleration is assumed to be directly proportional to the power associated with a given rate of fuel supply and inversely proportional to the speed of the train, then if follows that the applied acceleration is given by

$$u(x, v) = \frac{p}{v}$$

where $p = p(f)$ is the power per unit mass associated with fuel supply rate $f = f(x)$. For many trains, the graph $y = p(f)$ is convex. These trains are most efficient at the maximum rate of fuel supply, and so an idealised speed-holding strategy, using infinitesimal periods of alternate coast and maximum power, is theoretically more efficient than a true speed-holding strategy using an intermediate rate of fuel supply. This assertion is justified in Section 10.10. Thus, if $p(f)$ is convex, any level of acceleration can be implemented at maximum efficiency using infinitesimal periods of alternate coast and maximum power. Vehicles with peak efficiency at an intermediate rate of fuel supply are unable to do this. In practice, a small number of *coast–power* pairs is sufficient to obtain a good approximation to the idealised minimum cost strategy.

If the tractive power is directly proportional to the rate of fuel supply then $p(f) = Hf$ for some constant $H$, and a straightforward argument can be used to show that speed-holding, where feasible, is the most efficient driving mode. Consider the case where the train travels from $x = a$ to $x = b$ at speed $V$. From the equation of motion

$$v\frac{dv}{dx} = \frac{Hf(x)}{v} - r(v) + g(x)$$

it follows that

$$f(x) = \frac{V}{H}\left[r(V) - g(x)\right],$$

and hence the cost of this speed-holding segment is

$$J_0 = \int_a^b \frac{f(x)}{V} dx$$

$$= \frac{1}{H} \int_a^b [r(V) - g(x)] \, dx$$

$$= \frac{r(V)}{H} [b - a] + \frac{1}{H} [G(b) - G(a)]$$

where

$$G(x) = -\int g(x) dx.$$

Consider any other strategy that drives the train from speed $V$ at $x = a$ to speed $V$ at $x = b$, and for which the time taken is the same. Thus

$$\int_a^b \frac{1}{v} dx = \frac{b-a}{V}$$

and hence the cost of this segment is given by

$$J = \int_a^b \frac{f(x)}{v} dx$$

$$= \frac{1}{H} \int_a^b \left[ v \frac{dv}{dx} + r(v) - g(x) \right] dx$$

$$= \frac{1}{H} \int_a^b r(v) dx + \frac{1}{H} [G(b) - G(a)]$$

since

$$\int_a^b v \frac{dv}{dx} dx = \left[ \frac{1}{2} v^2 \right]_{x=a}^b = 0.$$

Because the graph $y = vr(v)$ is convex, a positive real number $\lambda$ can be found such that $vr(v) - Vr(V) \geq \lambda (v - V)$ . Therefore

$$J - J_0 = \frac{1}{H} \int_a^b [r(v) - r(V)]\, dx$$

$$\geq \frac{\lambda - r(V)}{H} \int_a^b \left(1 - \frac{V}{v}\right) dx$$

$$= 0.$$

Fluctuations in speed are therefore inefficient. If the speed of the train is too high, the resistance is increased. If the speed of the train is too low, the time constraint will be violated. Unnecessary braking at any stage of the journey will require excessive application of power at some other stage.

# CHAPTER 4
# CONSTRAINED OPTIMISATION – AN INTUITIVE VIEW

Both formulations of the train control problem can be solved using standard methods of constrained optimisation. This chapter shows how a standard optimisation problem can be formulated and then approximated by a linearised problem that defines necessary conditions for a solution to the original problem. The mathematical terminology is explained and illustrated with diagrams and examples. In particular, the Kuhn-Tucker conditions and the Pontryagin Principle are introduced.

## 4.1 Constrained Optimisation

The train control problem is a typical problem of constrained optimisation. The cost functional, which depends on the control function, must be minimised while certain constraints are satisfied. The equation of motion is a consequence of physical laws and appears as an equality constraint. The requirements that the train travel a certain distance within a certain time are additional restrictions we choose to impose, and can be written as inequality constraints.

The configuration of a physical system is normally described by a variable $x$, which may be a combination of control and state variables and can be thought of as an element in a vector space $X$. The *cost functional* $J = J(x)$ is a real number that depends on $x$, and must be minimised subject to certain constraints. In general these constraints are formulated as an *equality constraint* $F(x) = 0$ and an *inequality constraint* $G(x) \le 0$. We will suppose that $x = x_0$ is a local solution to our problem. That is $F(x_0) = 0$, $G(x_0) \le 0$ and $J(x) < J(x_0)$ for all feasible $x$ near $x_0$.

Rigorous explanations of the mathematical foundations for the techniques of constrained optimisation can be found in many standard references [17–25]. In this chapter we simply present an overview of the main principles.

We will illustrate the general optimisation problem with a simple example.

***Example 4.1***

*Minimise* $(x_1 - 2)^2 + x_2^2 + (x_3 - 4)^2$ *subject to the equality constraint* $x_1^2 + x_2^2 = 4 - x_3$ *and the inequality constraints* $x_1 \leq x_2,\ x_2 \geq 0,\ x_3 \geq 0$.

If we set

$$J(x) = (x_1 - 2)^2 + x_2^2 + (x_3 - 4)^2,$$

$$F(x) = x_1^2 + x_2^2 + x_3 - 4$$

and

$$G(x) = \begin{bmatrix} 1 & -1 & 0 \\ 0 & -1 & 0 \\ 0 & 0 & -1 \end{bmatrix} \begin{bmatrix} x_1 \\ x_2 \\ x_3 \end{bmatrix},$$

then the problem is in the standard form

*Minimise $J(x)$ subject to $F(x) = 0$ and $G(x) \leq 0$.*

The problem is illustrated in Figure 4-1. We wish to find the minimum distance between the point $(2, 0, 4)$ and the point $(x_1, x_2, x_3)$ in the dark shaded region on the parabolic surface. The problem is solved in Example 4.14.

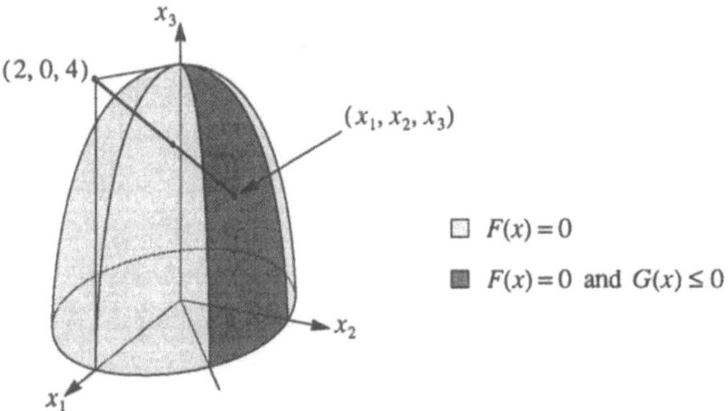


Figure 4-1: A constrained optimisation problem

The general constrained minimisation problem is illustrated in Figure 4-2.

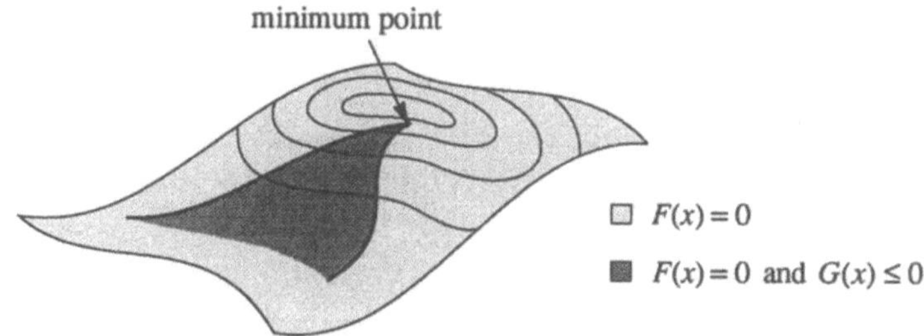


Figure 4-2: The general problem showing contour lines for the cost function

Necessary conditions for a minimum point are obtained by considering a corresponding linear problem. Let $x = x_0$ be the minimum point and let $\delta x = x - x_0$ denote the displacement from this point. Since $F(x_0) = 0$ the equation $F(x) = 0$ is replaced by a linear equation $A(\delta x) = 0$ which defines a tangent plane to approximate the surface $F(x) = 0$ near $x = x_0$. In a similar way, if we assume that $G(x_0) = 0$ the constraint $G(x) \leq 0$ can be replaced by a linear constraint $B(\delta x) \leq 0$ which defines a linearised cone to approximate the region $G(x) \leq 0$ near $x = x_0$. The intersection of the regions $A(\delta x) = 0$ and $B(\delta x) \leq 0$ forms a cone that is the projection of the cone $B(\delta x) \leq 0$ into the subspace $A(\delta x) = 0$. The linearisation procedure is shown in Figure 4-3.

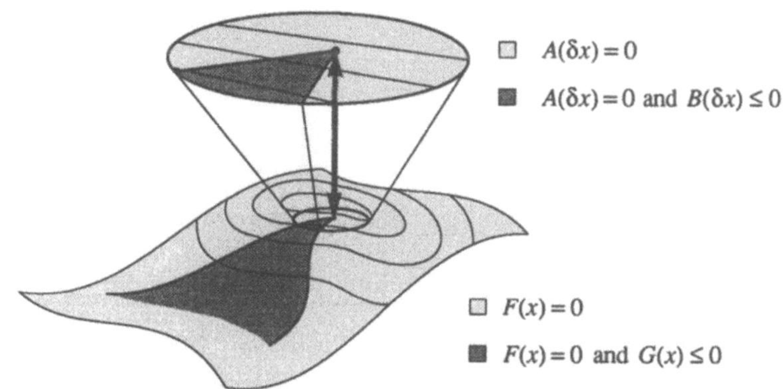


Figure 4-3: Linearising the general problem about the minimum point

## 4.2 Linearisation Techniques

Let $X$ and $Y$ be (Banach) vector spaces and let $M : X \to Y$ be a sufficiently smooth mapping from $X$ into $Y$. At any selected point $x_0$ we can use a standard limit procedure to define a *linearised mapping* $\nabla M : X \to Y$ which approximates the mapping $M$ near the point $x = x_0$. Let $\alpha > 0$ be a real number. For each increment $\delta x$ in $X$ we define the linear mapping $\nabla M [x_0] : X \to Y$ by setting

$$\nabla M [x_0] (\delta x) = \lim_{\alpha \to 0} \frac{1}{\alpha} [M(x_0 + \alpha \delta x) - M(x_0)] .$$

The set of all linear mappings from $X$ into $Y$ is denoted by $\mathcal{L}(X, Y)$. Thus we write $\nabla M [x_0] \in \mathcal{L}(X, Y)$.

If we regard the equation

$$M(x) = 0$$

as the equation of a surface in the space $X$ passing through the point $x = x_0$ where $M(x_0) = 0$ then the linearised approximation to this surface is given by

$$\nabla M [x_0] (\delta x) = 0$$

and is called the *tangent plane* to the surface $M(x) = 0$ at $x = x_0$.


***Example 4.2***

Let $X = \Re^3$, $Y = \Re^2$ and consider the mapping

$$M(x) = \begin{bmatrix} x_1^2 + (x_2 + 1)^2 + x_3^2 - 2 \\ x_1^2 + (x_2 - 1)^2 + x_3^2 - 2 \end{bmatrix}$$

The equation $M(x) = 0$ defines a circle of radius one centred at the origin in the plane $x_2 = 0$. The point

$$x_0 = (\frac{1}{\sqrt{2}}, 0, \frac{1}{\sqrt{2}})$$

lies on this circle. The linearised mapping at this point is given by

$$\nabla M [x_0] (\delta x) = \begin{bmatrix} \sqrt{2}\delta x_1 + 2\delta x_2 + \sqrt{2}\delta x_3 \\ \sqrt{2}\delta x_1 - 2\delta x_2 + \sqrt{2}\delta x_3 \end{bmatrix}$$

where $\delta x = x - x_0$. The equation

$$\nabla M [x_0] (\delta x) = 0$$

defines the tangent line $\delta x_1 + \delta x_3 = 0$ and $\delta x_2 = 0$ to the circle at this point.

$\square$


### Example 4.3

Suppose that the general problem

*Minimise $J(x)$ subject to $F(x) = 0$ and $G(x) \le 0$*

has a solution at the point $x = x_0$. Near the point $x_0$ we have

$$J(x_0 + \delta x) \approx J(x_0) + \nabla J [x_0] (\delta x),$$

$$F(x_0 + \delta x) \approx F(x_0) + \nabla F [x_0] (\delta x)$$

and

$$G(x_0 + \delta x) \approx G(x_0) + \nabla G [x_0] (\delta x).$$

If we define

$$f = \nabla J [x_0] \in \mathcal{L}(X, \Re)$$

$$A = \nabla F [x_0] \in \mathcal{L}(X, Y)$$

and

$$B = \nabla G [x_0] \in \mathcal{L}(X, Z)$$

and if we suppose $F(x_0) = 0$ and $G(x_0) = 0$ then the general problem can be replaced by the linearised problem

*Minimise $f(\delta x)$ subject to $A (\delta x) = 0$ and $B (\delta x) \le 0$*

where $\delta x = x - x_0$.

$\square$


The validity of our optimisation techniques relies on the following *Modified Linearisation Theorem*, which in turn depends on the *Implicit Function Theorem*. A careful statement and a proof of both theorems can be found in the book by Craven [18].

***Theorem 4.1 (Modified Linearisation Theorem)***

*If the problem*

$$\text{Minimise } J(x) \text{ subject to } F(x) = 0 \text{ and } G(x) \leq 0$$

*has a solution at $x = x_0$ then the linearised problem*

$$\text{Minimise } f(\delta x) \text{ subject to } A(\delta x) = 0 \text{ and } B(\delta x) \leq 0$$

*has a solution at $\delta x = 0$.*

$\square$

## 4.3 Hyperplanes and Half-Spaces

Let $X$ be a (Banach) vector space. The (Banach) vector space of all real valued linear functionals on $X$ is denoted by $X^* = \mathcal{L}(X, \mathfrak{R})$ and is called the *dual space* of $X$. For each $f \in X^*$ an equation of the form

$$f(x) = 1$$

defines a hyperplane in $X$. Every hyperplane in $X$ that does not pass through the origin is defined by an equation of this form.

### 4.3.1 Hyperplanes in Hilbert Space

If $X = H$ is a Hilbert space then $X$ has an inner product $<\cdot, \cdot>$ and an associated norm $\|\cdot\|$. Here the dual space $X^*$ can be identified with the original space $X$ in the following way.

Let $N$ be the hyperplane through the origin defined by $f(x) = 0$. This plane is called the *null space* of $f$. The orthogonal complement $N^\perp$ of $N$ in $H$ is the set of all vectors orthogonal to $N$. Thus

$$N^\perp = \{z| <z, x> = 0 \; \forall \; x \in N\}$$

If we choose any fixed point $z \in N^\perp$ with $f(z) \neq 0$ then for each $x \in H$ we note that

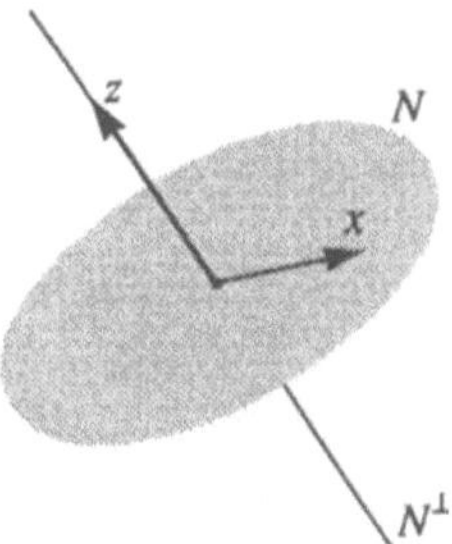

Figure 4-4: Orthogonal complement $N^{\perp}$ of a null space $N$

$$x - \frac{f(x)z}{f(z)} \in N$$

and hence

$$\left\langle z, x - \frac{f(x)z}{f(z)} \right\rangle = 0.$$

Therefore

$$f(x) = \left\langle \frac{f(z)z}{\|z\|^2}, x \right\rangle$$

and if $x_f \in X$ is defined by

$$x_f = \frac{f(z)z}{\|z\|^2}$$

then

$$f(x) = \langle x_f, x \rangle$$

for all $x \in H$.

For each $f \in H^*$ there is a uniquely defined vector $x_f \in H$. The vector $x_f$ is normal to $N$. We can now describe all hyperplanes $f(x) = 1$ in terms of the unique normal vector $x_f$. The length of $x_f$ is simply a scale factor for the corresponding linear functional.

***Theorem 4.2  (Riesz Theorem)***

*Let $X = H$ be a Hilbert space. For each $f \in H^*$ there is a uniquely defined vector $x_f \in H$ with*

$$f(x) = <x_f, x>$$

*for all $x \in H$.*

□

### 4.3.2  Hyperplanes in the Dual Space

We can also define hyperplanes in the dual space. One special way to do this is to use fixed elements of the original space. Thus for any fixed $x \in X$ we can define

$$\varphi_x(f) = f(x)$$

for each $f \in X^*$. The equation

$$\varphi_x(f) = 0$$

defines a hyperplane in $X^*$. It should be noted that in general there may be hyperplanes in $X^*$ that cannot be defined in this way.

### 4.3.3  Half-Spaces

Each hyperplane through the origin divides the space $X$ into two closed half-spaces $\{x | f(x) \geq 0\}$ and $\{x | f(x) \leq 0\}$.

***Example 4.4***

Let $X = \Re^2$ and let $f(x) = x_1 + x_2$. The set $S = \{x | f(x) \geq 0\}$ is a closed half-space in $\Re^2$.

□

## 4.4 Linear Mappings

Let $X$ and $Y$ be (Banach) vector spaces. A linear mapping $f \in \mathcal{L}(X, Y)$ has the property

$$f(\alpha_1 x_1 + \alpha_2 x_2) = \alpha_1 f(x_1) + \alpha_2 f(x_2)$$

for all $\alpha_1, \alpha_2 \in \mathfrak{R}$ and all $x_1, x_2 \in X$. Differentiation, integration and matrix multiplication are linear operations that are often used to define linear mappings.

### *Example 4.5*

The linear mappings used in the train control problems are usually constructed to provide a local approximation to some non-linear mapping. Consider the differential equation (1.2) used in Chapter 1 to describe the motion of a train. The speed of the train is denoted by $v = v(t)$ and satisfies the equation

$$Dv = u - r(v) \tag{4.1}$$

with $v(0) = v(T) = 0$ and $v(t) \geq 0$ for almost all $t \in [0, T]$. We have used an *operator notation* to denote the derivative $Dv$ of the function $v$. Thus $Dv(t) = v'(t)$. The control $u(t)$ is constrained by the inequality $|u(t)| \leq 1$ for all $t \in [0, T]$. The resistive acceleration $-r(v)$ is defined by the formula $r(v) = a + bv + cv^2$ where $a, b, c > 0$ are real constants.

To define this system precisely we let $L^\infty([0, T])$ denote the set of measurable and essentially bounded functions on the interval $[0, T]$ with norm

$$\|u\|_\infty = \operatorname*{ess.\ sup}_{t \in 0, T} |u(t)|$$

and let $C^{0,1}([0, T])$ denote the set of essentially Lipschitz functions on the interval $[0, T]$ with norm

$$\|v\| = \|v\|_\infty + \|Dv\|_\infty.$$

For each $v \in \mathcal{V} = C^{0,1}([0, T])$, equation (4.1) defines a corresponding control $u \in \mathcal{U} = L^\infty([0, T])$. We consider the set $\mathcal{F}$ of all such feasible pairs $(u, v) \in \mathcal{U} \times \mathcal{V}$ where $\|u\|_\infty \leq 1$ and $v(0) = v(T) = 0$.

To show that $\mathcal{F}$ is non-empty we can see that there a unique solution $v(t) = v_1(t)$ to the differential equation

$$Dv = 1 - r(v)$$

with $v(0) = 0$ in the region $t \geq 0$ and a unique solution $v(t) = v_2(t)$ to the differential equation

$$Dv = -1 - r(v)$$

with $v(T) = 0$ in the region $t \leq T$. There is also a unique point $t = \alpha$ where $v_1(\alpha) = v_2(\alpha)$. The pair $(u, v)$ defined by

$$u(t) = \begin{cases} 1 & 0 < t < \alpha \\ -1 & \alpha < t < T \end{cases}$$

and

$$v(t) = \begin{cases} v_1(t) & 0 < t < \alpha \\ v_2(t) & \alpha < t < T \end{cases}$$

satisfies the required conditions. The graph of $v$ is shown in Figure 4-5. In the original problem, this pair represents an *accelerate–brake* strategy.

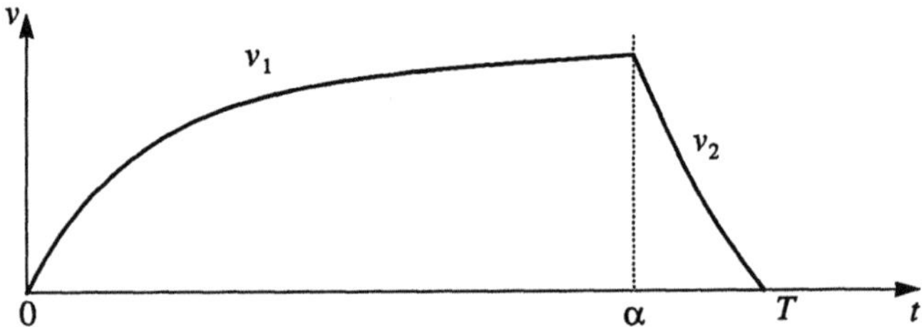

Figure 4-5: An *accelerate–brake* speed profile

We can find more feasible pairs by choosing $\beta \in [0, \alpha]$ and finding the unique solution $v(t) = v_3(t)$ to the differential equation

$$Dv = -r(v)$$

with $v(\beta) = v_1(\beta)$ in the region $t \geq \beta$. Provided $v_3(T) \geq 0$ there is a unique point $\gamma \in [\alpha, T]$ with $v_3(\gamma) = v_2(\gamma)$. The pair $(u, v)$ defined by

$$u(t) = \begin{cases} 1 & 0 < t < \beta \\ 0 & \beta < t < \gamma \\ -1 & \gamma < t < T \end{cases}$$

and

$$v(t) = \begin{cases} v_1(t) & 0 < t < \beta \\ v_3(t) & \beta < t < \gamma \\ v_2(t) & \gamma < t < T \end{cases}$$

A typical speed profile $v(t)$ of this type is shown in Figure 4-6. This pair represents a typical *accelerate–coast–brake* strategy.

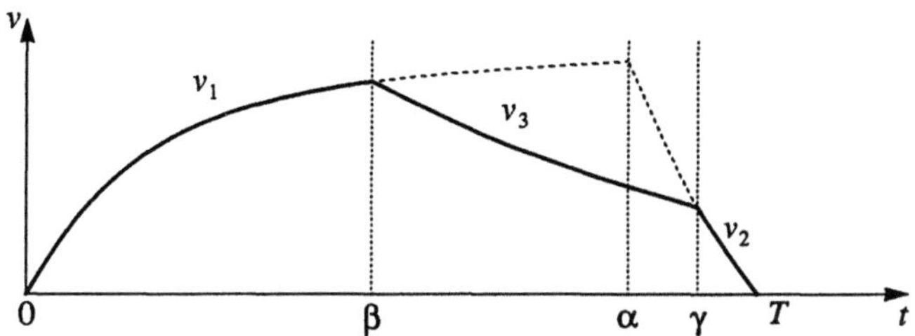


Figure 4-6: An *accelerate–coast–brake* speed profile

We define a non-linear mapping $E : \mathcal{U} \times \mathcal{V} \to \mathcal{U}$ by the formula

$$E(u, v) = Dv + R(v) - u$$

where

$$[R(v)](t) = r[v(t)]$$

for $t \in [0, T]$. The differential equation can now be written in the form

$$E(u, v) = 0.$$

We wish to replace the mapping $E(u, v)$ by a linear mapping near a point $(u_0, v_0)$. We have

$$
\nabla E\,[\,(u_0, v_0)\,]\,(\delta u, \delta v)
$$

$$
= \lim_{\alpha \to 0} \frac{1}{\alpha}\,[E(u_0 + \alpha \delta u,\, v_0 + \alpha \delta v) - E(u_0, v_0)]
$$

$$
= D(\delta v) + \lim_{\alpha \to 0} \frac{1}{\alpha}\,[R(v_0 + \alpha \delta v) - R(v_0)] - \delta u
$$

$$
= D(\delta v) + \nabla R\,[v_0]\,(\delta v) - \delta u
$$

where

$$
\{\nabla R\,[v_0]\,(\delta v)\}\,(t)
$$

$$
= \lim_{\alpha \to 0} \frac{1}{\alpha}\,\{r\,[v_0(t) + \alpha \delta v(t)] - r\,[v_0(t)]\,\}
$$

$$
= Dr\,[v_0(t)]\,\delta v(t)
$$

and hence

$$
\nabla R\,[v_0]\,(\delta v) = Dr(v_0)\delta v.
$$

Thus

$$
\nabla E\,[\,(u_0, v_0)\,]\,(\delta u, \delta v) = D(\delta v) + Dr(v_0)\delta v - \delta u.
$$

The differential equation

$$
\nabla E\,[\,(u_0, v_0)\,]\,(\delta u, \delta v) = 0
$$

is called the *variational equation* associated with the equation $E(u, v) = 0$ at $(u_0, v_0)$, and in this case can be written as

$$
D\delta v + Dr(v_0)\delta v = \delta u,
$$

or more commonly in the form

$$
\delta v'(t) + r'\,[v_0(t)]\,\delta v(t) = \delta u(t).
$$

$\square$


### Notation

In the case where $X_1, \ldots, X_n$ are (Banach) vector spaces and $X = X_1 \times \ldots \times X_n$ is the cartesian product (Banach) vector space then $X^* = X_1^* \times \ldots \times X_n^*$ is the dual space and for each $f = (f_1, \ldots, f_n) \in X^*$ and each $x = (x_1, \ldots, x_n) \in X$ we use the notation

$$f(x) = <f, x> = \sum_{j=1}^{n} <f_j, x_j> = \sum_{j=1}^{n} f_j(x_j).$$

We will write elements in cartesian product spaces as either rows or columns, whichever is most convenient.


## 4.5 The Adjoint Mapping

For each linear mapping $A \in \mathcal{L}(X, Y)$ there is an associated mapping $A^{\mathrm{T}} \in \mathcal{L}(Y^*, X^*)$ called the *adjoint mapping* defined by the equation

$$A^{\mathrm{T}}g = f$$

where $f(x) = g(Ax)$ for each $x \in X$. More information about adjoint mappings and some useful examples can be found in a number of textbooks on applied functional analysis [24, 26–29].


### *Example 4.6*

Let $X = Y = L_2([0, 1])$, where $L_2([0, 1])$ is the Hilbert space of square integrable functions on the interval $[0, 1]$. Let $A \in \mathcal{L}(X, Y)$ be defined by

$$Ax(t) = \int_0^1 [u(s - t) - s] x(s)ds$$

$$= -X(t) + \bar{X}$$

where $u(t)$ is the unit step function defined by

$$u(t) = \begin{cases} 0 & t < 0 \\ 1 & t > 0 \end{cases}$$

and where

$$X(t) = \int_0^t x(s)ds$$

and

$$\bar{X} = \int_0^1 X(t)\,dt.$$

The adjoint mapping is defined by $A^{\mathrm{T}} g = f$, where

$$
\begin{aligned}
A^{\mathrm{T}} g(x) &= g(Ax) \\
&= \ <y_g, Ax>_Y \qquad \text{for some } y_g \in Y \\
&= \int_0^1 y_g(t) \left\{ \int_0^1 [u(s-t) - s]\, x(s)\,ds \right\} dt \\
&= \int_0^1 \left\{ \int_0^1 [u(s-t) - s]\, y_g(t)\,dt \right\} x(s)\,ds \\
&= \int_0^1 x_f(s) x(s)\,ds
\end{aligned}
$$

where $x_f \in X$ is defined by

$$x_f(s) = \int_0^1 [u(s-t) - s]\, y_g(t)\,dt.$$

Thus

$$f(x) = \ <x_f, x>_X.$$

Because of the isomorphism between a Hilbert space and its dual it is usual to write

$$x_f = A^{\mathrm{T}}(y_g)$$

and to regard $A^{\mathrm{T}} \in \mathcal{L}(Y, X)$ as defined by the formula

$$<A^{\mathrm{T}} y, x>_X = \ <y, Ax>_Y.$$

Here

$$
\begin{aligned}
A^{\mathrm{T}} y(s) &= \int_0^1 [u(s-t) - s]\, y(t)\,dt \\
&= Y(s) - sY(1)
\end{aligned}
$$

where

$$Y(s) = \int_0^s y(t)\,dt.$$

$\square$

## 4.6 Convex Cones

A set $S \subseteq X$ is a *convex cone* if $\alpha S \subseteq S$ for all $\alpha \in \Re$ with $\alpha \geq 0$ and if $S + S \subseteq S$. We have used the convention that

$$\alpha S = \{\alpha x \mid x \in S\}$$

and

$$S + S = \{x_1 + x_2 \mid x_1, x_2 \in S\}.$$

The cone is *pointed* if $S \cap (-S) = \{0\}$. The intersection of half spaces in $X$ defines a *polyhedral convex cone*.

### *Example 4.7*

Let $C \in \Re^3$ be the convex cone defined by the intersection of the half-spaces $x_1 + x_3 \leq 0$ and $x_3 \geq 0$. The hyperplanes defining $C$ are shown in Figure 4-7.

$\square$

If a polyhedral cone is defined by the intersection of half-spaces and if the linear functionals defining the half-spaces span the dual space $X^*$ then the cone is pointed and the cone has a vertex at the origin. In $\Re^n$, a pointed cone is formed by the intersection of $n$ independent half-spaces.

### *Example 4.8*

Let $S \subseteq \Re^3$ be the cone defined by the intersection of the half-spaces $x_1 \geq 0$, $-x_1 + x_2 - x_3 \geq 0$ and $x_3 \geq 0$. The cone $S$ is pointed, as shown in Figure 4-8.

$\square$

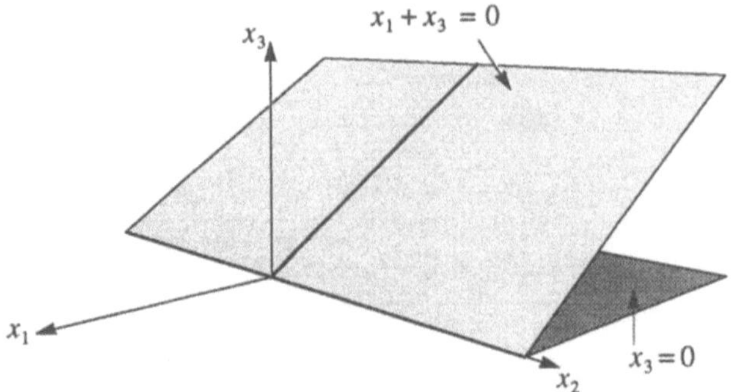


Figure 4-7: A cone defined by the intersection of two half-spaces

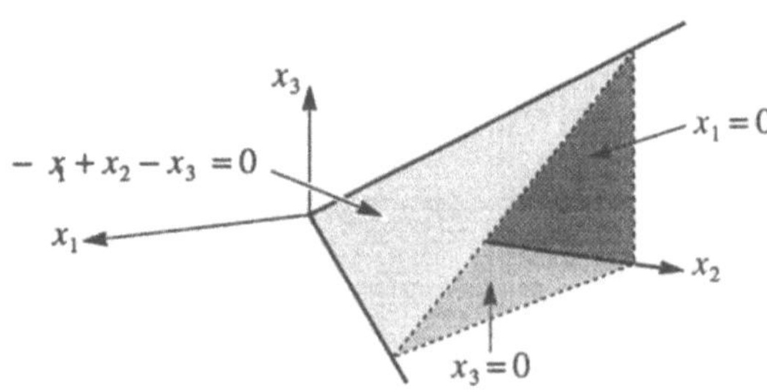


Figure 4-8: A pointed cone in $\mathfrak{R}^3$

In general the cone $S$ defined by $f_j(x) \geq 0$ for $j = 1, 2, \ldots, n$ is pointed if and only if $x = 0$ is the unique solution to the simultaneous equations $f_j(x) = 0$ for each $j = 1, 2, \ldots, n$.

### 4.6.1 Polar Cones

For each cone $S$ we define a corresponding *polar cone* $S^*$ in the dual space $X^*$ by

$$S^* = \{ f \mid f \in X^* \text{ and } f(x) \geq 0 \text{ for all } x \in S \} .$$

***Example 4.9***

In a Hilbert Space $H$ we can draw both cones $S$ and $S^*$ on the same diagram. If

$$S = \bigcap_{j=1}^{n} \{x \mid f_j(x) \geq 0\}$$

then the polar cone $S^*$ is given by

$$S^* = \{x \mid x = \sum_{j=1}^{n} \alpha_j x_j \text{ where } \alpha_1, \alpha_2, \ldots, \alpha_n \geq 0\}$$

where $x_j$ is normal to the hyperplane $P_j$ defined by $f_j(x) = 0$, and where $x_j \in H$ is chosen so that $f_j(x_j) > 0$.

In Figure 4-9, $S$ is formed by the intersection of the two half-spaces $f(x) \geq 0$ and $g(x) \geq 0$. The respective normals are denoted by $x_f$ and $x_g$.

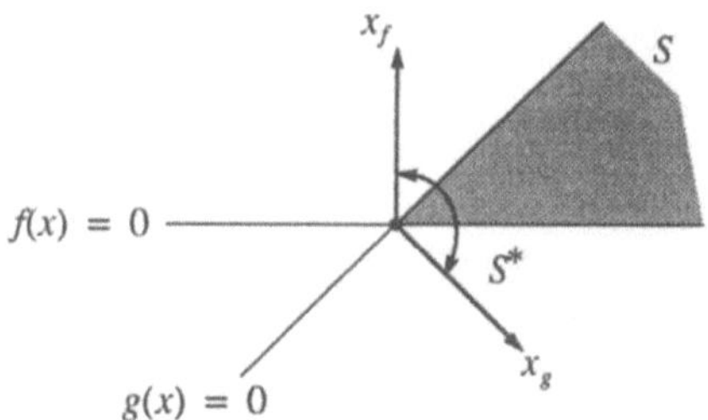


Figure 4-9: A cone $S$ and the corresponding polar cone $S^*$

$\square$

***Example 4.10***

Suppose that $P \subseteq \Re^3$ is defined by $P = \{x \mid x \in \Re^3 \text{ and } x_j \geq 0 \text{ for } j = 1, 2, 3\}$. Let $A \in L(\Re^3, \Re^2)$ be given by

$$A = (a_1, a_2, a_3)$$

where each $a_j \in \Re^2$. Let $S = A(P)$. Hence

$$S = \{y \mid y = x_1 a_1 + x_2 a_2 + x_3 a_3 \text{ for } x_1, x_2, x_3 \geq 0\}.$$

The polar cone is given by

$$S^* = \{g \mid g \in \mathfrak{R}^2 \text{ and } g(a_j) \geq 0 \text{ for } j = 1, 2, 3\} \,.$$

$\square$

If $S$ and $T$ are convex cones in $X$ then $S^* \cap T^* = (S + T)^*$.

### *Example 4.11*

Let $S$ and $T$ be given by

$$S = \{x \mid x \in \mathfrak{R}^2 \text{ with } x_1 \geq 0 \text{ and } -x_1 + 2x_2 \geq 0\}$$

and

$$T = \{x \mid x \in \mathfrak{R}^2 \text{ with } x_1 \geq 0 \text{ and } -x_1 - x_2 \geq 0\} \,.$$

The polar cones $S^*$ and $T^*$ are

$$S^* = \{x_f \mid x_f = x \in \mathfrak{R}^2 \text{ with } 2x_1 + x_2 \geq 0 \text{ and } x_2 \geq 0\}$$

and

$$T^* = \{x_f \mid x_f = x \in \mathfrak{R}^2 \text{ with } -x_1 + x_2 \geq 0 \text{ and } -x_2 \geq 0\}$$

The cones $S$, $S^*$, $T$, $T^*$, $S + T$ and $(S + T)^*$ are shown in Figure 4-10.

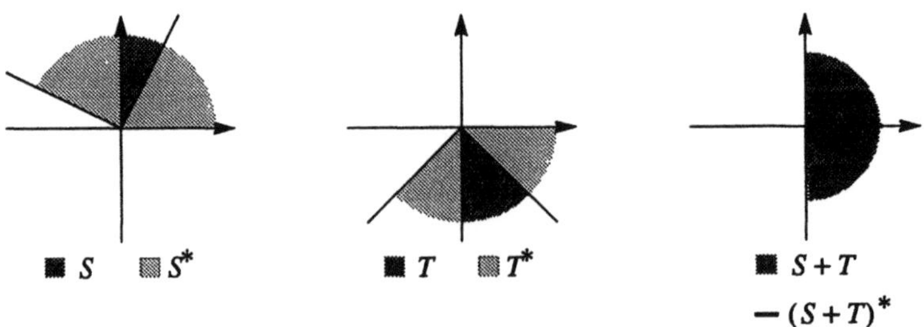


Figure 4-10: Intersection of polar cones

Note that $S^* \cap T^* = (S+T)^*$.

$\square$

## 4.6.2 Positivity

*Positivity* in general (Banach) vector spaces is defined using a designated cone $S$. Thus $x \in S \Leftrightarrow x \geq 0$ in $X$ and $f \in S^* \Leftrightarrow f \geq 0$ in $X^*$. If the cone is pointed then the inequalities $x \leq 0$ and $x \geq 0$ imply $x = 0$.

## 4.6.3 Cones in the Dual Space

In addition to the polar cones defined above, we can define cones directly in the dual space by considering the intersection of special half-spaces of the form

$$\varphi_x(f) \geq 0$$

where $\varphi_x(f) = f(x)$ for some fixed $x \in X$. In particular, let $x_1, ..., x_n \in X$ and define

$$P = \{f | f \in X^* \text{ and } \varphi_j(f) \geq 0 \text{ for each } j = 1, ...n\} \tag{4.2}$$

where $\varphi_j(f) = f(x_j)$ for each $j = 1, ..., n$. The *polar cone in X* for $P$ is denoted by $P^\#$ and is defined by

$$P^\# = \{x | x \in X \text{ and } f(x) \geq 0 \text{ for each } f \in P\}. \tag{4.3}$$

If $P$ and $Q$ are convex cones of the above form in $X^*$ then $P^\# \cap Q^\# = (P+Q)^\#$. Such cones also possess a special separation property.

## 4.6.4 The Cone Separation Theorem

*Theorem 4.3 (Cone Separation Theorem)*

*Let $P$ be defined by (4.2), and let $g \in X^*$ with $g \notin P$. There exists $x \in P^\#$ with $g(x) < 0$.*

***Proof***

Since $g \notin P$ we can find $k$ with $g(x_k) = \varphi_k(g) < 0$. But $x_k \in P^{\#}$ because for each $f \in P$ and each $j = 1, \ldots, n$ we have $f(x_j) = \varphi_j(f) \geq 0$.

$\square$

In Hilbert space the Separation Theorem can be illustrated with the diagram in Figure 4-11, where $g \notin P$ and $g(x_1) < 0$.

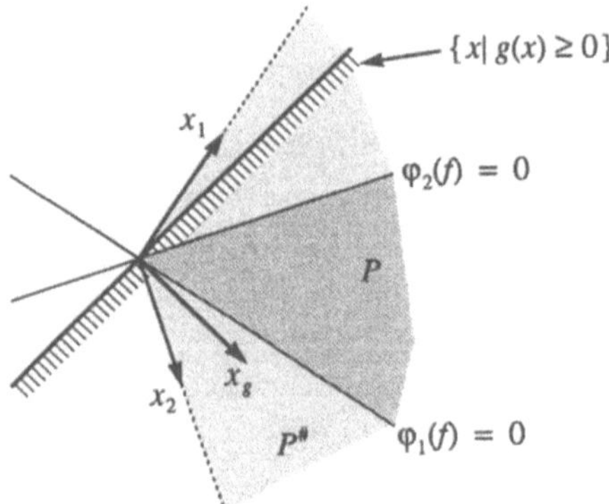


Figure 4-11: Cone Separation Theorem

### 4.6.5 The Cone Inclusion Theorem

***Theorem 4.4 (Cone Inclusion Theorem)***

*If $P$ and $Q$ are convex cones in $X^*$ with $P$ possessing the special separation property then $P^{\#} \subseteq Q^{\#} \Rightarrow P \supseteq Q$.*

***Proof***

If not, we can find $g \in Q$ and $g \notin P$. Since $g \notin P$ and $P$ has the separation property we can find $x \in P^{\#}$ with $g(x) < 0$. But $x \in P^{\#} \Rightarrow x \in Q^{\#}$ and hence $g(x) \geq 0$, which is a contradiction.

$\square$

### 4.6.6 Farkas Theorem for Convex Cones

***Theorem 4.5 (Farkas Theorem)***

*Let $X$ and $Y$ be (Banach) vector spaces and let $S \subseteq Y$ be a (closed) convex cone. Let $A \in L(X, Y)$ and $f \in X^*$. Provided the cone $A^T(S^*)$ possesses the special separation property we have the following equivalence:*

$$[Ax \in S \Rightarrow f(x) \geq 0] \iff [f \in A^T(S^*)]$$

***Outline of Proof***

The statement $Ax \in S$ is equivalent to $x \in [A^T(S^*)]^\#$ since

$$Ax \in S \iff g(Ax) \geq 0 \text{ for all } g \in S^*$$

$$\iff A^T g(x) \geq 0 \text{ for all } g \in S^*$$

$$\iff h(x) \geq 0 \text{ for all } h \in A^T(S^*)$$

$$\iff x \in [A^T(S^*)]^\#.$$

If we define

$$Q = \{q \mid q = \alpha f \text{ for some } \alpha \geq 0\}$$

then

$$Q^\# = \{x \mid x \in X \text{ and } f(x) \geq 0\}.$$

The condition $[Ax \in S \Rightarrow f(x) \geq 0]$ can now be rewritten as

$$[A^T(S^*)]^\# \subseteq Q^\#.$$

By the Cone Inclusion Theorem

$$Q \subseteq A^T(S^*).$$

Thus $f \in A^T(S^*)$. Therefore $f = A^T \lambda$ for some $\lambda \geq 0$.

$\square$

***Example 4.12***

We can illustrate Farkas Theorem in the case where $X = X^* = \Re^3$ and $Y = Y^* = \Re^2$. Let $A : \Re^3 \to \Re^2$ be defined by

$$A = \begin{bmatrix} 0 & 1 & -1 \\ 0 & 0 & 1 \end{bmatrix}$$

and let $S \subseteq \Re^2$ be the positive cone in $\Re^2$ defined by $\{y \mid y_1, y_2 \geq 0\}$. The cone

$$[A^\mathrm{T}(S^*)]^\#$$

is defined by the constraint $Ax \in S$ which implies $x_2 - x_3 \geq 0$ and $x_3 \geq 0$. If we write

$$p_1 = (0, \frac{1}{\sqrt{2}}, \frac{-1}{\sqrt{2}})$$

and

$$p_2 = (0, 0, 1)$$

then the constraint $Ax \in S$ can be written in the form $<p_1, x> \geq 0$ and $<p_2, x> \geq 0$. Note that $S^* = S$ and that the cone $A^\mathrm{T}(S^*)$ is given by

$$A^\mathrm{T}(S^*) = \{ (0, y_1, -y_1 + y_2) \mid y_1, y_2 \geq 0 \}$$
$$= \{x \mid x = y_1 \sqrt{2} p_1 + y_2 p_2 \text{ for } y_1, y_2 \geq 0 \}.$$

Both cones are shown in Figure 4-12.

For each linear functional $f \in X^*$ there is a uniquely defined vector $x_f \in X$ such that $f(x) = <x_f, x>$ for all $x \in X$. The linear functional $f$ is non-negative on the set $\{x \mid <p_1, x> \geq 0 \text{ and } <p_2, x> \geq 0\}$ if and only if $x_f = \alpha_1 p_1 + \alpha_2 p_2$ for $\alpha_1, \alpha_2 \geq 0$.

$\square$

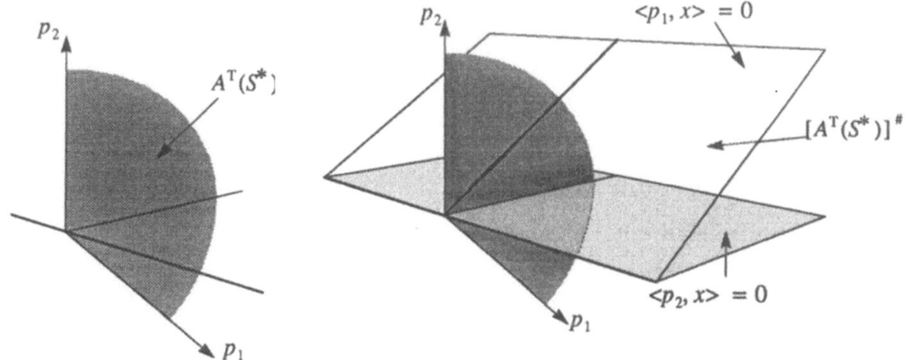


Figure 4-12: Example illustrating Farkas Theorem

## 4.7 The Optimisation Theorem

### *Theorem 4.6  (Optimisation Theorem)*

*Let $X$, $Y$ and $Z$ be (Banach) vector spaces and let $S = \{0\} \subseteq Y$ and $T \subseteq Z$ be (closed) convex cones. Let $A \in \mathcal{L}(X, Y)$, $B \in \mathcal{L}(X, Z)$ and $f \in X^*$. Provided the cone $A^{\mathrm{T}}(S^*) + B^{\mathrm{T}}(T^*)$ possesses the special separation property we have the following equivalence:*

$$[Ax \in S \text{ and } Bx \in T \Rightarrow f(x) \geq 0] \Leftrightarrow [f \in A^{\mathrm{T}}(S^*) + B^{\mathrm{T}}(T^*)]$$

### *Proof*

As above we let $Q = \{q \mid q = \alpha f \text{ for some } \alpha \geq 0\}$. We have

$$Ax \in S \Leftrightarrow x \in [A^{\mathrm{T}}(S^*)]^{\#}$$

$$Bx \in T \Leftrightarrow x \in [B^{\mathrm{T}}(T^*)]^{\#}$$

and hence

$$[A^{\mathrm{T}}(S^*)]^{\#} \cap [B^{\mathrm{T}}(T^*)]^{\#} \subseteq Q^{\#}.$$

Since

$$[A^{\mathrm{T}}(S^*)]^{\#} \cap [B^{\mathrm{T}}(T^*)]^{\#} = [A^{\mathrm{T}}(S^*) + B^{\mathrm{T}}(T^*)]^{\#}$$

the desired result follows from the Cone Inclusion Theorem. Note that

$$S = \{0\} \Rightarrow S^* = Y^*$$

and $Ax \in S$ is equivalent to $Ax = 0$.

$\square$

## 4.8 Linear Constrained Minimisation

Consider the linear constrained minimisation problem

$$\textit{Minimise } f(x) \textit{ subject to } Ax = 0 \textit{ and } Bx \leq 0$$

where $f \in X^*$, $A \in \mathcal{L}(X, Y)$ and $B \in \mathcal{L}(X, Z)$. Suppose that this problem has a solution at $x = 0$. Thus $Ax = 0$ and $Bx \leq 0$ implies $f(x) \geq 0$. We can find conditions for optimality if we let $T = \{z \mid z \leq 0\}$ and apply the Optimisation Theorem. We obtain

$$f \in A^{\mathrm{T}}(S^*) + B^{\mathrm{T}}(T^*)$$

which implies

$$f + A^{\mathrm{T}}\lambda + B^{\mathrm{T}}\mu = 0$$

where $\lambda \in S^*$ and $0 \leq \mu \in T^*$.

### Example 4.13

*Minimise* $-x_1 + 2x_2 + 4x_3$ *subject to* $x_3 = 0$, $-2x_1 - x_2 \leq 0$ *and* $2x_1 - x_2 \leq 0$.
   Let $X = \mathfrak{R}^3$, $Y = \mathfrak{R}$ and $Z = \mathfrak{R}^2$. Let

$$A = \begin{bmatrix} 0 & 0 & 1 \end{bmatrix}$$

and

$$B = \begin{bmatrix} -2 & -1 & 0 \\ 2 & -1 & 0 \end{bmatrix}$$

and define the convex cones $S \subseteq Y$ and $T \subseteq Z$ by

$$S = \{y | y \in \Re \text{ and } y = 0\}$$

and

$$T = \{z | z \in \Re^2 \text{ and } z_1, z_2 \leq 0\}.$$

We calculate

$$A^{\mathrm{T}}(S^*) = \{x | x \in \Re^3 \text{ and } x = \alpha_1 f_1\}$$

and

$$B^{\mathrm{T}}(T^*) = \{x | x \in \Re^3 \text{ and } x = \alpha_2 f_2 + \alpha_3 f_3 \text{ where } \alpha_2, \alpha_3 \leq 0\}$$

where

$$f_1 = \begin{bmatrix} 0 \\ 0 \\ 1 \end{bmatrix} \qquad f_2 = \begin{bmatrix} -2 \\ -1 \\ 0 \end{bmatrix} \qquad f_3 = \begin{bmatrix} 2 \\ -1 \\ 0 \end{bmatrix}.$$

The cost functional $f \in X^*$ given by $f(x) = -x_1 + 2x_2 + 4x_3$ can be represented in the form $f(x) = \langle x_f, x \rangle$ where

$$x_f = \begin{bmatrix} -1 \\ 2 \\ 4 \end{bmatrix}$$

and the condition

$$f \in A^{\mathrm{T}}(S^*) + B^{\mathrm{T}}(T^*)$$

can be written as

$$x_f = \begin{bmatrix} 0 \\ 0 \\ 1 \end{bmatrix} [4] + \begin{bmatrix} -2 & 2 \\ -1 & -1 \\ 0 & 0 \end{bmatrix} \begin{bmatrix} -\frac{5}{4} \\ -\frac{3}{4} \end{bmatrix}.$$

The calculations are illustrated in Figure 4-13.

$\square$

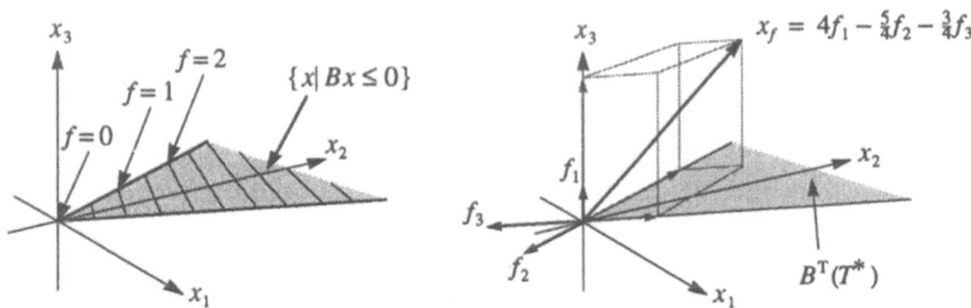



Figure 4-13: Optimality conditions for a constrained linear minimisation problem

## 4.9 The Kuhn-Tucker Conditions

### *Theorem 4.7 (Kuhn-Tucker Theorem)*

*Let $X$, $Y$ and $Z$ be (Banach) vector spaces and let $J : X \to R$, $F : X \to Y$ and $G : X \to Z$ be smooth functions. If the problem*

$$\text{Minimise } J(x) \text{ subject to } F(x) = 0 \text{ and } G(x) \le 0$$

*has a solution at $x = x_0$ then there exists $\lambda \in Y^*$ and $\mu \in Z^*$ with $\mu \ge 0$ such that*

$$\nabla J\,[x_0] + \nabla F\,[x_0]^{\mathrm{T}}\lambda + \nabla G\,[x_0]^{\mathrm{T}}\mu = 0 \qquad (4.4)$$

*and the transversality condition*

$$\mu(G(x_0)) = 0 \qquad (4.5)$$

*are satisfied.*

The conditions (4.4) and (4.5) are known as the *Kuhn-Tucker conditions*.

### *Proof*

Assume that $G(x_0) = 0$. From the Linearisation Theorem, the conditions $\nabla F\,[x_0]\,(\delta x) = 0$ and $\nabla G\,[x_0]\,(\delta x) \le 0$ imply that $\nabla J\,[x_0]\,(\delta x) \ge 0$. The result follows by applying the optimality conditions for a linear constrained minimisation problem.

$\square$

### *The Transversality Condition*

The *transversality condition*

$$\mu(G(x_0)) = 0$$

means that either $\mu = 0$ or else $G(x_0) = 0$. The case $\mu = 0$ allows $G(x_0) < 0$ and usually occurs when $x_0$ does not lie on the boundary of the region $G(x_0) \le 0$.


### *Example 4.14*

We reconsider the problem of Example 4.1:

> *Minimise* $(x_1 - 2)^2 + x_2^2 + (x_3 - 4)^2$ *subject to the equality constraint*
>
> $x_1^2 + x_2^2 = 4 - x_3$ *and the inequality constraints* $x_1 \le x_2$, $x_2 \ge 0$, $x_3 \ge 0$

Let the solution occur at $x = a$. Using the notation of Example 4.1 we find that the condition

$$\nabla J[a] + \nabla F[a]^T \lambda + \nabla B[a]^T \mu = 0$$

is given by

$$2\begin{bmatrix} a_1 - 2 \\ a_2 \\ a_3 - 4 \end{bmatrix} + \lambda \begin{bmatrix} 2a_1 \\ 2a_2 \\ 1 \end{bmatrix} + \mu_1 \begin{bmatrix} 1 \\ -1 \\ 0 \end{bmatrix} + \mu_2 \begin{bmatrix} 0 \\ -1 \\ 0 \end{bmatrix} + \mu_3 \begin{bmatrix} 0 \\ 0 \\ -1 \end{bmatrix} = 0. \tag{4.6}$$

From Figure 4-1 it is intuitively obvious that the solution lies on the surface $x_1 - x_2 = 0$. Thus we seek a solution of the form $a = (p, p, q)$ with $\mu_2 = \mu_3 = 0$. Equation (4.6) shows us that

$$\begin{aligned} (2\lambda + 2)p &= -\mu_1 + 4 \\ (2\lambda + 2)p &= \mu_1 \\ 2q &= -\lambda + 8 \end{aligned} \tag{4.7}$$

and the constraint $F(a) = 0$ gives

$$2p^2 + q - 4 = 0. \tag{4.8}$$

Solving (4.7) and (4.8) gives $a = (\frac{1}{2}, \frac{1}{2}, \frac{3}{2})$, $\lambda = 1$ and $\mu_1 = 2$.

$\square$

## 4.10  The Pontryagin Principle

Consider an optimal control problem in the following form.

***Optimal Control Problem***

*Minimise the cost functional*

$$J(u) = \int_0^T f_0(x(t), u(t))\,dt$$

*subject to the control constraint*

$$|u(t)| \le 1$$

*for all $t$ and an equation of motion*

$$x'(t) = f(x(t), u(t))$$

*with $x(0) = 0$ and additional constraints $g_1(x(T)) = 0$ and $g_2(x(T)) \le 0$ on the final
state $x(T)$.*

$\square$

In the above statement, $u \in U$ is the control variable and $x \in X$ is the state variable.
For each $t \in [0, T]$ we suppose that $u(t) \in \Re$ and $x(t) \in \Re^n$. The functions
$f_0 : X \times U \to \Re$ and $f : X \times U \to X$ are continuously differentiable. We define an
extra state variable $x_0$ by the state equation

$$x_0'(t) = f_0(x(t), u(t))$$

with $x_0(0) = 0$ so that the cost functional is given by $J(u) = x_0(T)$. We introduce the
extended state vector

$$\hat{x} = \begin{bmatrix} x_0 \\ x \end{bmatrix} \in \hat{X}$$

where $\hat{X} = \Re \times X$ is the extended state space. If we also define the extended function

$$\hat{f} = \begin{bmatrix} f_0 \\ f \end{bmatrix} : X \times U \to \hat{X}$$

then the extended state equations can be written in the form

$$\hat{x}'(t) = \hat{f}(x(t), u(t)).$$

The *Hamiltonian H* is defined by

$$H(x(t), \hat{z}(t), u(t)) = \hat{z}^{\mathrm{T}} \hat{x}'(t)$$

where $\hat{z} \in \hat{X}^*$ is the extended co-state vector. The co-state equations for $\hat{z} = \hat{z}(t)$ are given by

$$\frac{d\hat{z}}{dt} = -\frac{\partial H}{\partial \hat{x}}.$$

Since $H$ does not depend on $x_0$ these equations can be written in the form

$$\frac{dz_0}{dt} = 0$$

and

$$\frac{dz_i}{dt} = -\frac{\partial H}{\partial x_i}$$

for $i = 1, \ldots, n$. Suppose that this problem has a solution with optimal control $u^\dagger$ and corresponding optimal trajectories $x^\dagger$ and $\hat{z}^\dagger$. The *Pontryagin Principle* can now be stated as follows.


### Pontryagin Principle

*For a minimum of the Optimal Control Problem, the following conditions must hold:*

$$z_0^\dagger = -1,$$

$$H(x^\dagger, \hat{z}^\dagger, u^\dagger) = \sup_{u \in U} H(x^\dagger, \hat{z}^\dagger, u),$$

*and*

$$H(x^\dagger(t), \hat{z}^\dagger(t), u^\dagger(t)) = H_0$$

*for all $t \in [0, T]$ .*

$\square$

Proofs of the Pontryagin Principle can be found in many standard texts [18, 19, 21, 23, 24, 30]. A particular form of the Pontryagin Principle will be established and used later in the book.

# Analysis of the Mechanical Energy Model

---


## CHAPTER 5

# EXISTENCE OF AN OPTIMAL STRATEGY

---

In this chapter it will be shown that when the mechanical energy model is used to formulate the train control problem on level track in an appropriate function space, an optimal strategy does indeed exist. Furthermore, the existence of an optimal strategy will be established using a more general cost functional than that used in the original model.

This chapter was originally published in 1988 as a report to the School of Mathematics and Computer Studies at the South Australian Institute of Technology [31]. This chapter and the next two chapters were part of a systematic study to confirm the optimality of the *accelerate–hold–coast–brake* strategy for the mechanical energy model.

## 5.1 Introduction

In 1977-78 Milroy [1] considered the problem of driving a train from one station to the next along a level track in such a way that energy consumption is minimised. By applying the Pontryagin maximum principle Milroy obtained some basic strategies which he suggested as optimal strategies. Although his conjecture was supported by subsequent numerical calculations Milroy did not specify vector spaces for the various functions involved and hence was unable to provide an adequate mathematical justification for his suggested strategies. Since then, similar formulations of the train control problem have been considered by Kraft and Schnieder [32], by Asnis, Dmitruk and Osmolovskii [7] and by English, Moynihan and Boumeester [33].

## 5.2 Precise Formulation of the Train Control Problem

Let $\mathcal{B}$ denote the set of all real valued Borel measurable functions on the interval $[0, T]$. It is necessary to consider three subsets of $\mathcal{B}$ and the associated Banach spaces. $\mathcal{K}$ is the subset of all square integrable functions and if we define the scalar product

$$<k_1, k_2> = \int_0^T k_1(t) k_2(t) dt$$

and the corresponding norm

$$\|k\|_2 = <k, k>^{1/2}$$

then the associated space $K = (\mathcal{K}, <\cdot, \cdot>) = L^2([0, T])$ is a Hilbert space. $\mathcal{U}$ is the subset of all essentially bounded functions. If we define

$$\|u\|_\infty = \underset{t \in [0, T]}{\text{ess. sup}} |u(t)|$$

then $U = (\mathcal{U}, \|\cdot\|_\infty) = L^\infty([0, T])$ is a Banach space. $\mathcal{V}$ is the subset of all essentially Lipschitz functions and for these functions a suitable norm is given by

$$\|v\| = \|v\|_\infty + \|Dv\|_\infty$$

where $Dv$ denotes the derivative of $v$. In this case $V = (\mathcal{V}, \|\cdot\|) = C^{0,1}([0, T])$ is also a Banach space. For each $u \in \mathcal{U}$ it is easily seen that

$$\|u\|_2 \leq T^{1/2} \|u\|_\infty \tag{5.1}$$

and for each $v \in \mathcal{V}$ it is obvious that

$$\|v\|_\infty \leq \|v\| \tag{5.2}$$

Because of (5.1) and (5.2) it follows that $\mathcal{V} \subseteq \mathcal{U} \subseteq \mathcal{K}$. We will define a real valued cost functional $J$ on $\mathcal{U} \times \mathcal{V}$ by the formula

$$J(u, v) = \int_0^T p[u(t)] \, q[v(t)] \, dt \tag{5.3}$$

where $p : \mathfrak{R} \to \mathfrak{R}$ and $q : \mathfrak{R} \to \mathfrak{R}$ are functions with the following properties. The function $p$ is convex with $p(u) = 0$ when $u \leq 0$ and with $p(u) > 0$ when $u > 0$. It can now be shown that for each bounded interval $[-b, b]$ there is a finite positive real number $P_b$ with

$$|p(u + \delta u) - p(u)| \leq P_b |\delta u|$$

whenever $u, u + \delta u \in [-b, b]$. The function $q$ satisfies the requirements that $q(0) = 0$ and that $q(v)$ is strictly increasing when $v > 0$. We will also assume that for each bounded interval $[-b, b]$ we can find a finite positive number $Q_b$ with

$$|q(v + \delta v) - q(v)| \le Q_b |\delta v|$$

whenever $v, v + \delta v \in [-b, b]$. It can now be seen that $J(u, v)$ is convex in $u$ for each fixed $v$ and also that

$$|J(u + \delta u, v + \delta v) - J(u, v)| \le m_b \, (\|\delta u\|_2 + \|\delta v\|_2) \tag{5.4}$$

where $m_b = b P_b Q_b T^{1/2}$ whenever $\|u\|_\infty$, $\|u + \delta u\|_\infty$, $\|v\|_\infty$ and $\|v + \delta v\|_\infty$ are less than or equal to $b$. We wish to minimise $J(u, v)$ over all $u \in \mathcal{U}$ and $v \in \mathcal{V}$ subject to certain constraints. The primary constraint has the form

$$D(v) = u - R(v) \tag{5.5}$$

where $D : \mathcal{V} \to \mathcal{U}$ is the linear operator defined by

$$[D(v)] \, (t) = v'(t)$$

almost everywhere and where $R : \mathcal{V} \to \mathcal{U}$ is defined by $[R(v)] \, (t) = r \, [v(t)]$. Note that $\|D(v)\|_\infty \le \|v\|$. We will assume that the resistance function $r : \mathfrak{R} \to \mathfrak{R}$ is a function with $r(0) = 0$, $r(v)$ strictly increasing for $v \ge 0$ and with $r(v) \downarrow r_0 > 0$ when $v \downarrow 0$. Thus $r(v)$ has a jump discontinuity from the right at $v = 0$. We will also assume that there is a function $\bar{r} : \mathfrak{R} \to \mathfrak{R}$ with $\bar{r}(v) = r(v)$ for $v > 0$ and such that for each bounded interval $[-b, b]$ there is a finite positive number $R_b$ with

$$|\bar{r}(v + \delta v) - \bar{r}(v)| \le R_b |\delta v|$$

whenever $v, v + \delta v \in [-b, b]$. The remaining constraints will be written in the form $v \in C$ where $C$ is the subset of $\mathcal{V}$ specified by the conditions

$$v(0) = v(T) = 0, \tag{5.6}$$

$$v(t) \ge 0 \tag{5.7}$$

for all $t$,

$$\int_0^T v(t) dt = X \tag{5.8}$$

and finally

$$|v'(t) + r\,[v(t)]| \le 1 \tag{5.9}$$

almost everywhere.

It is now possible to specify the problem precisely.

### Problem Formulation

*Minimise the cost functional*

$$J(u, v) \;=\; \int_0^T p\,[u(t)]\, q\,[v(t)]\, dt$$

*over all* $(u, v) \in \mathcal{U} \times \mathcal{V}$ *subject to the constraints* $D(v) = u - R(v)$ *and* $v \in C$.

☐

Alternatively we could write the problem in the following way.

### Equivalent Problem Formulation

*Minimize the cost functional* $J_0(v) = J\,[D(v) + R(v), v]$ *over all* $v \in C$.

☐

Although formulation of the problem is now complete it is convenient to return to a discussion of the jump discontinuity in the resistance function. While this discontinuity has practical significance we intend to show that it is not an essential part of the solution procedure. Consider what happens when we replace the resistance function $r : \mathfrak{R} \to \mathfrak{R}$ by the modified resistance function $\bar{r} : \mathfrak{R} \to \mathfrak{R}$ described above. We define $\bar{r} : \mathcal{V} \to \mathcal{U}$ by the formula $[\bar{R}(v)]\,(t) = \bar{r}\,[v(t)]$ and note that when $v \ge 0$ the condition $\bar{r}(v) \ne r(v)$ implies $v = 0$ and hence $q(v) = 0$. Therefore when $v \in C$ we have

$$J\,[D(v) + \bar{R}(v), v]$$

$$= \int_0^T p\,\{v'(t) + \bar{r}\,[v(t)]\,\}\, q\,[v(t)]\, dt$$

$$= \int_0^T p\,\{v'(t) + r\,[v(t)]\,\}\, q\,[v(t)]\, dt$$

$$= J\,[D(v) + R(v), v]$$

and hence we can see that the cost of the journey can be calculated using $J[D(v) + \bar{R}(v), v]$ instead of $J[D(v) + R(v), v]$. For convenience we will use the modified resistance function but will revert to the original notation.

## 5.3  Sufficient Conditions for a Feasible Strategy

To establish the existence of an optimal strategy it is necessary to show that there is at least one feasible strategy. Thus we must find a condition that ensures that $C$ is non-empty. Let $V > 0$ be the unique point with $r(V) = 1$ and define $t_1 : [0, V) \rightarrow [0, \infty)$ by the formula

$$t_1(v) = \int_0^v \frac{1}{[1 - r(w)]} dw.$$

We know that there exists a constant $R_V$ such that

$$r(V) - r(w) \le R_V (V - w)$$

for all $w \in [0, V]$ and hence

$$1 - r(w) \le R_V (V - w)$$

for all $w \in [0, V]$. Therefore

$$t_1(v) \ge \frac{1}{R_v} \ln\left[\frac{V}{V - v}\right]$$

which shows that $t_1(v) \uparrow \infty$ as $t \uparrow V$. If

$$v_1 : [0, \infty) \rightarrow [0, V)$$

is the inverse function of $t_1$ then

$$v_1'(t) = 1 - r[v_1(t)]$$

almost everywhere and $v_1(t) \uparrow V$ as $t \rightarrow \infty$. We also define

$$t_{-1} : [0, V] \rightarrow [0, t_{-1}(0)]$$

by the formula

$$t_{-1}(v) = \int_{v}^{V} \frac{1}{[1 + r(w)]} dw$$

and let

$$v_{-1} : [0, t_{-1}(0)] \rightarrow [0, V]$$

be the inverse function of $t_{-1}$ in which case it can be seen that

$$v_{-1}'(t) = -1 - r[v_{-1}(t)] .$$

Because $v_{-1}'(t) \leq -1$ it follows that $v_{-1}[t - T + t_{-1}(0)] \geq T - t$. Hence $\tau \in [0, T]$ can be chosen such that $v_1(\tau) = v_{-1}[\tau - T + t_{-1}(0)]$. Now it is possible to define a composite function $v^{\dagger} : [0, T] \rightarrow [0, v_1(\tau)]$ by setting

$$v^{\dagger}(t) = \begin{cases} v_1(t) & 0 \leq t \leq \tau \\ v_{-1}[t - T + t_{-1}(0)] & \tau \leq t \leq T \end{cases}$$

and we claim that $C$ is non-empty provided

$$X^{\dagger} = \int_{0}^{T} v^{\dagger}(t) dt \geq X.$$

The graph $v = v^{\dagger}(t)$ is shown in Figure 5-1.

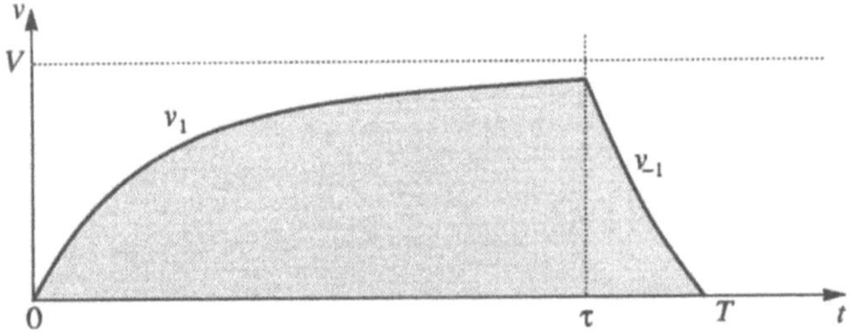


Figure 5-1: The graph $v = v^{\dagger}(t)$. The area under the curve is $X^{\dagger}$.

For each $w \in [0, v_1(\tau)]$ we can construct another composite function $(v^{\dagger} \wedge w) : [0, T] \rightarrow [0, w]$ by setting

$$(v^{\dagger} \wedge w)(t) = \min[v^{\dagger}(t), w] .$$

If $X(w)$ is defined by

$$X(w) = \int_0^T (v^\dagger \wedge w)\,(t)\,dt$$

then $X(w) = 0$, $X[v_1(\tau)] = X^\dagger \geq X$ and $X(w)$ is continuous and strictly increasing for $w \in [0, v_1(\tau)]$. Thus we can choose $w_0$ with $X(w_0) = X$ and hence $(v^\dagger \wedge w_0) \in C$. The graph $v = (v^\dagger \wedge w)\,(t)$ is shown in Figure 5-2.

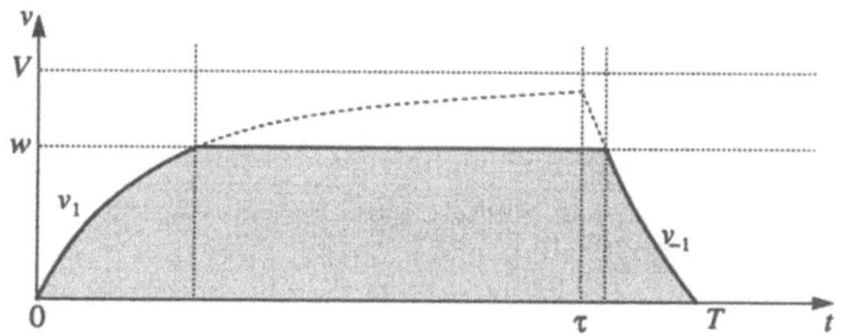


Figure 5-2: The graph $v = (v^\dagger \wedge w)\,(t)$. The area under the curve is $X(w)$.

## 5.4  Existence of an Optimal Strategy

Assume that the set $C$ defined by (5.6)–(5.9) is non-empty. When $v \in C$ it is clear that $v'(t) \leq 1 - r\,[v(t)] \leq 1$ almost everywhere and hence $0 \leq v(t) \leq T$. Thus $|v'(t)| \leq 1 + r(T)$ almost everywhere and consequently

$$\|v\| = \|v\|_\infty + \|D(v)\|_\infty \leq T + [1 + r(T)].$$

It follows that $C$ is bounded in $V$. Let $\{v_n\}$ be any sequence in $C$. Because $|v_n'(t)| \leq 1 + r(T)$ almost everywhere we can use the *Arzela-Ascoli Theorem* [25] to deduce that $\{v_n\}$ has a uniformly convergent subsequence. For convenience we will denote this subsequence by $\{v_n\}$. Thus we can find $v_0 \in V$ with $\|v_n - v_0\|_\infty \to 0$ as $n \to \infty$. Because

$$\|D(v_n)\|_2 \leq T^{1/2}\|D(v_n)\|_\infty \leq T^{1/2}[1 + r(T)]$$

the sequence $\{D(v_n)\}$ is bounded in the Hilbert space $K$ and therefore has a weakly convergent subsequence in $K$. Once again it is convenient to continue to use the notation $\{v_n\}$ for this more restrictive subsequence. Thus we have $y_0 \in K$ such that

$\{D(v_n)\}$ converges weakly to $y_0$ in $K$ and such that $\|v_n - v_0\|_\infty \to 0$ as $n \to \infty$. Let $\mathcal{F} = C_0^1[\,(0, T)\,]$ denote the set of all real valued smooth functions $\varphi$ on the real line with support $\mathrm{spt}(\varphi)$ satisfying

$$\mathrm{spt}(\varphi) \subseteq (0, T) \,.$$

That is, for each $\varphi \in \mathcal{F}$ there is a closed interval $[a, b] \subseteq (0, T)$ such that $\varphi(t) = 0$ for all $t \notin [a, b]$. It is well known that $\mathcal{F}$ is dense in $K$. In other words, for each $\varepsilon > 0$ and each $k \in K$ we can find $\varphi \in \mathcal{F}$ such that $\|k - \varphi\|_2 < \varepsilon$. We note that

$$\int_0^T [y_0(t) - v_0'(t)]\, \varphi(t)\, dt$$

$$= \lim_{n \to \infty} \int_0^T [v_n'(t) - v_0'(t)]\, \varphi(t)\, dt$$

$$= (-1) \lim_{n \to \infty} \int_0^T [v_n(t) - v_0(t)]\, \varphi'(t)\, dt$$

$$= 0$$

for all $\varphi \in \mathcal{F}$. Thus $y_0(t) = v_0'(t)$ almost everywhere and hence $\{D(v_n)\}$ converges weakly to $D(v_0)$ in $K$. It is easy to see that $v_0(0) = v_0(T) = 0$,

$$\int_0^T v_0(t)\, dt = X$$

and $v_0(t) \geq 0$. In addition, for any real valued, bounded and measurable function $\psi$ on $[0, T]$ with

$$\int_0^T |\psi(t)|\, dt \leq 1$$

we see that

$$\int_0^T \{v_n'(t) + r[v_n(t)]\}\, \psi(t)\, dt \leq 1$$

for all $n$ and by taking the limit as $n \to \infty$ we have

$$\int_0^T \{ v_0'(t) + r\,[v_0(t)] \}\, \psi(t)\,dt \le 1.$$

Here we have used the fact that $\psi \in K$ in conjunction with the weak convergence of $\{D(v_n)\}$ to $D(v_0)$ in $K$. Because this inequality holds for all such $\psi$ we must have $|v_0'(t) + r\,[v_0(t)]| \le 1$ almost everywhere. Thus $v_0 \in C$. We have in fact proved the following two lemmas.

## Lemma 5.1

*If $\{v_n\} \subseteq C$ then there exists a subsequence (again denoted by $\{v_n\}$ ) and $v_0 \in C$ such that $\|v_n - v_0\|_2 \to 0$ as $n \to \infty$.*

$\square$

## Lemma 5.2

*If $\{u_n\} \subseteq D(C)$ then there exists a subsequence (again denoted by $\{u_n\}$ ) and $y_0 \in D(C)$ such that*

$$\int_0^T [u_n(t) - y_0(t)]\, k(t)\,dt \to 0$$

*as $n \to \infty$ for all $k \in K$. Furthermore, if $u_n = D(v_n)$ where $\{v_n\} \subseteq C$ and $v_0$ is defined by Lemma 5.1 then $y_0 = D(v_0)$.*

$\square$

Now we define

$$J_0(v) = J\,[D(v) + R(v),\, v]$$

for all $v \in \mathcal{V}$. Clearly, $J_0(v) \ge 0$ when $v \in C$. Choose $\{v_n\} \in C$ such that

$$J_0(v_n) \to J_0 = \inf_{v \in C} J_0(v)$$

as $n \to \infty$. Using Lemmas 5.1 and 5.2 we can assume that the sequence has been chosen in such a way that for some $v_0 \in C$ we have $\|v_n - v_0\|_2 \to 0$ and

$$\int_0^T [v_n{}'(t) - v_0{}'(t)]\, k(t)\, dt \to 0$$

for all $k \in K$ as $n \to \infty$. Because $\{D(v_n)\}$ converges weakly to $D(v_n)$ in $K$ we can use *Mazur's Theorem* [24, 25] to choose a strictly increasing sequence $\{n_j\}$ of positive integers and real numbers $\gamma_n$ with

$$\gamma_n \geq 0 \quad \text{and} \quad \sum_{n = n_j + 1}^{n_{j+1}} \gamma_n = 1 \tag{5.10}$$

and

$$w_j = \sum_{n = n_j + 1}^{n_{j+1}} \gamma_n v_n \tag{5.11}$$

and such that $\|D(w_j) - D(v_0)\|_2 \to 0$ as $j \to \infty$. If we define

$$\varepsilon_j = \sup_{m,\, n\, >\, n_j} \|v_m - v_n\|_2$$

then $\varepsilon_j \to 0$ as $j \to \infty$. We can also see that for $m > n_j$ we have

$$\|w_j - v_m\|_2 = \left\| \sum_{n = n_j + 1}^{n_{j+1}} \gamma_n (v_n - v_m) \right\|_2 \leq \varepsilon_j.$$

Now from the convexity of $J$ we know that

$$\begin{aligned}
J_0(w_j) &= J[D(w_j) + R(w_j),\, w_j] \\
&= J\left[ \sum_{n = n_j + 1}^{n_{j+1}} \gamma_n \{D(v_n) + R(w_j)\},\, w_j \right] \\
&\leq \sum_{n = n_j + 1}^{n_{j+1}} \gamma_n J[D(v_n) + R(w_j),\, w_j].
\end{aligned}$$

We can find a positive real number $b$ for which $\|D(v_n) + R(w_j)\|_\infty$, $\|D(v_n) + R(v_n)\|_\infty$, $\|R(w_j)\|_\infty$ and $\|v_n\|_\infty$ are less than or equal to $b$. Hence for each $n > n_j$ we can use (5.4) to see that

$$|J [D(v_n) + R(w_j), w_j] - J_o(v_n)|$$
$$= |J [D(v_n) + R(w_j), w_j] - J [D(v_n) + R(v_n), v_n]|$$
$$\leq m_b (\|R(w_j) - R(v_n)\|_2 + \|w_j - v_n\|_2)$$
$$\leq m_b (R_b + 1) \varepsilon_j.$$

Since $J_0(v_n) \to J_0$ as $n \to \infty$ we deduce that

$$\limsup_{j \to \infty} J_0(w_j)$$
$$\leq \limsup_{j \to \infty} \sum_{n = n_j + 1}^{n_{j+1}} \gamma_n J [D(v_n) + R(w_j), w_j]$$
$$\leq \limsup_{j \to \infty} \sum_{n = n_j + 1}^{n_{j+1}} \gamma_n \{|J [D(v_n) + R(w_j), w_j] - J_0(v_n)| + J_o(v_n)\}$$
$$\leq \limsup_{j \to \infty} \sum_{n = n_j + 1}^{n_{j+1}} \gamma_n \{m_b (R_b + 1) \varepsilon_j + J_0(v_n)\}$$
$$\leq \limsup_{j \to \infty} \left\{ m_b (R_b + 1) \varepsilon_j + \sum_{n = n_j + 1}^{n_{j+1}} \gamma_n J_0(v_n) \right\}$$
$$= J_0.$$

On the other hand

$$|J_0(w_j) - J_0(v_0)|$$
$$= |J [D(w_j) + R(w_j), w_j] - J [D(v_0) + R(v_0), v_0]|$$
$$= |J [D(w_j) + R(w_j), w_j] - J [D(v_0) + R(w_j), w_j]|$$
$$\quad + |J [D(v_0) + R(w_j), w_j] - J [D(v_0) + R(v_0), v_0]|$$
$$\leq (m_b \|D(w_j) - D(v_0)\|_2 + m_b (R_b + 1) \|w_j - v_0\|_2)$$
$$\to 0$$

as $j \to \infty$. If follows that $J_0 = J_0(v_0)$ and hence the infimum is achieved for some $v_0 \in C$.

## 5.5 Conclusions

We have shown that when the train control problem is formulated in an appropriate way we can guarantee the existence of an optimal strategy. In the next two chapters we will show that the Pontryagin principle can be applied to find the necessary conditions on the optimal strategy, and that these conditions can then be used to find the precise nature of the strategy.

## 5.6 Appendix—Convergence in Banach Spaces

Let $X$ be a Banach space and let $X^*$ be the dual space. If we define a norm on $X^*$ by the formula

$$\|f\| = \sup_{x \in X, \, \|x\| \leq 1} |f(x)|$$

for each $f \in X^*$ then $X^*$ is also a Banach space. There are two commonly used forms of convergence in Banach space.

***Strong Convergence***

We say that the sequence $\{x_n\}_{n=1, 2, \dots} \subseteq X$ converges *strongly* to $x \in X$ if $\|x_n - x\| \to 0$ as $n \to \infty$. A necessary and sufficient condition for strong convergence is that $\{x_n\}$ be a Cauchy sequence in $X$. Thus the sequence $\{x_n\}$ converges strongly if and only if for each $\varepsilon > 0$ we can find $N = N(\varepsilon)$ such that

$$\|x_m - x_n\| < \varepsilon$$

for all $m, n > N$.

***Example 5.1***

Let $\{v_n\}_{n=1, 2, \dots}$ be a sequence of functions $v_n : 0, T \to \Re$ such that there exists $C, K \in \Re$ such that $|v_n(t)| \leq C$ and $|v_n(s) - v_n(t)| \leq K|s - t|$ for all $s, t \in [0, T]$ and all $n = 1, 2, \dots$. If we let $U = L^\infty([0, T])$ denote the Banach space of all Borel measurable functions on $[0, T]$ with the uniform norm

$$\|f\| = \underset{t \in [0, T]}{\text{ess. sup}} |f(t)|$$

then $v_n \in U$ and $\|v_n\| \le C$ for all $n = 1, 2, \ldots$. We will show that there exists $v \in U$ and a subsequence $\{v_{n(j)}\}_{j=1, 2, \ldots}$ such that $\|v_{n(j)} - v\| \to 0$ as $j \to \infty$. Let $\{r_k\}_{k=1, 2, \ldots}$ denote the set of all rational numbers in $[0, T]$. The set

$$\{v_n(r_1)\}_{n=1, 2, \ldots} \subseteq [-C, C]$$

is a bounded sequence and hence contains a convergent subsequence

$$\{v_{n(1, j)}(r_1)\}_{j=1, 2, \ldots},$$

where $n(1, 1) < n(1, 2) < \ldots$. The set

$$\{v_{n(1, j)}(r_2)\}_{j=1, 2, \ldots} \subseteq [-C, C]$$

is a also a bounded sequence and hence contains a convergent subsequence

$$\{v_{n(2, j)}(r_2)\}_{j=1, 2, \ldots}$$

where $n(2, 1) = n(1, 1)$ and $n(2, 1) < n(2, 2) < \ldots$. Note that $\{v_{n(2, j)}(t)\}$ converges for all $t \in \{r_1, r_2\}$. By continuing in this way we can find a subsequence

$$\{v_{n(j)}\}_{j=1, 2, \ldots} \subseteq U$$

with $n(j) = n(j, j)$ for all $j = 1, 2, \ldots$ and such that $\{v_{n(j)}(t)\}$ converges for all $t \in \{r_k\}_{k=1, 2, \ldots}$.

We can now show that there exists $v \in U$ such that the subsequence $\{v_{n(j)}\}$ converges strongly to $v$. Let $\varepsilon > 0$. Choose $M \in Z_+$ sufficiently large to ensure that for each $t \in [0, T]$ we can find $k \in \{1, 2, \ldots, M\}$ with $|t - r_k| < \varepsilon/3K$. Now choose $N \in Z_+$ such that

$$\left| v_{n(i)}(r_k) - v_{n(j)}(r_k) \right| < \frac{\varepsilon}{3}$$

for each $i, j > N$ and each $k \in \{1, 2, \ldots, M\}$. It follows that

$$\left| v_{n(i)}(t) - v_{n(j)}(t) \right|$$
$$\le \left| v_{n(i)}(t) - v_{n(i)}(r_k) \right| + \left| v_{n(i)}(r_k) - v_{n(j)}(r_k) \right| + \left| v_{n(j)}(r_k) - v_{n(j)}(t) \right|$$
$$\le K|t - r_k| + \frac{\varepsilon}{3} + K|t - r_k|$$
$$\le \varepsilon$$

for each $i, j > N$. Thus $\|v_{n(i)} - v_{n(j)}\| < \varepsilon$ for all $i, j > N$. This proves that $\{v_{n(j)}\}$ is a Cauchy sequence in $U$.

$\square$

### Weak Convergence

We say that the sequence $\{x_n\}_{n = 1, 2, \ldots} \subseteq X$ converges *weakly* to $x \in X$ if $|f(x_n) - f(x)| \to 0$ as $n \to \infty$ for each $f \in X^*$.

### Example 5.2

Let $H = L_2([0, 1])$ be the Hilbert space of square integrable functions $f : [0, 1] \to \Re$ with scalar product

$$<f, g> = \int_0^1 f(t) g(t) dt$$

and norm

$$\|f\| = <f, f>^{1/2}$$

We will show that the sequence $\{\sin n\pi t\}_{n = 1, 2, \ldots}$ converges weakly to $0$ in $H$. By Theorem 4.2, each linear functional on $H$ can be represented in the form of a scalar product. Thus each $\varphi \in H^*$ can be represented in the form $\varphi = \varphi_g$ where $g \in H$ and $\varphi_g(f) = <g, f>$ for all $f \in H$. We show that

$$\varphi_g(\sin n\pi t) = \int_0^1 g(t) \sin n\pi t \, dt$$
$$\to 0$$

as $n \to \infty$ for each $g \in H$. For each $\varepsilon > 0$ we can find a step function

$$s(t) = \sum_{j=0}^{N-1} s_j \left[ u(t - t_{j+1}) - u(t - t_j) \right]$$

such that $\|s - g\| < \varepsilon$. Note that $u : \Re \to \Re$ is the unit step function defined by

$$u(t) = \begin{cases} 0 & t < 0 \\ 1 & t > 0 \end{cases}.$$

Since

$$\left| \varphi_g(\sin n\pi t) - \int_0^1 s(t) \sin n\pi t\, dt \right| < \varepsilon$$

and since

$$\int_0^1 s(t) \sin n\pi t\, dt = \sum_{j=0}^{N-1} s_j \int_{t_j}^{t_{j+1}} \sin n\pi t\, dt$$

$$= \sum_{j=0}^{N-1} s_j \frac{\cos n\pi t_j - \cos n\pi t_{j+1}}{n\pi}$$

$$< \frac{2}{n\pi} \sum_{j=0}^{N-1} s_j$$

$$\rightarrow 0$$

as $n \rightarrow \infty$ it follows that $\varphi_g(\sin n\pi t) \rightarrow 0$ as $n \rightarrow \infty$.

$\square$

## CHAPTER 6

# NECESSARY CONDITIONS FOR AN OPTIMAL STRATEGY

With an appropriate formulation of the train control problem, we have already shown that an optimal driving strategy exists. In this chapter, necessary conditions of the Fritz-John type will be obtained for the optimal strategy, and these conditions will be used to find a Hamiltonian function and to demonstrate the validity of the Pontryagin Principle for this problem. This chapter was originally published in 1988 as a report to the School of Mathematics and Computer Studies at the South Australian Institute of Technology [34]. The methods used are an extension of the methods developed by Craven [18].

## 6.1 An Equivalent Formulation of the Train Control Problem

We use the same notation as we used in Chapter 5. In particular we consider the two basic subsets $\mathcal{U}$ and $\mathcal{V}$ of the set $\mathcal{B}$ and the associated Banach spaces $U$ and $V$. It is convenient to introduce the subset $\mathcal{V}_0$ defined by

$$\mathcal{V}_0 = \{ v | v \in \mathcal{V} \text{ with } v(0) = v(T) = 0 \}$$

and the subspace $V_0 = (\mathcal{V}_0, \| \cdot \|)$ which is again a Banach space. We consider the real valued cost functional $J$ on $\mathcal{U} \times \mathcal{V}$ defined in (5.3) by the formula

$$J(u, v) = \int_0^T p\,[u(t)]\, q\,[v(t)]\, dt$$

where $p : \mathfrak{R} \to \mathfrak{R}$ and $q : \mathfrak{R} \to \mathfrak{R}$ are functions with the properties described previously. It can now be seen that $J(u, v)$ is convex in $u$ for each fixed $v$ and also that for each $(u, v) \in U \times V$ the (partial) *Frechet derivative* $J_v\,[\,(u, v)\,] \in V^*$ given by

$$J_v\,[\,(u, v)\,]\,(\delta v) = \lim_{\alpha \to 0} \frac{1}{\alpha}\,[J(u, v + \alpha \delta v) - J(u, v)]$$

is well defined with

$$J_v \left[\, (u, v)\, \right] (\delta v) \;=\; \int_0^l p\, [u(t)]\, q'\, [v(t)]\, \delta v(t)\, dt \;.$$

(6.1)

We wish to minimize $J(u, v)$ over all $u \in \mathcal{U}$ and $v \in \mathcal{V}_0$ subject to the equality constraints (5.5) and (5.8). If we define $\mathcal{W} = \mathfrak{R} \times \mathcal{U}$ and the operator $E: \mathcal{U} \times \mathcal{V} \to \mathcal{W}$ by the formula

$$E(u, v) \;=\; \left[ \int_0^T v(t)\, dt - X,\; D(v) + R(v) - u \right]$$

then the equation

$$E(u, v) \;=\; 0$$

incorporates both equality constraints. There are also the two inequality constraints (5.7) and (5.9). The latter will now be written in the form

$$|u(t)| \le 1$$

for almost all $t$. We define a positive cone $M \subseteq \mathcal{U}$ by setting

$$M \;=\; \{\, u\,|\, u \in \mathcal{U} \text{ and } u(t) \ge 0 \text{ almost everywhere} \,\} \;.$$

If we define $F: \mathcal{U} \to \mathcal{U}$ and $G: \mathcal{V} \to \mathcal{U}$ by the formulae

$$\begin{aligned}
[F(u)]\,(t) \;&=\; f\,[u(t)] \\
&=\; [u(t)]^2 - 1
\end{aligned}$$

(6.2)

and

$$\begin{aligned}
[G(v)]\,(t) \;&=\; g\,[v(t)] \\
&=\; -v(t)
\end{aligned}$$

(6.3)

where $f: \mathfrak{R} \to \mathfrak{R}$ and $g: \mathfrak{R} \to \mathfrak{R}$ are defined by

$$f(u) \;=\; u^2 - 1$$

(6.4)

and

$$g(v) \;=\; -v$$

(6.5)

then these constraints can be written as $F(u) \leq 0$ and $G(v) \leq 0$ respectively. The problem formulation in Chapter 5 can now be rewritten as follows.

### Problem (P)

*Minimise the cost functional*

$$J(u, v) = \int_0^T p\,[u(t)]\, q\,[v(t)]\, dt$$

*over all* $(u, v) \in \mathcal{U} \times \mathcal{V}_0$ *subject to the equality constraint* $E(u, v) = 0$ *and the inequality constraints* $F(u) \leq 0$ *and* $G(v) \leq 0$.

$\square$

We can define a norm on $\mathcal{W} = \mathfrak{R} \times \mathcal{U}$ by the formula

$$\| (s, u) \|_W = |s| + \| u \|_\infty$$

in which case $W = (\mathcal{W}, \| \cdot \|_W)$ is a Banach space. Because the modified resistance function $r(v)$ is continuously differentiable, it is clear that for each $(u, v) \in U \times V$ the (partial) Frechet derivatives $E_u\,[\,(u, v)\,]$ and $E_v\,[\,(u, v)\,]$ given by

$$E_u\,[\,(u, v)\,]\,(\delta u) = \lim_{\alpha \to 0} \frac{1}{\alpha}\,[E(u + \alpha\delta u, v) - E(u, v)]$$

and

$$E_v\,[\,(u, v)\,]\,(\delta v) = \lim_{\beta \to 0} \frac{1}{\beta}\,[E(u, v + \beta\delta v) - E(u, v)]$$

are well defined with

$$E_u\,[\,(u, v)\,]\,(\delta u) = (0, -\delta u) \tag{6.6}$$

and

$$E_v\,[\,(u, v)\,]\,(\delta v) = \left[ \int_0^T \delta v(t)dt\,,\ D(\delta v) + Dr(v)\delta v \right]. \tag{6.7}$$

## 6.2 Necessary Conditions for an Optimal Strategy

This section follows closely the methods of Craven [18]. The main result is a modified statement of the Pontryagin Principle. Details of the modification process are non-trivial, and only the basic alternative theorem can be regarded as entirely standard. For convenience, a brief outline of the proof of this theorem has been included. A more comprehensive account can be found in the book by Craven.

In the following discussion it will often be necessary to use the notion of a cartesian product. If $X = (\mathcal{X}, \|\cdot\|_X)$ and $Y = (\mathcal{Y}, \|\cdot\|_Y)$ are Banach spaces then for each $(x, y) \in \mathcal{X} \times \mathcal{Y}$ we will define the norm $\|(x, y)\|_{X \times Y} = \|x\|_X + \|y\|_Y$ and note that the space $X \times Y = (\mathcal{X} \times \mathcal{Y}, \|\cdot\|_{X \times Y})$ is also a Banach space. We will consider necessary conditions for the solution of Problem (P).

Throughout this section we will assume that the point $(u_0, v_0) \in \mathcal{U} \times \mathcal{V}_0$ is fixed and we define a set $Q = Q(u_0, v_0) \subseteq \mathcal{W}$ by

$$Q = \{ (\delta r, \delta u) \mid \alpha \delta r \geq J(u_0 + \alpha \delta u, v_0) - J(u_0, v_0) \text{ for some } \alpha > 0 \}$$

### Lemma 6.1

*The set $Q$ defined above is a convex cone.*

### Proof

Let $(\delta r, \delta u) \in Q$. Thus for some $\alpha > 0$ we have

$$\alpha \delta r \geq J(u_0 + \alpha \delta u, v_0) - J(u_0, v_0).$$

For all $\beta > 0$ it follows that

$$(\frac{\alpha}{\beta}) \beta \delta r \geq J\left[ u_0 + (\frac{\alpha}{\beta}) \beta \delta u, v_0 \right] - J(u_0, v_0)$$

and hence $\beta (\delta r, \delta u) \in Q$. On the other hand if $(\delta r_i, \delta u_i) \in Q$ for $i = 1, 2$ we can find $\alpha_i > 0$ such that

$$\alpha_i \delta r_i \geq J(u_0 + \alpha_i \delta u_i, v_0) - J(u_0, v_0)$$

and hence

$$\frac{\alpha_1 \alpha_2}{\alpha_1 + \alpha_2} [\delta r_1 + \delta r_2]$$

$$= \frac{\alpha_2}{\alpha_1 + \alpha_2} \alpha_1 \delta r_1 + \frac{\alpha_1}{\alpha_1 + \alpha_2} \alpha_2 \delta r_2$$

$$\geq \frac{\alpha_2}{\alpha_1 + \alpha_2} [J(u_0 + \alpha_1 \delta u_1, v_0) - J(u_0, v_0)]$$

$$+ \frac{\alpha_1}{\alpha_1 + \alpha_2} [J(u_0 + \alpha_2 \delta u_2, v_0) - J(u_0, v_0)]$$

$$\geq J\left(u_0 + \frac{\alpha_1 \alpha_2}{\alpha_1 + \alpha_2} [\delta u_1 + \delta u_2], v_0\right) - J(u_0, v_0)$$

and therefore $(\delta r_1 + \delta r_2, \delta u_1 + \delta u_2) \in Q$.

$\square$

We will write $(\delta r, \delta u) \geq 0$ when $(\delta r, \delta u) \in Q$ and $(\delta r, \delta u) > 0$ when $(\delta r, \delta u) \in \text{int } Q$. The interior of $Q$, denoted by $\text{int } Q$, is clearly non-empty, and in fact $(\delta r, \delta u) > 0$ is equivalent to $\alpha \delta r > J(u_0 + \alpha \delta u, v_0) - J(u_0, v_0)$ for some $\alpha > 0$.

We will now show that the equation $E(u, v) = 0$ is *locally solvable* at the point $(u_0, v_0)$. That is, we must show that we can find $\delta > 0$ such that whenever the direction $(\delta u, \delta v)$ satisfies

$$E_u [\, (u_0, v_0) \,] (\delta u) + E_v [\, (u_0, v_0) \,] (\delta v) = 0$$

and $\| (\delta u, \delta v) \|_{U \times V} < \delta$ there exists a solution

$$(u, v) = (u_0, v_0) + \alpha (\delta u, \delta v) + (\rho(\alpha), \sigma(\alpha))$$

to the equation $E(u, v) = 0$, valid for sufficiently small $\alpha > 0$, where

$$\frac{1}{\alpha} \| (\rho [\alpha], \sigma [\alpha]) \|_{U \times V} \to 0$$

as $\alpha \downarrow 0$. By the Implicit Function Theorem [18] it is sufficient to show that the Frechet derivative

$$\nabla E [\, (u, v) \,] = (E_u [u, v], E_v [u, v])$$

is continuous in $(u, v)$ and that the linear mapping $\nabla E [\, (u_0, v_0) \,]$ maps $U \times V_0$ onto $W$. Thus we must show that for each $(\delta r, \delta u) \in W$ we can find $(\delta u_0, \delta v_0) \in U \times V_0$ with

$$\nabla E\,[\,(u_0, v_0)\,]\,(\delta u_0, \delta v_0) \;=\; (\delta r, \delta u)\,.$$

Thus we need to find $(\delta u_0, \delta v_0)$ satisfying the equation

$$\delta v_0{}'(t) + r'\,[v_0(t)]\,\delta v_0(t) - \delta u_0(t) \;=\; \delta u(t)$$

and such that

$$\int_0^T \delta v_0(t)\,dt \;=\; \delta r.$$

If we introduce the integrating factor

$$K(t) \;=\; \exp\left\{\int_0^t r'\,[v_0(s)]\,ds\right\}$$

then the required solution is given by

$$(\delta u_0(t),\, \delta v_0(t)) \;=\; (-\,\delta u(t) + \frac{k'(t)}{K(t)}\,,\ \frac{k(t)}{K(t)})$$

where the function $k(t)$ is defined by

$$k(t) \;=\; \frac{3\delta r}{2T^3}\,[T^2 - (T - 2t)^2]\,K(t).$$

Hence the mapping $\nabla E\,[\,(u_0, v_0)\,]$ is onto and the equation $E(u, v) = 0$ is locally solvable at $(u_0, v_0)\,.$

We will now associate the Problem (P) with the following linearised problem, which will subsequently be referred to as Problem (L).


### Linearised Problem (L)

*Find*

$$(\beta,\, \gamma,\, \delta u,\, \delta v) \in \Re \times \Re \times U \times V_0$$

*such that*

$$\nabla E\,[\,(u_0, v_0)\,]\,(\delta u, \delta v) \;=\; 0,$$

$$(-J_v [u_0, v_0] (\delta v), \delta u) > 0,$$

$$(-1) [\beta F(u_0) + \nabla F [u_0] (\delta u)] > 0,$$

*and*

$$(-1) [\gamma G(v_0) + \nabla G [v_0] (\delta v)] > 0.$$

$\square$

If we define linear mappings

$$A : \Re \times \Re \times U \times V_0 \to W$$

and

$$B : \Re \times \Re \times U \times V_0 \to W \times U \times U$$

by setting

$$A(\beta, \gamma, \delta u, \delta v) = \nabla E [ (u_0, v_0) ] (\delta u, \delta v)$$
$$= E_u [ (u_0, v_0) ] (\delta u) + E_v [ (u_0, v_0) ] (\delta v)$$

and

$$B(\beta, \gamma, \delta u, \delta v) = \begin{pmatrix} -J_v [ (u_0, v_0) ] (\delta v) \\ \delta u \\ -[\beta F(u_0) + \nabla F [u_0] (\delta u)] \\ -[\gamma G(v_0) + \nabla G [v_0] (\delta v)] \end{pmatrix}$$

then we can write Problem (L) in a more convenient form.

**Equivalent Linearised Problem (L)**

*Find*

$$(\beta, \gamma, \delta u, \delta v) \in \Re \times \Re \times U \times V_0$$

*such that*

$$A(\beta, \gamma, \delta u, \delta v) = 0$$

*and*

$$B(\beta, \gamma, \delta u, \delta v) > 0.$$

$\square$

From the above definition we see that the map $A$ can be represented in the form

$$A = (0, 0, E_u [\, (u_0, v_0)\,], E_v [\, (u_0, v_0)\,])$$
$$\in \mathcal{L}(\Re \times \Re \times U \times V_0, W)$$

Similarly, the map $B$ can be represented by an element

$$B = \begin{bmatrix} (0, 0, 0, -J_v [\, (u_0, v_0)\,]) \\ (0, 0, I, 0) \\ (-F(u_0), 0, -\nabla F [u_0], 0) \\ (0, -G(v_0), 0, -\nabla G [v_0]) \end{bmatrix}$$
$$\in \mathcal{L}(\Re \times \Re \times U \times V_0, \Re) \times \mathcal{L}(\Re \times \Re \times U \times V_0, U)$$
$$\times \mathcal{L}(\Re \times \Re \times U \times V_0, U) \times \mathcal{L}(\Re \times \Re \times U \times V_0, U).$$

The following theorem modifies a result in [18].

### Theorem 6.2  (Linearisation Theorem)

*If $(u_0, v_0)$ is a local minimum for Problem (P) then Problem (L) has no solution.*

### Proof

Suppose that Problem (L) has a solution $(\beta, \gamma, \delta u, \delta v)$. For $\kappa > 0$ and sufficiently large

$$(\delta u_1, \delta v_1) = \frac{(\delta u, \delta v)}{\kappa}$$

satisfies $\| (\delta u_1, \delta v_1) \|_{U \times V} < \delta$ and since $\nabla E [\, (u_0, v_0)\,] (\delta u_1, \delta v_1) = 0$ it follows that $E(u, v) = 0$ has a solution

$$(u, v) = (u_0, v_0) + \alpha (\delta u_1, \delta v_1) + (\rho(\alpha), \sigma(\alpha))$$

for sufficiently small $\alpha > 0$. Then

$$J(u, v) - J(u_0, v_0)$$
$$= \{J(u_0 + \alpha\delta u_1 + \rho(\alpha), v_0 + \alpha\delta v_1 + \sigma(\alpha)) - J(u_0 + \alpha\delta u_1 + \rho(\alpha), v_0)\}$$
$$+ \{J(u_0 + \alpha\delta u_1 + \rho(\alpha), v_0) - J(u_0, v_0)\}$$
$$= J_v[(u_0 + \alpha\delta u_1 + \rho(\alpha), v_0)](\alpha\delta v_1 + \sigma(\alpha)) + \theta_1(\alpha)$$
$$+ \{J(u_0 + \alpha\delta u_1 + \rho(\alpha), v_0) - J(u_0, v_0)\}$$

where

$$\frac{\theta_1(\alpha)}{\alpha} \to 0$$

as $\alpha \downarrow 0$. Thus

$$J(u, v) - J(u_0, v_0)$$
$$= J_v[(u_0, v_0)](\alpha\delta v_1) + \{J(u_0 + \alpha\delta u_1, v_0) - J(u_0, v_0)\} + \theta_2(\alpha)$$

where

$$\frac{\theta_2(\alpha)}{\alpha} \to 0$$

as $\alpha \downarrow 0$. We have used the continuity of $J_v[(u, v)]$ in $(u, v)$ and the fact that convexity of $J(u, v)$ in $u$ implies that $J(u, v)$ is locally Lipschitz in $u$. Now since

$$(-J_v[(u_0, v_0)](\delta v_1), \delta u_1) > 0$$

we must have some $\alpha_0 > 0$ and some $\delta_0 > 0$ with

$$-J_v[(u_0, v_0)](\alpha_0 \delta v_1) = [J(u_0 + \alpha_0\delta u_1, v_0) - J(u_0, v_0)] + \delta_0.$$

For all $\alpha$ with $0 < \alpha \leq \alpha_0$ it follows that

$$-J_v[(u_0, v_0)](\alpha\delta v_1) \geq [J(u_0 + \alpha\delta u_1, v_0) - J(u_0, v_0)] + \frac{\delta_0}{\alpha_0}\alpha$$

and hence

$$J(u, v) - J(u_0, v_0) \leq -\frac{\delta_0}{\alpha_0}\alpha + \theta_2(\alpha)$$

Thus the left hand side will be negative for $\alpha > 0$ and sufficiently small. On the other hand

$$F(u_0 + \alpha \delta u_1 + \rho(\alpha))$$

$$= F(u_0) + \nabla F[u_0](\alpha \delta u_1 + \rho(\alpha)) + \theta_3(\alpha)$$

$$= (1 - \frac{\alpha \beta}{\kappa}) F(u_0) + \frac{\alpha}{\kappa} [\beta F(u_0) + \nabla F[u_0](\delta u)] + \theta_4(\alpha)$$

where

$$\frac{\|\theta_3(\alpha)\|_\infty}{\alpha} \to 0$$

and

$$\frac{\|\theta_4(\alpha)\|_\infty}{\alpha} \to 0$$

as $\alpha \downarrow 0$. Because

$$F(u_0) \leq 0,$$

$$\beta F(u_0) + \nabla F[u_0](\delta u) < 0$$

and

$$\frac{\|\theta_4(\alpha)\|_\infty}{\alpha} \to 0$$

as $\alpha \downarrow 0$ we can see that

$$F(u_0 + \alpha \delta u_1 + \rho(\alpha)) < 0$$

for $\alpha > 0$ and sufficiently small. A similar argument shows that

$$G(v_0 + \alpha \delta v_1 + \sigma(\alpha)) < 0$$

if $\alpha > 0$ is sufficiently small. Thus the minimum of Problem (P) is contradicted.

$\square$

We have indicated previously that if $P$ is a convex cone in a Banach space $Z$ we can define a pre-order on $Z$ by setting $z \geq 0$ if $Z \in P$ and $z > 0$ if $z \in \text{int } P$. There is a corresponding pre-order defined on the dual space $Z^*$ via the polar cone

$$P^\# = \{\pi^* | \pi^* \in Z^* \text{ and } \pi^*(z) \geq 0 \text{ when } z \geq 0\}.$$

When $\pi^* \in P^\#$ we write $\pi^* \geq 0$ and when $\pi^* \in \text{int } P^\#$ we write $\pi^* > 0$. For convenience we will now state a basic separation theorem. A more general version of this theorem is proved elsewhere [24, 25].


### Theorem 6.3 *(Mazur's Theorem)*

*Let $K$ be a convex set with non-empty interior in a real normed linear vector space $Z$. Let $S$ be a subspace in $Z$ containing no interior points of $K$. There is an element $f \in S^\perp$ such that $f(z) > 0$ for all $z \in \text{int } K$.*


### Outline of Proof

The *Hahn-Banach Theorem* [24, 25] can be used to show that there exists a hyperplane $P$ that contains $S$ and that contains no interior points of $K$. Hence there exists a linear functional $f \in P^\perp \subseteq S^\perp \subseteq Z^*$ such that

$$P = \{z \mid z \in Z \text{ and } f(z) = 0\}$$

and such that

$$\text{int } K \subseteq \{z \mid z \in Z \text{ and } f(z) > 0\}.$$

If $Z$ is a Hilbert space there is a uniquely defined vector $z_f \in Z$ such that $f(z) = \, <(z_f, z)>$, and $P^\perp$ is the orthogonal complement of $P$ in $Z$. The theorem is illustrated for Hilbert space in Figure 6-1.

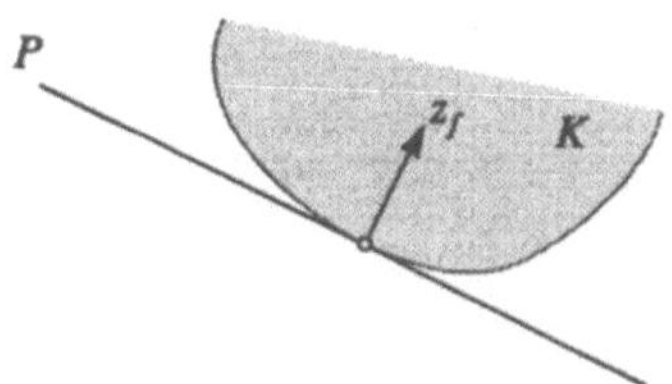


Figure 6-1: Mazur's Theorem in Hilbert space with $z_f \in P^\perp$ and $K \subseteq \{z \mid <z_f, z> \, > 0\}$


$\square$


We can use Theorem 6.3 to prove a basic alternative theorem. A more general version of this theorem can be found in [18].

### Theorem 6.4  (Basic Alternative Theorem)

*Let $Z$ be a Banach space with a pre-order and let $Z_0$ be a subspace of $Z$. If we can find $z \in Z$ with $z > 0$ then exactly one of the following two systems has a solution:*

*System 1:*     $z_0 \in Z_0$ *and* $z_0 > 0$

*System 2:*     $\pi^* \in Z_0^{\perp}$ *with* $\pi^* \geq 0$ *and* $\pi^* \neq 0$

Note that $Z_0^{\perp} = \{\zeta^* \,|\, \zeta^* \in Z^* \text{ and } \zeta^*(z) = 0 \text{ for all } z \in Z_0\}$ .

### Proof

The set

$$P_0 = \{z \,|\, z + z_0 > 0 \text{ for some } z_0 \in Z_0\}$$

is open and convex. We have either $0 \in P_0$ or $0 \notin P_0$.

The condition $0 \in P_0$ implies $z_0 > 0$ for some $z_0 \in Z_0$, and hence System 1 has a solution. If $0 \notin P_0$ we can use Theorem 6.3 with $S = Z_0$ to deduce that there exists a non-zero element $\pi^* \in Z^*$ with $\pi^*(z) \geq 0$ for all $z \in P_0$, and hence System 2 has a solution.

$\square$

The next theorem is an appropriate modification of a well known result.

### Theorem 6.5  (Fritz-John Theorem)

*A necessary condition for Problem (P) to attain a local minimum at $(u_0, v_0)$ is given by the Fritz-John Conditions*

$$\lambda^* E_u [\,(u_0, v_0)\,] + \mu^* \nabla F[u_0] = \eta^* \tag{6.8}$$

$$\tau J_v [\,(u_0, v_0)\,] + \lambda^* E_v [\,(u_0, v_0)\,] + v^* \nabla G[v_0] = 0 \tag{6.9}$$

$$\mu^* F(u_0) = 0 \tag{6.10}$$

$$v^* G(v_0) = 0 \tag{6.11}$$

*where*

$$\lambda^* = (\sigma, \xi^*) \in \Re \times U^* = W^*, \tag{6.12}$$

$$\pi^* = ((\tau, \eta^*), \mu^*, v^*)$$
$$\in (\Re \times U^*) \times U^* \times U^* = W^* \times U^* \times U^* \tag{6.13}$$

*with*

$$\pi^* \geq 0 \tag{6.14}$$

*and*

$$\pi^* \neq 0. \tag{6.15}$$

## *Proof*

Let Problem (P) attain a local minimum at $(u_0, v_0)$. Let $Z = W \times U \times U$ and let $A : \Re \times \Re \times U \times V_0 \to W$ and $B : \Re \times \Re \times U \times V_0 \to Z$ be the linear maps defined previously. Now $Z_0$ is the subspace of $Z$ defined by the image under $B$ of the null space of $A$. That is,

$$Z_0 = B\,[\mathcal{N}(A)] .$$

If we let $P = Q \times M \times M$ be the positive cone in $Z$ then, by Theorem 6.2, Problem (L) has no solution. That is, $A\,(\beta, \gamma, \delta u, \delta v) = 0$ and $B\,(\beta, \gamma, \delta u, \delta v) > 0$ has no solution. Equivalently we may say there is no solution to the system

*System 1:* $\quad \delta z \in Z_0^{\perp}$ *and* $\delta z > 0$

Since we know that int $P$ is non-empty and hence that there is some $\delta z \in Z$ with $\delta z > 0$ we can now use Theorem 6.4 to find a solution to the system

*System 2:* $\quad \pi^* \in Z_0^{\perp}$ *with* $\pi^* \geq 0$ *and* $\pi^* \neq 0.$

Now for all $\delta z \in Z_0$ we have $\pi^*(\delta z) = 0$. Therefore for all $(\beta, \gamma, \delta u, \delta v) \in \mathcal{N}(A)$ it follows that

$$(B^T \pi^*)\,(\beta, \gamma, \delta u, \delta v) = \pi^*\,[B(\beta, \gamma, \delta u, \delta v)] = 0$$

Hence

$$B^T \pi^* \in [\mathcal{N}(A)]^{\perp}.$$

Let $\mathcal{R}(A)$ denote the range of $A$. Since $\mathcal{R}(A) = W$ is closed we know that $[\mathcal{N}(A)]^{\perp} = \mathcal{R}(A^T)$. This result is easily proved [24]. Thus there exists some $\lambda^* \in W^*$ such that

$$B^T \pi^* = A^T \lambda^*. \tag{6.16}$$

If we write

$$\pi^* = [\,(\tau, \eta^*)\,, \mu^*, \nu^*\,]$$

then

$$B^T \pi^* (\beta, \gamma, \delta u, \delta v)$$
$$= \pi^* \,[B(\beta, \gamma, \delta u, \delta v)]$$
$$= (\,(\tau, \eta^*)\,, u^*, \nu^*) \begin{bmatrix} (-J_v\,[\,(u_0, v_0)\,]\,(\delta v), \delta u) \\ -\beta F(u_0) - \nabla F\,[\,(u_0, v_0)\,]\,(\delta u) \\ -\gamma G(v_0) - \nabla G\,[\,(u_0, v_0)\,]\,(\delta v) \end{bmatrix}$$
$$= -\tau J_v\,[\,(u_0, v_0)\,]\,(\delta v) + \eta^* \delta u - \mu^* \{\beta F(u_0) + \nabla F\,[\,(u_0, v_0)\,]\,(\delta u)\}$$
$$\qquad - \nu^* \{\gamma G(v_0) + \nabla G\,[\,(u_0, v_0)\,]\,(\delta v)\}$$

and hence

$$B^T \pi^* = \begin{bmatrix} -\mu^* F(u_0) \\ -\nu^* G(v_0) \\ \eta^* + \mu^* \nabla F\,[\,(u_0, v_0)\,] \\ -\tau J_v\,[\,(u_0, v_0)\,] - \nu^* \nabla G\,[\,(u_0, v_0)\,] \end{bmatrix}.$$

Similarly

$$A^T \lambda^* (\beta, \gamma, \delta u, \delta v)$$
$$= \lambda^* \,[A(\beta, \gamma, \delta u, \delta v)]$$
$$= \lambda^* \nabla E\,[\,(u_0, v_0)\,]\,(\delta u, \delta v)$$
$$= \lambda^* \,[E_u\,[\,(u_0, v_0)\,]\,(\delta u) + E_v\,[\,(u_0, v_0)\,]\,(\delta v)]$$

and hence

$$A^T \lambda^* = \begin{bmatrix} 0 \\ 0 \\ \lambda^* E_u \left[ (u_0, v_0) \right] \\ \lambda^* E_v \left[ (u_0, v_0) \right] \end{bmatrix}.$$

Therefore (6.16) is equivalent to the Fritz-John equations (6.8)–(6.11). Note that $\pi^* \geq 0$ becomes $(\tau, \eta^*) \geq 0$, $\mu^* \geq 0$ and $v^* \geq 0$ because $P^\# = Q^\# \times M^\# \times M^\#$.

$\square$

### *Comment on Theorem 6.5*

Since $\lambda^* \in W^* = \Re \times U^*$ we can write $\lambda^* = (\sigma, \xi^*)$ where $\sigma \in \Re$ and $\xi^* \in U^*$. Using (6.1), (6.6) and (6.7) we can rewrite (6.8) and (6.9) in the form

$$-\xi^* \delta u + \mu^* \nabla F \left[ u_0 \right] = \eta^*$$

and

$$\tau \int_0^T p \left[ u_0(t) \right] q' \left[ v_0(t) \right] \delta v(t) dt + \sigma \int_0^T \delta v(t) dt$$

$$+ \xi^* (D(\delta v) + D r(v_0) \delta v) + v^* \nabla G \left[ v_0 \right] (\delta v) = 0.$$

When $\tau > 0$ we can divide through by $\tau$ in (6.9) and the Fritz-John equations reduce to the more usual Kuhn-Tucker equations.

$\square$

We define an *abstract Hamiltonian* $\mathcal{H} : U \to \Re$ by setting

$$\mathcal{H}(u) = -\tau J(u, v_0) - \lambda^* E(u, v_0)$$

$$= -\tau J(u, v_0) - \sigma \int_0^T v_0(t) dt - \xi^* (D(v_0) + R(v_0) - u)$$

where $(u_0, v_0)$, $((\tau, \eta^*), \mu^*, v^*)$ and $\lambda^* = (\sigma, \xi^*)$ are defined by the Fritz-John equations (6.8)–(6.15). The next result shows that the Hamiltonian is related to our original problem. This theorem is a form of the Pontryagin Principle.

**Theorem 6.6  (Abstract Pontryagin Principle)**

*If $F(u_0) + \nabla F[u_0](\alpha \delta u) \leq 0$ for all $\alpha > 0$ and sufficiently small and if*

$$\lambda^* E_u(u_0, v_0) + \mu^* \nabla F[u_0] = \eta^*$$

*and*

$$\mu^* F(u_0) = 0$$

*where $(\tau, \eta^*) \geq 0$ and $\mu^* \geq 0$ then $\mathcal{H}(u)$ has a quasi-maximum at $u_0$ in the direction $\delta u$. That is*

$$\mathcal{H}(u_0 + \alpha \delta u) - \mathcal{H}(u_0) \leq \theta_1(\alpha)$$

*where*

$$\frac{\theta_1(\alpha)}{\alpha} \to 0$$

*as $\alpha \downarrow 0$.*

**Proof**

$$\mathcal{H}(u_0 + \alpha \delta u) - \mathcal{H}(u_0)$$

$$= -\tau [J(u_0 + \alpha \delta u, v_0) - J(u_0, v_0)] - \lambda^* [E(u_0 + \alpha \delta u, v_0) - E(u_0, v_0)]$$

$$= -\tau [J(u_0 + \alpha \delta u, v_0) - J(u_0, v_0)] - \lambda^* E_u[(u_0, v_0)](\alpha \delta u) + \theta_1(\alpha)$$

where

$$\frac{\theta_1(\alpha)}{\alpha} \to 0$$

as $\alpha \downarrow 0$. Using (6.8) we can therefore write

$$\mathcal{H}(u_0 + \alpha \delta u) - \mathcal{H}(u_0)$$

$$= -\tau [J(u_0 + \alpha \delta u, v_0) - J(u_0, v_0)] - (\eta^* - \mu^* \nabla F[u_0])(\alpha \delta u) + \theta_1(\alpha)$$

$$= -(\tau, \eta^*) \left[ \frac{J(u_0 + \alpha\delta u, v_0) - J(u_0, v_0)}{\alpha\delta u} \right]$$

$$+ \mu^*(F(u_0) + \nabla F[u_0](\alpha\delta u)) + \theta_1(\alpha)$$

$$\leq \theta_1(\alpha)$$

since $(\tau, \eta^*) \geq 0$, $\mu^* \geq 0$,

$$(J(u_0 + \alpha\delta u, v_0) - (J(u_0, v_0), \alpha\delta u)) \geq 0$$

and

$$F(u_0) + \nabla F[u_0](\alpha\delta u) \leq 0.$$

$\square$

### A Note on Convexity

Let $S$ be a convex cone in $Y$. A function $F : X \to Y$ is *S-convex* if

$$\alpha F(p) + (1 - \alpha) F(q) - F(\alpha p + (1 - \alpha) q) \in S$$

for all $p, q \in X$ and $0 < \alpha < 1$.

We can apply this definition to Problem (P) in the following way. If $F : U \to U$ is $M$-convex, $F(u_0) \leq 0$ and $F(u_0 + \alpha_0 \delta u) \leq 0$ for some $\alpha_0 > 0$ then $\delta u$ is a *feasible direction*. That is, $F(u_0 + \alpha\delta u) \leq 0$ for all $\alpha$ with $0 \leq \alpha \leq \alpha_0$. We can use the $M$-convexity of $F$ to show that

$$F(u_0) + \nabla F[u_0](\alpha\delta u) \leq 0$$

for all $\alpha$ with $0 < \alpha \leq \alpha_0$. On the other hand, if $F(u_0) < 0$ then all directions are feasible.

$\square$

## 6.3  The Adjoint Differential Equation and the Pontryagin Principle

This section elaborates on the results of the previous section. We suppose, as before, that $(u_0, v_0)$ is a minimum of Problem (P). Thus for each increment $\delta v \in V_0$ the adjoint equation (6.9) gives us

$$[\tau J_v [\,(u_0, v_0)\,] + \lambda^* E_v [\,(u_0, v_0)\,] + v^* \nabla G [v_0]\,](\delta v) = 0.$$

Thus

$$\tau \int_0^T p\,[u_0(t)]\,q'\,[v_0(t)]\,\delta v(t)dt$$

$$+ \left\{ \sigma \int_0^T \delta v(t)dt + \xi^*\,(D(\delta v) + Dr(v_0)\delta v) \right\}$$

$$+ v^*\,(Dg(v_0)\delta v) = 0$$

where we have used (6.3) and (6.5) and written $\lambda^* = (\sigma, \xi^*)$. For convenience we will rewrite this equation in the form

$$\int_0^T j(t)\,\delta v(t)dt + \xi^*\,(D(\delta v) + k\delta v) + v^*\,(m\delta v) = 0 \tag{6.17}$$

where

$$j(t) = \tau p\,[u_0(t)]\,q'\,[v_0(t)] + \sigma,$$

$$k(t) = r'\,[v_0(t)]$$

and

$$m(t) = g'\,[v_0(t)]\,.$$

We wish to show that the linear functionals $\xi^*, v^* \in U^*$ can be represented by appropriate integrals.

If we define

$$K(t) = \exp\left[\int_0^t k(s)ds\right] \tag{6.18}$$

and let

$$\zeta(t) = K(t)\int_0^t \frac{j(s)}{K(s)}ds,$$

from which it follows that

$$\zeta'(t) - k(t)\zeta(t) = j(t) \tag{6.19}$$

almost everywhere, then we can define a corresponding element $\zeta^* \in U^*$ by the formula

$$\zeta^*(u) = \int_0^T \zeta(t)u(t)dt$$

for all $u \in U$. Now because $\delta v(0) = \delta v(T) = 0$ we can use integration by parts to show that (6.17) can be rewritten as

$$\zeta^*(D(\delta v) + k\delta v) = \xi^*(D(\delta v) + k\delta v) + v^*(m\delta v). \tag{6.20}$$

To further simplify these results we note that the equation

$$\delta v'(t) + k(t)\delta v(t) = \delta u(t) \tag{6.21}$$

with $\delta v(0) = \delta v(T) = 0$ can be solved for $\delta v \in V_0$ whenever $\delta u \in U$ and $\delta u$ belongs to the null space of $K$. Therefore if

$$K^*(\delta u) = \int_0^T K(t)\delta u(t)dt = 0 \tag{6.22}$$

the solution to (6.21) is given by

$$\delta v(t) = \frac{1}{K(t)}\int_0^t K(s)\delta u(s)ds.$$

Equation (6.20) can now be rewritten in the form

$$\zeta^*(\delta u) = \xi^*(\delta u) + v^*(m\delta v).$$

Let us define a finitely additive measure $\psi$ on the Borel subsets of $[0, T]$ by the formula $\psi(B) = v^*(\chi_B)$ where $\chi_B$, defined by

$$\chi_B(t) = \begin{cases} 1 & t \in B \\ 0 & t \notin B \end{cases},$$

is the characteristic function of $B$. Now we can write

$$v^*(u) = \int_{[0, T]} u(t)\psi(dt)$$

for all $u \in U$ where the integral is a *Radon integral*. The Radon integral is discussed in Section 6.5. Since $v^* \geq 0$ implies $\psi$ is non-negative, we can define a monotone increasing function $w : [0, T] \to \Re$ by the formula

$$w(t) = \psi([0, t)).$$

As shown in Section 6.5, it follows that when $v \in V$ and satisfies $v(T) = 0$ we can use formal integration by parts to obtain

$$v^*(v) = -\int_0^T w(t)v'(t)dt$$

where the integral is now an ordinary Lebesgue integral. If $m \in V$ we have

$$v^*(m\delta v) = -\int_0^T [m'(t)\delta v(t) + m(t)\delta v'(t)]\, w(t)dt$$

$$= -\int_0^T \left\{ \frac{m'(t) - k(t)m(t)}{K(t)} \int_0^t K(s)\delta u(s)ds + m(t)\delta u(t) \right\} w(t)dt.$$

Reversing the order of integration gives

$$v^*(m\delta v) = -\int_0^T \left\{ m(t)w(t) + K(t)\int_t^T \frac{[m'(s) - k(s)m(s)]\, w(s)}{K(s)} ds \right\} \delta u(t)dt$$

$$= - \int_0^T \rho(t)\delta u(t)dt$$

$$= - \rho^*(\delta u)$$

where

$$\rho(t) = m(t)w(t) + K(t)\int_t^T \frac{[m'(s) - k(s)m(s)]\, w(s)}{K(s)}ds \qquad (6.23)$$

and where $\rho^* \in U^*$ is defined by

$$\rho^*(u) = \int_0^T \rho(t)u(t)dt.$$

Since $w(t)$ is monotone increasing it is differentiable almost everywhere and hence from (6.23) it is easy to show that

$$\rho'(t) - k(t)\rho(t) = m(t)w'(t). \qquad (6.24)$$

Thus we now have

$$\xi^*(\delta u) = [\zeta^* + \rho^*](\delta u)$$

for all $\delta u \in U$ satisfying (6.22). Clearly

$$\delta u_0 = \delta u - \frac{K^*(\delta u)}{K^*(K)}K$$

satisfies

$$K^*(\delta u_0) = 0$$

for all $\delta u \in U$. Thus $\delta u_0$ lies in the null space of $K^*$. Therefore

$$\xi^*(\delta u) = \xi^*(\delta u - \frac{K^*(\delta u)}{K^*(K)}K) + \frac{K^*(\delta u)}{K^*(K)}\xi^*(K)$$

$$= \xi^*(\delta u_0) + \frac{K^*(\delta u)}{K^*(K)}\xi^*(K)$$

$$= [\zeta^* + \rho^*]\,(\delta u_0) + \frac{K^*(\delta u)}{K^*(K)}\,\xi^*(K)$$

and hence

$$\xi^*(\delta u) = \left[\zeta^* + \rho^* + \frac{[\xi^* - \zeta^* - \rho^*]\,(K)}{K^*(K)}\,K^*\right](\delta u).$$

Now if we define

$$\xi(t) = \zeta(t) + \rho(t) + \frac{[\xi^* - \zeta^* - \rho^*]\,(K)}{K^*(K)}\,K(t) \tag{6.25}$$

we can represent $\xi^*(\delta u)$ by the integral

$$\xi^*(\delta u) = \int_0^T \xi(t)\delta u(t)dt.$$

for all $\delta u \in U$. From (6.18), (6.19), (6.24) and (6.25) we can see that

$$\xi'(t) - k(t)\xi(t) = j(t) + m(t)w'(t) \tag{6.26}$$

for almost all $t \in [0, T]$. The equation (6.26) is called the *adjoint differential equation* and in terms of the original functions it becomes

$$\xi'(t) - r'\,[v_0(t)]\,\xi(t) = \tau p\,[u_0(t)]\,q'\,[v_0(t)] + g'\,[v_0(t)]\,w'(t). \tag{6.27}$$

Because of the integral representation established above the abstract Hamiltonian can now be written as an integral. We have

$$\begin{aligned}
\mathcal{H}(u) &= -\tau J(u, v_0) - \lambda^* E(u, v_0) \\
&= -\tau \int_0^T p\,[u(t)]\,q\,[v_0(t)]\,dt - \sigma\left(\int_0^T v_0(t)dt - X\right) \\
&\quad - \int_0^T \xi(t)\,\{v_0'(t) + r\,[v_0(t)] - u(t)\}\,dt
\end{aligned}$$

where $\xi(t)$ is determined by the adjoint differential equation. Thus

$$\mathcal{H}(u) = \int_0^T H\,[u(t),\,t]\,dt + H_0$$

where $H_0 \in \mathfrak{R}$ and $H : \mathfrak{R} \times \mathfrak{R} \to \mathfrak{R}$ is defined by

$$H(u, t) = -\tau p(u)q\,[v_0(t)] - \sigma v_0(t) + \xi(t)\,\{u - r\,[v_0(t)]\,\}.$$

We call $H(u, t)$ the pointwise Hamiltonian function or simply the *Hamiltonian* function. We define

$$H_u\,[\,(u, t)\,]\,(\delta u) = \lim_{\alpha \downarrow 0} \frac{[H(u + \alpha \delta u, t) - H(u, t)]}{\alpha}$$

and say that $H(u, t)$ is *pseudo-concave* at $u_0$ if

$$[H(u_0 + \delta u, t) > H(u_0, t)] \;\Rightarrow\; [H_u(u_0, t)(\delta u) > 0]\,.$$

It is certainly true in our case that the convexity of $p(u)$ implies that $H(u, t)$ is pseudoconcave at almost all points. If the function $F : U \to U$ is defined by an expression of the form $[F(u)]\,(t) = f\,[u(t)]$ where $f : \mathfrak{R} \to \mathfrak{R}$ is convex as in (6.2) and (6.4) then $F(u)$ is $M$-convex and we have the following version of the Pontryagin Principle.

### Theorem 6.7  (Pontryagin Principle)

*Let $H(u, t)$ be pseudoconcave in $u$ at $u_0$ for each $t$ and let $F(u)$ be $M$-convex with $F(u_0) \leq 0$. If $\mathcal{H}(u)$ has a quasi-maximum at $u_0$ in every feasible direction $\delta u$ then*

$$H\,[u_0(t),\,t] \geq H\,[u_0(t) + \delta u(t),\,t]$$

*for almost all $t$.*

### Proof

Let

$$A = \{t\,|\,t \in \mathfrak{R} \text{ and } H\,[u_0(t) + \delta u(t),\,t] > H\,[u_0(t),\,t]\,\}$$

and suppose that $A$ has positive measure. Now for $t \in A$ we have

$$H\left[u_0(t) + \alpha\delta u(t), t\right] - H\left[u_0(t), t\right]$$
$$= \alpha H_u\left[u_0(t), t\right]\delta u(t) + \theta_\alpha(t)$$
$$= \alpha\delta(t) + \theta_\alpha(t)$$

where $\delta(t) > 0$ and

$$\frac{\theta_\alpha(t)}{\alpha} \to 0$$

as $\alpha \downarrow 0$. Define

$$A_\varepsilon = \left\{ t \,\middle|\, t \in A \text{ and } \frac{|\theta_\alpha(t)|}{\alpha} < \frac{\delta(t)}{2} \text{ for } 0 < \alpha < \varepsilon \right\}.$$

Clearly

$$A = \lim_{\varepsilon \downarrow 0} A_\varepsilon$$

and hence we can find $\varepsilon > 0$ with $A_\varepsilon$ having positive measure. If we let

$$\delta u_\varepsilon(t) = \begin{cases} \delta u(t) & t \in A_\varepsilon \\ 0 & \text{otherwise} \end{cases}$$

then we note that $\|\delta u_\varepsilon\|_\infty \le \|\delta u\|_\infty$ and furthermore that for all $\alpha$ with $0 < \alpha \le \varepsilon$ we have

$$\mathcal{H}(u_0 + \alpha\delta u_\varepsilon) - \mathcal{H}(u_0)$$
$$= \int_{A_\varepsilon} \left\{ H\left[u_0(t) + \alpha\delta u(t), t\right] - H\left[u_0(t), t\right] \right\} dt$$
$$= \int_{A_\varepsilon} \left[ \alpha\delta(t) + \theta_\alpha(t) \right] dt$$
$$\ge \int_{A_\varepsilon} \frac{\alpha\delta(t)}{2} dt$$
$$= \frac{\alpha\delta}{2}$$

where

$$\delta = \int_{A_\varepsilon} \delta(t)\,dt > 0.$$

Since

$$[F(u_0 + \alpha\delta u_\varepsilon)]\,(t)$$

$$= f(u_0(t) + \alpha\delta u_\varepsilon(t))$$

$$= \begin{cases} f(u_0(t) + \alpha\delta u(t)) & t \in A_\varepsilon \\ f(u_0(t)) & t \notin A_\varepsilon \end{cases}$$

$$= \begin{cases} [F(u_0 + \alpha\delta u)]\,(t) & t \in A_\varepsilon \\ [F(u_0)]\,(t) & t \notin A_\varepsilon \end{cases}$$

it is easy to see that $\delta u_\varepsilon$ is a feasible direction at $u_0$ and hence we have contradicted the quasi-maximum of $\mathcal{H}(u)$ at $u_0$.

$\square$

## 6.4 Conclusions

We have shown that when the train control problem is formulated in an appropriate way the optimal strategy coincides with a local maximum for the pointwise Hamiltonian function. Thus we have established the validity of the Pontryagin Principle as a necessary condition for an optimal strategy. In the next chapter it will be shown that this principle can be used to find the precise nature of the optimal strategy.

## 6.5 Appendix—The Radon Integral

Representation of the elements of the dual space $L^\infty([0, T])^*$ can be achieved using an integral based on a finitely additive measure. In certain cases these integrals can be replaced by Lebesgue integrals. Let $v^* \in L^\infty([0, T])^*$ and for any Borel subset $B$ of $[0, T]$ define

$$\psi(B) = v^*(\chi_B) \tag{6.28}$$

where $\chi_B$ is the characteristic function of $B$. Since

$$|v^*(\chi_B)| \leq \|v^*\|\,\|\chi_B\|_\infty$$

it is easy to see that the measure $\psi$ has the following properties:

1.  if $B_1, B_2$ are disjoint then $\psi(B_1 \cup B_2) = \psi(B_1) + \psi(B_2)$,

2.  $|\psi(B)| \leq \|v^*\|$ for all $B$, and

3.  if $B$ has Lebesgue measure zero then $\psi(B) = 0$.

Now for any $u \in L^\infty([0, T])$ and each non-negative integer $N$ we consider a partition of the intervals $[0, \|u\|_\infty + 1)$ into a finite system of disjoint half open intervals

$$A_n(N) = [n\varepsilon, (n+1)\varepsilon)$$

for each $n = 0, 1, ..., N$ where

$$\varepsilon = \varepsilon(N) = \frac{\|u\|_\infty + 1}{N+1}.$$

If we set

$$B_n(N) = \{t \mid t \in [0, T] \text{ and } u(t) \in A_n(N)\}$$

then for any points $\alpha_n \in A_n = A_n(N)$ we have

$$\left\| u - \sum_{n=0}^{N} \alpha_n \chi_n \right\|_\infty \leq \varepsilon$$

where $\chi_n$ is the characteristic function of $B_n = B_n(N)$. Thus it follows that

$$\left| v^*(u) - \sum_{n=0}^{N} \alpha_n \psi(B_n) \right| \leq \|v^*\| \varepsilon.$$

Now if we let $N \to \infty$ and hence $\varepsilon = \varepsilon(N) \to 0$ we obtain

$$v^*(u) = \lim_{N \to \infty} \sum_{n=0}^{N} \alpha_n \psi(B_n)$$

independently of the choice of $\alpha_n$. The limit on the right hand side is called the *Radon integral* of $u$ with respect to $\psi$. We write

$$v^*(u) = \int_{[0, T]} u(t) \psi(dt).$$

In the case where $v^* \geq 0$ we know that $v^*(u) \geq 0$ whenever $u \geq 0$ and we can define a monotone increasing function $w$ on $[0, T]$ by setting

$$w(t) = v^*(\chi_{[0,T)}) = \psi([0, t)).\tag{6.29}$$

Clearly $w(0) = 0$.


## Theorem 6.8

*If $v \in V = C^{0,1}([0, T])$ and $v^* \in U^* = L^\infty([0, T])^*$ then we have*

$$v^*(v) = \int_{[0, T]} v(t)\psi(dt)$$

$$= v(T)w(T) - \int_0^T w(t)v'(t)dt$$

*where $\psi$ and $w$ are defined above by (6.28) and (6.29).*


## Proof

To begin we assume that $v(t)$ is strictly monotone increasing with $v'(t) > \delta$ for some $\delta > 0$. Clearly

$$B_n(N) = [b_n(N), b_{n+1}(N))$$

where $b_n(N)$ is defined by $v[b_n(N)] = v(0) + n\varepsilon(N)$ for each $n = 0, 1, ..., N+1$, and so

$$\int_{[0, T]} v(t)\psi(dt)$$

$$= \lim_{N \to \infty} \sum_{n=0}^{N} [v(0) + n\varepsilon(N)] \, \psi[B_n(N)]$$

$$= \lim_{N \to \infty} \sum_{n=0}^{N} [v(0) + n\varepsilon(N)] \, \{w[b_{n+1}(N)] - w[b_n(N)]\}.$$

Therefore summation by parts gives

$$\int_{[0,\,T]} v(t)\psi(dt)$$

$$= \lim_{N \to \infty}\ \{\ [v(0) + n\varepsilon(N)]\, w\,[b_n(N)]\,\big|_{n=0}^{N+1} - \varepsilon(N)\sum_{n=0}^{N} w\,[b_{n+1}(N)]\ \}\ .$$

However,

$$\varepsilon(N)\ =\ \Delta v\,[b_n(N)]$$

$$=\ (v\,[b_{n+1}(N)] - v\,[b_n(N)]\,)$$

and because $v'(t) > \delta$ it follows that

$$\big|b_{n+1}(N) - b_n(N)\big|\ <\ \frac{\varepsilon(N)}{\delta}\,.$$

Thus $\Delta b_n(N) \to 0$ as $N \to \infty$ and hence

$$\int_{[0,\,T]} v(t)\psi(dt)\ =\ v(T)w(T) - \lim_{N \to \infty}\sum_{n=0}^{N} w\,[b_{n+1}(N)]\,\Delta v\,[b_n(N)]\,.$$

With regard to the right hand side, however, we can see that

$$\sum_{n=0}^{N} w\,[b_{n+1}(N)]\,\Delta v\,[b_n(N)]$$

$$=\ \sum_{n=0}^{N} w\,[b_{n+1}(N)]\int_{b_n(N)}^{b_{n+1}(N)} v'(t)\,dt$$

$$\geq\ \sum_{n=0}^{N}\int_{b_n(N)}^{b_{n+1}(N)} w(t)v'(t)\,dt$$

$$=\ \int_{0}^{T} w(t)v'(t)\,dt\,.$$

A similar argument gives

$$\sum_{n=0}^{N} w\,[b_n(N)]\,\Delta v\,[b_n(N)] \leq \int_{0}^{T} w(t)v'(t)\,dt\,.$$

On the other hand it is also clear that

$$\sum_{n=0}^{N} \{ w[b_{n+1}(N)] - w[b_n(N)] \} \Delta v[b_n(N)]$$

$$= \varepsilon(N) \sum_{n=0}^{N} \Delta w[b_n(N)]$$

$$= \varepsilon(N) w(T)$$

$$\rightarrow 0$$

as $N \rightarrow \infty$. Thus we have finally

$$\int_{[0,T]} v(t)\psi(dt) = v(T)w(T) - \int_0^T w(t)v'(t)dt.$$

In general, by writing $v_1(t) = v(t) + kt$ for sufficiently large $k$ we can ensure that $v_1'(t) > \delta$ for some $\delta > 0$. The result now follows for $v_1$ and consequently for $v$ we have

$$\int_{[0,T]} v(t)\psi(dt)$$

$$= \int_{[0,T]} [v_1(t) - kt]\, \psi(dt)$$

$$= \int_{[0,T]} v_1(t)\psi(dt) - k\int_{[0,T]} t\psi(dt)$$

$$= \left[ v_1(T)w(T) - \int_0^T w(t)v_1'(t)dt \right] + k\left[ Tw(T) - \int_0^T w(t)dt \right]$$

$$= v(T)w(T) - \int_0^T w(t)v'(t)dt.$$

$\square$

# CHAPTER 7
# DETERMINATION OF OPTIMAL DRIVING STRATEGIES

In this chapter the Pontryagin Principle will be used to find the nature of the optimal strategy for the mechanical energy model with the cost functional

$$J(u, v) = \int_0^T p\,[u(t)]\,q\,[v(t)]\,dt$$

when the function $p : \Re \to \Re$ is piecewise linear. Using the adjoint differential equation and the Hamiltonian function, we show that the optimal strategy uses piecewise constant acceleration, and that only certain distinct values of $u$ should be used. Furthermore, the acceleration decreases as the journey progresses. Using this information to reformulate the problem we find key equations that determine the precise speeds at which the acceleration should be changed. The results are illustrated with an example that highlights some deficiencies in the mechanical energy model.

This chapter was originally published in 1988 as a report to the School of Mathematics and Computer Studies at the South Australian Institute of Technology [35] and later in the *Journal of the Australian Mathematical Society Series B* [6].

## 7.1  A Special Case of the Train Control Problem

We again consider the real valued cost functional $J$ on $\mathcal{U} \times \mathcal{V}$ given by the formula

$$J(u, v) = \int_0^T p\,[u(t)]\,q\,[v(t)]\,dt$$

where $p : \Re \to \Re$ and $q : \Re \to \Re$ are known functions with the properties described in Chapter 5. In addition, we will assume that the function $p$ is piecewise linear. More precisely we will assume that there exist points $0 = u_1 < u_2 < ... < u_n = 1$ such that

$$p(u) = \begin{cases} 0 & u < 0 \\ a_j(u - u_j) + b_j & u_j \leq u < u_{j+1} \\ a_n(u - 1) + b_n & 1 \leq u \end{cases}$$

where $0 < a_1 < a_2 < ... < a_n$, $b_1 = 0$ and $\Delta b_j = a_j \Delta u_j$. We will also assume that the function $q$ is continuously differentiable when $v > 0$. We will need some additional assumptions on $p$ and $q$, and these will be introduced in the next section. For the moment we observe that we wish to minimise $J(u, v)$ over all $u \in \mathcal{U}$ and $v \in \mathcal{V}_0$ subject to the constraints

$$\int_0^T v(t)\,dt = L$$

and

$$v'(t) = u(t) - r\,[v(t)]$$

where $r : \mathfrak{R} \to \mathfrak{R}$ is the resistance function specified in Chapter 5. There are also two inequality constraints, $|u(t)| \leq 1$ and $v(t) \geq 0$, which must be observed.

## 7.2  The Nature of the Optimal Strategy

We mentioned earlier that additional assumptions on $p$ and $q$ would be introduced. To this purpose we define functions $p_j : \mathfrak{R} \to \mathfrak{R}$ for each $j = 1, 2, ..., n$ by setting

$$p_j(v) = a_j\,[r(v) - u_j] + b_j$$

and define the function $(p_j q) : \mathfrak{R} \to \mathfrak{R}$ by the formula

$$(p_j q)\,(v) = p_j(v)q(v).$$

We will assume that the functions $(p_1 q)$ and $(p_n q)$ are strictly convex. It is now easy to establish that $(p_j q)$ is strictly convex for $j = 1, 2, ..., n$. With these additional assumptions we can show that the adjoint differential equation

$$\xi'(t) - r'\,[v_0(t)]\,\xi(t) = \tau p\,[u_0(t)]\,q'\,[v_0(t)] + \sigma - w'(t) \tag{7.1}$$

and the Hamiltonian function

$$H(u, t) = \tau p(u)q\,[v_0(t)] + \sigma v_0(t) + \xi(t)\,\{r\,[v_0(t)] - u\}$$

can be used to determine the nature of the optimal strategy. In this context $\tau \in \mathfrak{R}$ and $\sigma \in \mathfrak{R}$ are adjoint variables with $w : \mathfrak{R} \to \mathfrak{R}$ and $\xi : \mathfrak{R} \to \mathfrak{R}$ are functions used when representing the adjoint variables by appropriate integrals. The function $v_0$ is the optimal speed of the train and the function $u_0$ is the optimal applied acceleration.

From the previous chapter we know that the optimal acceleration $u_0$ can be obtained by minimising $H(u, t)$ over all $u \in I = [-1, 1]$. We must therefore obtain one of the following basic situations:

(C1):    $\tau a_{j-1} q\,[v_0(t)] - \xi(t) < 0$    and    $\tau a_j q\,[v_0(t)] - \xi(t) > 0$    for    some $j = 2, 3, ..., n$, in which case

$$H(u_j, t) = \min_{u \in I} H(u, t);$$

(C2):    $\tau a_j q\,[v_0(t)] - \xi(t) = 0$ for some $j = 1, 2, ..., n$, in which case

$$H(u^*, t) = \min_{u \in I} H(u, t)$$

for all $u^* \in [u_j, u_{j+1}]$ ;

(C3):    $\xi(t) = 0$ in which case

$$H(u^*, t) = \min_{u \in I} H(u, t)$$

for all $u^* \in [-1, 0]$ ;

(C4):    $\xi(t) < 0$ in which case

$$H(-1, t) = \min_{u \in I} H(u, t).$$

We will now state several results which allow us to develop the overall structure of the optimal strategy. For convenience we will state them as a sequence of lemmas.

### Lemma 7.1

*Let $t_1, t_2 \in \mathfrak{R}$ with $0 \le t_1 < t_2$. If $v_0(t) > 0$ for all $t \in (t_1, t_2)$ then $w(t)$ is constant in this interval.*

*Proof*

From Theorem 6.5 we know that

$$v^*(v_0) = 0.$$

That is,

$$\int_{[0,T]} v_0(t)w(dt) = 0$$

where $w(t)$ is increasing and where the integral is the Radon integral described in Section 6.5. This condition derives from the constraint $v_0(t) \geq 0$. It follows that

$$\begin{aligned}
0 &= \int_{[0,T]} v_0(t)w(dt) \\
&\geq \int_{[t_1 + \delta, t_2 - \delta]} v_0(t)w(dt) \\
&\geq \varepsilon \left[ w(t_2 - \delta) - w(t_1 + \delta) \right]
\end{aligned}$$

for some $\delta > 0$ and an associated $\varepsilon > 0$. Since $w(t)$ is increasing it must be constant. $\square$


### Lemma 7.2

*Let $t_1, t_2 \in \Re$ with $0 \leq t_1 < t_2$. If $\tau a_{j-1} q\,[v_0(t)] - \xi(t) < 0$ and $\tau a_j q\,[v_0(t)] - \xi(t) > 0$ for some $j = 2, 3, \ldots, n$ and all $t \in (t_1, t_2)$, and if $u_0(t_1) = u_j > r\,[v_0(t_1)]$ then $u_0(t) = u_j > r\,[v_0(t)]$ for all $t \in [t_1, t_2]$.*


*Proof*

Suppose there is some value $t_3 \in (t_1, t_2]$ with $u_j = r\,[v_0(t_3)]$. For each $t \in (t_1, t_3)$ we have

$$t = t_1 + \int_{v_0(t_1)}^{v_0(t)} \frac{1}{[u_j - r(v)]}\,dv.$$

From the *Mean Value Theorem* [29] we can find $c_v$ with $v < c_v < v_0(t_3)$ such that

$$r\,[v_0(t_3)] - r(v) = r'(c_v)\,[v_0(t_3) - v]$$

and hence

$$t > t_1 + \varepsilon \int_{v_0(t_1)}^{v_0(t)} \frac{1}{[v_0(t_3) - v]}\, dv$$

where $\varepsilon > 0$ is a lower bound for $1/r'(v)$ on the interval $[v_0(t_1), v_0(t_3)]$. Clearly the right hand side approaches infinity as $t \uparrow t_3$. Thus $t_3$ cannot be finite.

$\square$

It is convenient to relate the optimality conditions to the function $(p_j q) : \Re \to \Re$. Let $t_1, t_2 \in \Re$ with $0 \leq t_1 < t_2$, and suppose that $u_j \leq u_0(t) < u_{j+1}$ for some $j = 2, 3, \ldots, n$ and all $t \in (t_1, t_2)$. Then

$$p\,[u_0(t)] = a_j\,[u_0(t) - u_j] + b_j$$

and

$$\begin{aligned}
\tau\,(p_j q)\,'\,[v_0(t)] + \sigma \\
= \tau\,\{p_j'\,[v_0(t)]\,q\,[v_0(t)] + p_j\,[v_0(t)]\,q'\,[v_0(t)]\,\} + \sigma \\
= \tau\,\{a_j r'\,[v_0(t)]\,q\,[v_0(t)] + [a_j\,(r\,[v_0(t)] - u_j) + b_j]\,q'\,[v_0(t)]\,\} + \sigma
\end{aligned}$$

and since $w'(t) = 0$ the adjoint equation (7.1) and the equation of motion (5.5) show us that

$$\begin{aligned}
\tau\,(p_j q)\,'\,[v_0(t)] + \sigma = \xi'(t) - \tau a_j q'\,[v_0(t)]\,v_0'(t) \\
+ r'\,[v_0(t)]\,\{\tau a_j q\,[v_0(t)] - \xi(t)\}\,.
\end{aligned} \tag{7.2}$$

When $r'\,[v_0(t)] > 0$ we obtain

$$\tau\,(p_j q)\,'\,[v_0(t)] + \sigma > \xi'(t) - \tau a_j q'\,[v_0(t)]\,v_0'(t) \tag{7.3}$$

if $\tau a_j q\,[v_0(t)] - \xi(t) > 0$ and

$$\tau\,(p_j q)\,'\,[v_0(t)] + \sigma < \xi'(t) - \tau a_j q'\,[v_0(t)]\,v_0'(t) \tag{7.4}$$

if $\tau a_j q\,[v_0(t)] - \xi(t) < 0$. When $\tau a_j q\,[v_0(t)] - \xi(t) = 0$ equation (7.2) reduces to

$$\tau\,(p_j q)\,'\,[v_0(t)] + \sigma = \xi'(t) - \tau a_j q'\,[v_0(t)]\,v_0'(t). \tag{7.5}$$

We will refer to (7.2)–(7.5) in the following lemmas.

### Lemma 7.3

*Let $t_1, t_2 \in \Re$ with $0 \le t_1 < t_2$. If $\tau a_j q\,[v_0(t)] - \xi(t) = 0$ for some $j = 1, 2, \ldots, n$ and all $t \in (t_1, t_2)$ then $v_0(t) = v_0(t_1)$ and $u_0(t) = r\,[v_0(t_1)]$ for all such $t$. This can only occur if $u_j \le r\,[v_0(t_1)] \le u_{J+1}$.*

### Proof

Since $\tau a_j q\,[v_0(t)] - \xi(t) = 0$ on $(t_1, t_2)$ it follows that

$$\tau a_j q'\,[v_0(t)]\,v_0'(t) - \xi'(t) = 0$$

on this interval. Using (7.5) this condition becomes

$$\tau (p_j q)'\,[v_0(t)] + \sigma = 0$$

and since $(p_j q)$ is strictly convex there is at most one value $V$ such that $\tau (p_j q)'(V) + \sigma = 0$. Thus $v_0(t) = V$ for all $t \in (t_1, t_2)$.

$\square$

### Lemma 7.4

*Let $t_1, t_2, t_3 \in \Re$ with $0 \le t_1 < t_2 < t_3$. If $\tau a_{j-1} q\,[v_0(t)] - \xi(t) < 0$ and $\tau a_j q\,[v_0(t)] - \xi(t) > 0$ for some $j = 2, 3, \ldots, n$ and all $t \in (t_1, t_2)$ and if $u_0(t) = u_j > r\,[v_0(t_1)]$ then it is not possible to have $\tau a_j q\,[v_0(t)] - \xi(t) = 0$ for all $t \in (t_1, t_3)$.*

### Proof

Suppose the contrary. From Lemma 7.2 we have $u_j > r\,[v_0(t_2)]$ and from Lemma 7.3 we have $u_j \le r\,[v_0(t_2)]$ which is a contradiction.

$\square$

## *Lemma 7.5*

*Let $t_1, t_2, t_3 \in \Re$ with $0 \le t_1 < t_2 < t_3$. If $\tau a_{j-1} q \left[ v_0(t) \right] - \xi(t) < 0$ and $\tau a_j q \left[ v_0(t) \right] - \xi(t) > 0$ for some $j = 2, 3, \ldots, n$ and all $t \in (t_1, t_2)$ and if $u_0(t) = u_j > r \left[ v_0(t_1) \right]$ on this interval then it is not possible to have $\tau a_j q \left[ v_0(t) \right] - \xi(t) < 0$ and $\tau a_{j+1} q \left[ v_0(t) \right] - \xi(t) > 0$ for all $t \in (t_2, t_3)$.*

## *Proof*

Suppose the contrary. From Lemma 7.2 it is clear that $u_j > r \left[ v_0(t_2) \right]$ and since $u_{j+1} > u_j > r \left[ v_0(t_2) \right]$ it is also clear from Lemma 7.2 that $u_{j+1} > r \left[ v_0(t_3) \right]$. Thus $v_0(t)$ increases on $(t_1, t_3)$. We also know that $\tau a_j q \left[ v_0(t) \right] - \xi(t)$ is positive on $(t_1, t_2)$ and negative on $(t_2, t_3)$. From (7.4) and the convexity of $(p_j q)$ it follows that

$$\tau a_j q' \left[ v_0(t) \right] v_0'(t) - \xi'(t) \;<\; (-1) \left\{ \tau (p_j q)' \left[ v_0(t) \right] + \sigma \right.$$
$$\le \; (-1) \left\{ \tau (p_j q)' \left[ v_0(t_2) \right] + \sigma \right\}$$

for all $t \in (t_2, t_3)$. Since $\tau a_j q' \left[ v_0(t_2) \right] - \xi'(t_2) = 0$ equation (7.5) gives

$$\tau a_j q' \left[ v_0(t) \right] - \xi'(t) < \tau a_j q' \left[ v_0(t_2) \right] - \xi'(t_2)$$

for all $t \in (t_2, t_3)$. Now we know that $\tau a_j q \left[ v_0(t) \right] - \xi(t)$ is decreasing at $t = t_2$ and hence

$$\tau a_j q' \left[ v_0(t_2) \right] v_0'(t_2) - \xi'(t_2) \le 0.$$

Thus we have shown that $\tau a_j q' \left[ v_0(t) \right] - \xi'(t)$ is negative and decreasing throughout the interval $(t_2, t_3)$. Let us suppose that the optimal strategy changes at $t = t_3$. Let $t_4 \in \Re$ with $t_3 < t_4$. From Lemma 7.4 we can see that it is not possible to have $\tau a_{j+1} q \left[ v_0(t) \right] - \xi(t) = 0$ for all $t \in (t_3, t_4)$ and so the only possible change would require $\tau a_{j+1} q \left[ v_0(t) \right] - \xi(t) < 0$ at $t \in (t_3, t_4)$. A continuation of this argument will show that such a journey can never terminate.

$\square$

Although the above results do not constitute a complete determination of the overall structure of the optimal strategy they do indicate a procedure that can be used. For example, Lemma 7.2 shows us that if Condition (C1) is valid on a time interval $[t_1, t_2]$ and if the optimal acceleration $u_0(t) = u_j$ for some $j = 2, 3, \ldots, n$ exceeds the frictional resistance $r \left[ v_0(t) \right]$ at the beginning of the time interval $[t_1, t_2]$, then it exceeds the frictional resistance at the end of the interval as well. Lemma 7.4 shows us that at $t = t_2$ the condition

$$\tau a_{j-1} q \left[ v_0(t_2) \right] - \xi(t_2) = 0$$

must be true. Thus on a next time interval $(t_2, t_3)$ we must have either

$$\tau a_{j-1} q \left[ v_0(t) \right] - \xi(t) = 0$$

or else we have both

$$\tau a_{j-1} q \left[ v_0(t) \right] - \xi(t) < 0$$

and

$$\tau a_j q \left[ v_0(t) \right] - \xi(t) > 0$$

throughout the interval. In the former case Condition (C2) holds and we can apply Lemma 7.2 with $j$ replaced by $j-1$ to see that a constant acceleration

$$u_0(t) = r \left[ v_0(t) \right] = r(V) \leq u_j$$

is applied throughout the interval. In the latter case Condition (C1) again applies throughout the interval, but with $j$ replaced by $j-1$ and acceleration $u_0(t) = u_{j-1}$. Lemma 7.5 shows that it is not possible for the acceleration to be given by $u_0(t) = u_j$ on a particular time interval and then to be given by $u_0(t) = u_{j+1}$ on the next time interval. In summary, the optimal acceleration must progressively decrease with the passage of time.

### The Optimal Strategy

From the above arguments, it follows that an optimal strategy must consist of

1. an acceleration phase, during which time $u_0(t) = 1 \;\; \to \;\; u_0(t) = u_{n-1} \;\; \to \;\; u_0(t) = u_{n-2} \to \ldots \to \;\; u_0(t) = u_{n-k+1}$;

2. a speed-hold phase, with $u_0(t) = r \left[ v_0(t) \right]$;

3. a coast phase, during which time $u_0(t) = u_{n-k} \;\; \to \;\; u_0(t) = u_{n-k-1} \to \ldots \to \;\; u_0(t) = 0$ and the speed decreases gradually; and

4. a brake phase, with $u_0(t) = -1$.

It is possible to have an optimal strategy in which certain stages are omitted.

## 7.3 The Complete Solution

In this section it is convenient to retain the notation of the original paper [6], in which $\alpha_1, \alpha_2, \ldots, \alpha_k$ denote the durations of the time intervals during the acceleration phase, $\beta$ denotes the duration of the speed-hold phase, $\gamma_1, \gamma_2, \ldots, \gamma_{n-k}$ denote the durations of the time intervals during the coast phase, and $\delta$ denotes the duration of the brake phase. For algebraic convenience we let $\alpha_0 = 0$ and $\gamma_0 = 0$.

We will consider an optimal strategy with the acceleration phase defined by

$$u_0(t) = u_{n-i} \text{ for } t \in \left( \sum_{r=0}^{i} \alpha_r, \sum_{r=0}^{i+1} \alpha_r \right)$$

and each $i = 0, 1, 2, \ldots, k-1$. We will define the speeds $V_0 < V_1 < \ldots < V_k$ by the formulae

$$V_i = v_0 \left( \sum_{r=0}^{i} \alpha_r \right)$$

for each $i = 0, 1, \ldots, k$. We assume that $\alpha_r \geq 0$ for all $r = 1, 2, \ldots, k$. The speed-hold phase will be given by

$$u_0(t) = r(V_k) \text{ for } t \in \left( \sum_{r=0}^{k} \alpha_r, \sum_{r=0}^{k} \alpha_r + \beta \right)$$

where $\beta \geq 0$. The coast phase is specified by

$$u_0(t) = u_{n-(k+j)} \text{ for } t \in \left( \sum_{r=0}^{k} \alpha_r + \beta + \sum_{s=0}^{j} \gamma_s, \sum_{r=0}^{k} \alpha_r + \beta + \sum_{s=0}^{j+1} \gamma_s \right)$$

and each $j = 0, 1, \ldots, n-k-1$. We define the speeds $V_k > V_{k+1} > \ldots > V_n \geq 0$ by the formulae

$$V_{k+j} = v_0 \left( \sum_{r=0}^{k} \alpha_r + \beta + \sum_{s=0}^{j} \gamma_s \right)$$

for each $j = 0, 1, \ldots, n-k$. We assume that $\gamma_s \geq 0$ for all $s = 1, 2, \ldots, n-k$. Finally, the brake phase is defined by

$$u_0(t) = -1 \text{ for } t \in \left( \sum_{r=0}^{k} \alpha_r + \beta + \sum_{s=0}^{n-k} \gamma_s, \sum_{r=0}^{k} \alpha_r + \beta + \sum_{s=0}^{n-k} \gamma_s + \delta \right)$$

where the speed $V_{n+1}$ is defined by

$$V_{n+1} = v_0\left(\sum_{r=0}^{k}\alpha_r + \beta + \sum_{s=0}^{j}\gamma_s + \delta\right).$$

Note the constraints on the initial speed $V_0 = 0$ and the final speed $V_{n+1} = 0$. The actual functions used in the synthesis of $v_0(t)$ are defined in the following way. For each $i = 0, 1, ..., k-1$ we define $t_i : [0, V_i^*) \rightarrow [0, \infty)$ by the formula

$$t_i(v) = \int_0^v \frac{1}{u_{n-i} - r(w)}\,dw$$

where $V_i^*$ is the unique point with $r(V_i^*) = u_{n-i}$. We can choose a constant $r_i$ such that

$$u_{n-i} - r(w) \le r_i(V_i^* - w)$$

for all $w \in [0, V_i^*)$ . Thus

$$t_i(v) \ge \frac{1}{r_i} \ln\left[1 - \frac{v}{V_i^*}\right]$$

and hence $t_i(v) \uparrow \infty$ as $v \uparrow V_i^*$. Now if we define $v_i^* : [0, \infty) \rightarrow [0, V_i^*)$ as the inverse function of $t_i$ then it is easy to see that

$$v_i^{*\prime}(t) = u_{n-i} - r[v_i^*(t)]$$

Furthermore it is clear that $v_i^*(t) \uparrow V_i^*$ as $t \rightarrow \infty$. Since $0 \le V_i < V_i^*$ we can define the acceleration phase of the optimal strategy by the formulae

$$v_0(t) = v_i^*\left(t - \sum_{r=0}^{i}\alpha_r + t_i[V_i]\right) \text{ for } t \in \left(\sum_{r=0}^{i}\alpha_r, \sum_{r=0}^{i+1}\alpha_r\right)$$

for each $i = 0, 1, ..., k-1$. Incidentally, it is clear that

$$V_{i+1} = v_i^*(\alpha_{i+1} + t_i[V_i]).$$

For the speed-hold phase of the optimal strategy it is clear that

$$v_0(t) = V_k \text{ for } t \in \left(\sum_{r=0}^{k}\alpha_r, \sum_{r=0}^{k+1}\alpha_r + \beta\right)$$

Now for each $j = 0, 1, ..., n - k - 2$ we define

$$t_{k+j} : (V_{k+j}{}^*, V_k] \rightarrow [0, \infty)$$

by the formula

$$t_{k+j}(v) = \int_v^{V_k} \frac{1}{r(w) - u_{n-k-j}} \, dw$$

where $V_{k+j}{}^*$ is the unique point with $r(V_{k+j}{}^*) = u_{n-k-j}$. It is necessary to assume at this stage that $u_2 > r_0$. That is, the lowest non-zero level of acceleration is assumed sufficient to overcome the initial resistance to motion. We can choose a constant $r_k$ so that

$$r(w) - u_{n-k-j} \le r_k (w - V_{k+j}{}^*)$$

for all $w \in (V_{k+j}{}^*, V_k]$ and so

$$t_{k+j}(v) \ge \frac{1}{r_k} \ln \left[ \frac{V_k - V_{k+j}{}^*}{v - V_{k+j}{}^*} \right]$$

and hence $t_{k+j}(v) \uparrow \infty$ as $v \downarrow V_{k+j}{}^*$. Now we define

$$v_{k+j}{}^* : [0, \infty) \rightarrow (V_{k+j}{}^*, V_k]$$

as the inverse function of $t_{k+j}$ and it is easy to see that

$$v_{k+j}{}^{*'}(t) = u_{n-k-j} - r [v_{k+j}{}^*(t)] .$$

Furthermore it is clear that $v_{k+j}{}^*(t) \downarrow V_{k+j}{}^*$ as $t \rightarrow \infty$. Since $V_k \ge V_{k+j} > V_{k+j}{}^*$ we can define all stages except the final stage for the coast phase of the optimal strategy by the formulae

$$v_0(t) = v_{k+j}{}^*(t - \left[ \sum_{r=0}^{k} \alpha_r + \beta + \sum_{s=0}^{j} \gamma_s \right] + t_{k+j}[V_{k+j}])$$

for

$$t \in \left( \sum_{r=0}^{k} \alpha_r + \beta + \sum_{s=0}^{j} \gamma_s , \sum_{r=0}^{k} \alpha_r + \beta + \sum_{s=0}^{j+1} \gamma_s \right)$$

for each $j = 0, 1, \ldots, n - k - 2$. For the final stage of the coast phase we begin by defining $t_{n-1} : [0, V_k] \rightarrow [0, t_{n-1}(0)]$ by the formula

$$t_{n-1}(v) = \int_v^{V_k} \frac{1}{r(w)} \, dw$$

Since $r(w) \geq r_0 > 0$ it follows that

$$t_{n-1}(v) \leq \frac{(V_k - v)}{r_0}.$$

Now the inverse function $v_{n-1}^* : [0, t_{n-1}(0)] \rightarrow [0, V_k]$ satisfies the differential equation

$$v_{n-1}^{*\prime}(t) = (-1) \, r \, [v_{n-1}^*(t)]$$

and since $V_k \geq V_{n-1} > 0$ we can define the final stage of the coast phase by the formula

$$v_0(t) = v_{n-1}^* \left( t - \left[ \sum_{r=0}^{k} \alpha_r + \beta + \sum_{s=0}^{n-k-1} \gamma_s \right] + t_{n-1}[V_{n-1}] \right)$$

for

$$t \in \left( \sum_{r=0}^{k} \alpha_r + \beta + \sum_{s=0}^{n-k-1} \gamma_s, \; \sum_{r=0}^{k} \alpha_r + \beta + \sum_{s=0}^{n-k} \gamma_s \right).$$

Incidentally, we can now see that for all $j = 0, 1, \ldots, n - k - 1$ we have $V_{k+j+1} = v_{k+j}^* (\gamma_{j+1} + t_{k+j}[V_{k+j}])$. For the brake phase of the optimal strategy we can begin by defining $t_n : [0, V_k] \rightarrow [0, t_n(0)]$ by the formula

$$t_n(v) = \int_v^{V_k} \frac{1}{r(w) + 1} \, dw.$$

Since $r(w) + 1 \geq r_0 + 1 > 0$ it follows that

$$t_n(v) \leq \frac{V_k - v}{r_0 + 1}.$$

Now the inverse function $v_n^* : [0, t_n(0)] \rightarrow [0, V_k]$ satisfies the differential equation

$$v_n^{*\prime}(t) = -1 - r\,[v_n^*(t)]\,,$$

and since $V_k \geq V_n \geq 0$ we can define the brake phase of the optimal strategy by the formula

$$v_0(t) = v_n^*(t - \left[ \sum_{r=0}^{k} \alpha_r + \beta + \sum_{s=0}^{n-k} \gamma_s \right] + t_n[V_n])$$

for

$$t \in \left( \sum_{r=0}^{k} \alpha_r + \beta + \sum_{s=0}^{n-k} \gamma_s,\ \sum_{r=0}^{k} \alpha_r + \beta + \sum_{s=0}^{n-k} \gamma_s + \delta \right)$$

where $\delta$ is chosen so that $V_{n+1} = v_n^*(\delta + t_n[V_n]) = 0$. Hence $\delta = t_n(0) - t_n(V_n)$ depends on $\alpha$, $\beta$ and $\gamma$.

Now that we can describe the basic format of the optimal strategy it is possible to state the problem in a simplified form. The cost of the optimal strategy is given by

$$J_0(\alpha, \beta, \gamma) = \sum_{i=0}^{k-1} \int_0^{\alpha_{i+1}} p(u_{n-i})q\,[v_i^*(\tau + t_i[V_i])]\,d\tau$$

$$+ p\,[r(V_k)]\,q(V_k)\beta$$

$$+ \sum_{j=0}^{n-k-1} \int_0^{\gamma_{j+1}} p(u_{n-k-j})q\,[v_{k+j}^*(\tau + t_{k+j}[V_{k+j}])]\,d\tau$$

and the distance travelled during the optimal strategy is given by

$$x(\alpha, \beta, \gamma) = \sum_{i=0}^{k-1} \int_0^{\alpha_{i+1}} v_i^*(\tau + t_i[V_i])\,d\tau + V_k\beta$$

$$+ \sum_{j=0}^{n-k-1} \int_0^{\gamma_{j+1}} v_{k+j}^*(\tau + t_{k+j}[V_{k+j}])\,d\tau$$

$$+ \int_0^{t_n(0) - t_n(V_n)} v_n^*(\tau + t_n[V_n])\,d\tau.$$

We can therefore consider the original problem in the following form.

### Simplified Problem

*Minimise $J_0(\alpha, \beta, \gamma)$ subject to the equality constraint $x(\alpha, \beta, \gamma) = X$ and the inequality constraints $\alpha \geq 0$, $\beta \geq 0$, $\gamma \geq 0$ and*

$$\sum_{r=1}^{k} \alpha_r + \beta + \sum_{s=1}^{n-k} \gamma_s + t_n(0) - t_n(V_n) \leq T.$$

□

Thus we form a Lagrangean function

$$\mathcal{J}_0(\alpha, \beta, \gamma; \lambda, \mu, \nu, \eta, \xi)$$

$$= J_0(\alpha, \beta, \gamma) + \lambda \left[X - x(\alpha, \beta, \gamma)\right] - \left[\sum_{r=1}^{k} \mu_r \alpha_r + \nu\beta + \sum_{s=1}^{n-k} \eta_s \gamma_s\right]$$

$$+ \xi \left[\sum_{r=1}^{k} \alpha_r + \beta + \sum_{s=1}^{n-k} \gamma_s + t_n(0) - t_n(V_n) - T\right].$$

The necessary conditions for optimality are now the Kuhn-Tucker equations

$$\frac{\partial \mathcal{J}_0}{\partial \alpha} = 0, \frac{\partial \mathcal{J}_0}{\partial \beta} = 0, \frac{\partial \mathcal{J}_0}{\partial \gamma} = 0$$

with $\lambda \left[X - x(\alpha, \beta, \gamma)\right] = 0$, $\mu_r \alpha_r = 0$, $\nu\beta = 0$, $\eta_s \gamma_s = 0$, and

$$\xi \left[\sum_{r=1}^{k} \alpha_r + \beta + \sum_{s=1}^{n-k} \gamma_s + t_n(0) - t_n(V_n) - T\right] = 0$$

with $\mu_r \geq 0$, $\nu \geq 0$, $\eta_s \geq 0$ and $\xi \geq 0$ for all $r$ and $s$.

To calculate the above derivatives efficiently we must first establish some convenient formulae. When $m < r + 1$ we have

$$\frac{\partial V_{r+1}}{\partial \alpha_m} = \frac{\partial}{\partial \alpha_m} \{v_r^* (\alpha_{r+1} + t_r[V_r])\}$$

$$= v_r^{*\prime}(\alpha_{r+1} + t_r[V_r])t_r'(V_r) \frac{\partial V_r}{\partial \alpha_m}$$

$$= \frac{u_{n-r} - r(V_{r+1})}{u_{n-r} - r(V_r)} \frac{\partial V_r}{\partial \alpha_m},$$

while we also have

$$\frac{\partial V_m}{\partial \alpha_m} = \frac{\partial}{\partial \alpha_m} \{ v_{m-1}^* (\alpha_m + t_{m-1}[V_{m-1}]) \}$$

$$= v_{m-1}^{*\prime}(\alpha_m + t_{m-1}[V_{m-1}])$$

$$= u_{n-m+1} - r(V_m).$$

Hence by induction we obtain the general formula

$$\frac{\partial V_i}{\partial \alpha_m} = \prod_{r=m}^{i-1} \frac{u_{n-r+1} - r(V_r)}{u_{n-r} - r(V_r)} [u_{n-i+1} - r(V_i)]$$

whenever $m < i$. We can extend this result by noting that

$$\frac{\partial V_{k+s+1}}{\partial \alpha_m} = \frac{\partial}{\partial \alpha_m} \{ v_{k+s}^* (\gamma_{s+1} + t_{k+s}[V_{k+s}]) \}$$

$$= v_{k+s}^{*\prime}(\gamma_{s+1} + t_{k+s}[V_{k+s}]) t_{k+s}^{\prime}(V_{k+s}) \frac{\partial}{\partial \alpha_m}(V_{k+s})$$

$$= \frac{u_{n-k-s} - r(V_{k+s+1})}{u_{n-k-s} - r(V_{k+s})} \frac{\partial V_{k+s}}{\partial \alpha_m},$$

from which it follows by induction that

$$\frac{\partial V_{k+j}}{\partial \alpha_m} = \prod_{s=0}^{j-1} \frac{u_{n-k-s} - r(V_{k+s+1})}{u_{n-k-s} - r(V_{k+s})} \frac{\partial V_k}{\partial \alpha_m}$$

for each $j = 1, 2, ..., n - k$. Now the previous result can be used to give

$$\frac{\partial V_{k+j}}{\partial \alpha_m} = \prod_{s=0}^{j-1} \frac{u_{n-k-s} - r(V_{k+s+1})}{u_{n-k-s} - r(V_{k+s})} \prod_{r=m}^{k-1} \frac{u_{n-r+1} - r(V_r)}{u_{n-r} - r(V_r)} [u_{n-k+1} - r(V_k)]$$

$$= \prod_{r=m}^{k+j-1} \frac{u_{n-r+1} - r(V_r)}{u_{n-r} - r(V_r)} [u_{n-k-j+1} - r(V_{k+j})].$$

In similar fashion when $l < s + 1$ we can see that

$$\frac{\partial V_{k+s+1}}{\partial \gamma_l} = \frac{\partial}{\partial \gamma_l}\{v_{k+s}{}^*(\gamma_{s+1}+t_{k+s}+t_{k+s}[V_{k+s}])\}$$

$$= v_{k+s}{}^{*\prime}(\gamma_{s+1}+t_{k+s}[V_{k+s}])t_{k+s}{}'(V_{k+s})\frac{\partial V_{k+s}}{\partial \gamma_l}$$

$$= \frac{u_{n-k-s}-r(V_{k+s+1})}{u_{n-k-s}-r(V_{k+s})}\frac{\partial V_{k+s}}{\partial \gamma_l}.$$

Since it is also clear that

$$\frac{\partial V_{k+l}}{\partial \gamma_l} = \frac{\partial}{\partial \gamma_l}\{v_{k+l-1}{}^*(\gamma_l+t_{k+l-1}[V_{k+l-1}])\}$$

$$= v_{k+l-1}{}^{*\prime}(\gamma_l+t_{k+l-1}[V_{k+l-1}])$$

$$= u_{n-k+l+1}-r(V_{k+l}),$$

we can again use induction to obtain the general formula

$$\frac{\partial V_{k+j}}{\partial \gamma_l} = \prod_{s=l}^{j-1}\frac{u_{n-k-s+1}-r(V_{k+s})}{u_{n-k-s}-r(V_{k+s})}[u_{n-k-j+1}-r(V_{k+j})]$$

whenever $l < j$. By applying these formulae in the calculation of the appropriate partial derivatives we obtain a suitable form for the Kuhn-Tucker equations. In fact we have

$$\frac{\partial \mathcal{J}_0}{\partial \alpha_m} = p(u_{n-m+1})q[v_{m-1}{}^*(\alpha_m+t_{m-1}[V_{m-1}])]$$

$$+ \sum_{i=m}^{k-1}\int_0^{\alpha_{i+1}}p(u_{n-i})q'[v_i{}^*(\tau+t_i[V_i])]$$

$$\times v_i{}^{*\prime}(\tau+t_i[V_i])t_i{}'(V_i)\frac{\partial V_i}{\partial \alpha_m}d\tau$$

$$+ \left\{p'[r(V_k)]r'(V_k)q(V_k)+p[r(V_k)]q'(V_k)\right\}\frac{\partial V_k}{\partial \alpha_m}\beta$$

$$+ \sum_{j=0}^{n-k-1} \int_0^{\gamma_{j+1}} p(u_{n-k-j}) q' \left[ v_{k+j}{}^* (\tau + t_{k+j}[V_{k+j}]) \right]$$

$$\times v_{k+j}{}^{*\prime}(\tau + t_{k+j}[V_{k+j}]) t_{k+j}(V_{k+j}) \frac{\partial V_{k+j}}{\partial \alpha_m} d\tau$$

$$- \lambda \left\{ v_{m-1}{}^*(\alpha_m + t_{m-1}[V_{m-1}]) \right.$$

$$+ \sum_{i=m}^{k-1} \int_0^{\alpha_{i+1}} v_i{}^{*\prime}(\tau + t_i[V_i]) t_i{}'(V_i) \frac{\partial V_i}{\partial \alpha_m} d\tau + \frac{\partial V_k}{\partial \alpha_m}\beta$$

$$+ \sum_{j=0}^{n-k-1} \int_0^{\gamma_{j+1}} v_{k+j}{}^{*\prime}(\tau + t_{k+j}[V_{k+j}]) t_{k+j}{}'(V_{k+j}) \frac{\partial V_{k+j}}{\partial \alpha_m} d\tau$$

$$\left. - v_n{}^* (t_n[V_n]) t_n{}'(V_n) \frac{\partial V_n}{\partial \alpha_m} \right\}$$

$$- \mu_m + \xi \left[ 1 - t_n{}'(V_n) \frac{\partial V_n}{\partial \alpha_m} \right] = 0.$$

This equation can be rewritten in the simplified form

$$p(u_{n-m+1}) q(V_m) + \sum_{i=m}^{n-1} p(u_{n-i}) [q(V_{i+1}) - q(V_i)] \prod_{r=m}^{i} \frac{u_{n-r+1} - r(V_r)}{u_{n-r} - r(V_r)}$$

$$+ (p_{n-k} q)'(V_k) \prod_{r=m}^{k-1} \frac{u_{n-r+1} - r(V_r)}{u_{n-r} - r(V_r)} [u_{n-k+1} - r(V_k)] \beta$$

$$- \lambda \left\{ V_m + \sum_{i=m}^{n} [V_{i+1} - V_i] \prod_{r=m}^{i} \frac{u_{n-r+1} - r(V_r)}{u_{n-r} - r(V_r)} \right.$$

$$\left. + \prod_{r=m}^{k-1} \frac{u_{n-r+1} - r(V_r)}{u_{n-r} - r(V_r)} [u_{n-k+1} - r(V_k)] \beta \right\}$$

$$- \mu_m + \xi \left[ 1 - \prod_{r=m}^{n} \frac{u_{n-r+1} - r(V_r)}{u_{n-r} - r(V_r)} \right] = 0 \qquad (7.6)$$

for each $m$ with $1 \leq m \leq k$. A much less complicated calculation now gives

$$\frac{\partial \mathcal{I}_0}{\partial \beta} = p\,[r(V_k)]\,q(V_k) - \lambda V_k - \nu + \xi = 0$$

from which we obtain the equation

$$(p_{n-k}q)(V_k) - \lambda V_k - \nu + \xi = 0. \tag{7.7}$$

Finally we calculate

$$
\begin{aligned}
\frac{\partial \mathcal{I}_0}{\partial \gamma_1} =\ & p(u_{n-k-l+1})q\,[v_{k-l-1}{}^*\,(\gamma_l + t_{k+l-1}\,[V_{k+l-1}])\,] \\[2mm]
& + \sum_{j=l}^{n-k-1} \int_0^{\gamma_{j+1}} p(u_{n-k-j})q'\,[v_{k+j}{}^*\,(\tau + t_{k+j}\,[V_{k+j}])\,] \\[2mm]
& \qquad\qquad \times v_{k+j}{}^{*\prime}(\tau + t_{k+j}\,[V_{k+j}])\,t_{k+j}{}'(V_{k+j})\frac{\partial V_{k+j}}{\partial \gamma_l}\,d\tau \\[3mm]
& - \lambda \Bigg\{ v_{k+l-1}{}^*\,(\gamma_l + t_{k+l-1}\,[V_{k+l-1}]) \\[2mm]
& \qquad + \sum_{j=l}^{n-k-1} \int_0^{\gamma_{j+1}} v_{k+j}{}^{*\prime}(\tau + t_{k+j}\,[V_{k+j}])\,t_{k+j}{}'(V_{k+j})\frac{\partial V_{k+j}}{\partial \gamma_l}\,d\tau \\[2mm]
& \qquad - v_n{}^*\,(t_n\,[V_n])\,t_n{}'(V_n)\frac{\partial V_n}{\partial \gamma_l} \Bigg\} \\[2mm]
& - \eta_l + \xi\left[1 - t_n{}'(V_n)\frac{\partial V_n}{\partial \gamma_l}\right] = 0
\end{aligned}
$$

and note that this equation can be rewritten in the form

$$
\begin{aligned}
& p(u_{n-k-l+1})q(V_{k+l}) \\[2mm]
& + \sum_{j=l}^{n-k-1} p(u_{n-k-j})\,[q(V_{k+j+1}) - q(V_{k+j})]\prod_{s=1}^{j}\frac{u_{n-k-s+1} - r(V_{k+s})}{u_{n-k-s} - r(V_{k+s})} \\[2mm]
& - \lambda \left\{ V_{k+l} + \sum_{j=l}^{n-k}[V_{k+j+1} - V_{k+j}]\prod_{s=1}^{j}\frac{u_{n-k-s+1} - r(V_{k+s})}{u_{n-k-s} - r(V_{k+s})} \right\} \\[2mm]
& - \eta_l + \xi\left[1 - \prod_{s=1}^{n-k}\frac{u_{n-k-s+1} - r(V_{k+s})}{u_{n-k-s} - r(V_{k+s})}\right] = 0 \tag{7.8}
\end{aligned}
$$

for each $l$ with $1 \leq l \leq n - k$. For convenience in the above formulae we have used the notation $u_0 = -1$ wherever necessary. In order to solve (7.6)–(7.8) we assume to begin with that $\alpha$, $\beta$ and $\gamma$ are positive. Thus we must have $\mu = 0$, $\nu = 0$, and $\eta = 0$. If we take (7.8) with $l = n - k$ it follows that

$$\xi = \lambda V_n.$$

Thus $\xi = 0$ only if $\lambda = 0$ or $V_n = 0$ and from (7.7) it is clear that $\lambda = 0$ is not possible. If we put $l = n - k - 1$ in (7.8) we obtain

$$(p_1 q)(V_{n-1}) - \lambda [V_{n-1} - V_n] = 0$$

and if we subsequently use an inductive argument on (7.8) then it follows that

$$(p_j q)(V_{n-j}) - \lambda [V_{n-j} - V_n] = 0$$

for each $j = 1, 2, ..., n - k - 1$. From (7.7) we get

$$(p_{n-k} q)(V_k) - \lambda [V_k - V_n] = 0.$$

If these result are used in conjunction with (7.6) in the case $m = k$ then it can be seen that

$$(p_{n-k} q)'(V_k) = \lambda. \tag{7.9}$$

With $m = k - 1$ in (7.6) we find that

$$(p_{n-k+1} q)(V_{k-1}) - \lambda [V_{k-1} - V_n] = 0$$

and once again an inductive argument, applied this time to (7.6), can be used to deduce that

$$(p_{n-k+i} q)(V_{k-i}) - \lambda [V_{k-i} - V_n] = 0$$

for each $i = 1, 2, ..., k - 1$. These results can be written collectively in the form

$$\frac{(p_{j-1} q)(V_{n-j+1}) - (p_j q)(V_{n-j})}{V_{n-j+1} - V_{n-j}} = \lambda \tag{7.10}$$

for each $j = 1, 2, ..., n - 1$. In this form (7.10) compares in nature with (7.9). If we have either $\alpha_m = 0$ for some $m$ or $\gamma_l = 0$ for some $l$ then the corresponding stages are simply omitted. The fundamental nature of the results is however the same and the same formulae apply. The notation must be adjusted to allow for re-numbering of the appropriate stages but if we imagine that the whole problem is simply reworked with the null stages omitted it is easy to see that the same analysis will again apply. Similar comments can be made when $\beta = 0$ and the speed-hold phase is omitted. This time however the results are changed to the extent that (7.9) will be deleted.

For a strategy that contains a speed-hold phase, it can now be seen that the maximum speed $\overline{V}$ achieved during the speed-hold phase is sufficient to determine the parameter $\lambda$ and the speeds $V_1, V_2, ..., V_n$. For each selected configuration of null stages and a given maximum speed $\overline{V}$ the time allowed for the speed-hold phase can be adjusted to achieve the appropriate value of $x(\overline{V}, \beta)$, namely $x(\overline{V}, \beta) = X$. Of course for some selected configurations it is possible there may be a violation of the time constraint. When both the distance and time constraints can be satisfied the optimal type strategy is feasible and the cost can be calculated.

## 7.4  Examples

### *Example 7.1  An Optimal Strategy*

In this example, to obtain realistic values for acceleration, we use $u \in [-0.2, 0.1]$ instead of $u \in [-1, 1]$. This does not change the nature of the analysis. We use

$$p(u) = \begin{cases} 0 & u < 0 \\ 5u & 0 \le u < 0.02 \\ 10\,(u - 0.02) + 0.1 & 0.02 \le u < 0.08 \\ 15\,(u - 0.08) + 0.7 & 0.08 \le u \le 0.1 \end{cases}$$

$$q(v) = v$$

and

$$r(v) = 0.015 + 0.00003v + 0.000006v^2.$$

The desired journey distance is $X = 150000$ and desired journey time is $T = 9000$. Distance is measured in metres, and time in seconds. The hold speed is $\overline{V} = 19.536$. The control changes occur at speeds $V_1 = 1.296$, $V_2 = 8.690$, $V_3 = \overline{V} = 19.536$ and $V_4 = 4.380$. The Lagrange multiplier has the value $\lambda = 0.115211$, and the cost of the strategy is $J = 13315$. The speed profile is shown in Figure 7-1. The distance travelled is represented by the area under the curve.

$\square$

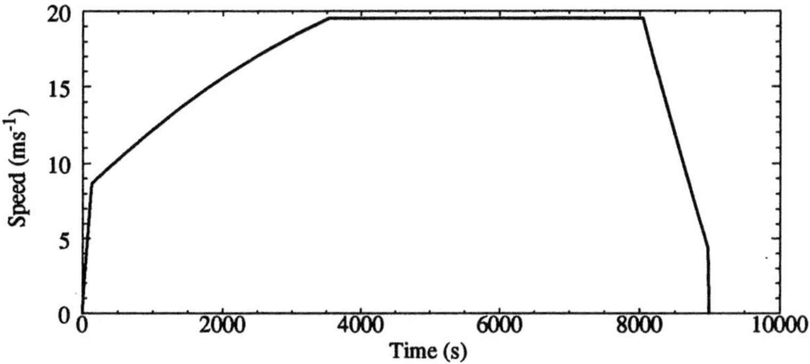

Figure 7-1: Speed profile for optimal strategy of Example 7.1

### *Example 7.2   An Optimal Strategy with Restricted Control*

This example uses the same parameters as Example 7.1, but omits the control phases $u(t) = 0.08$ and $u(t) = 0.02$. The hold speed is $\overline{V} = 17.607$. The control changes occur at speeds $V_1 = \overline{V} = 17.607$ and $V_2 = 3.457$. The Lagrange multiplier has the value $\lambda = 0.108184$. The cost of the strategy is $J = 13935$. The speed profile is shown in Figure 7-2.

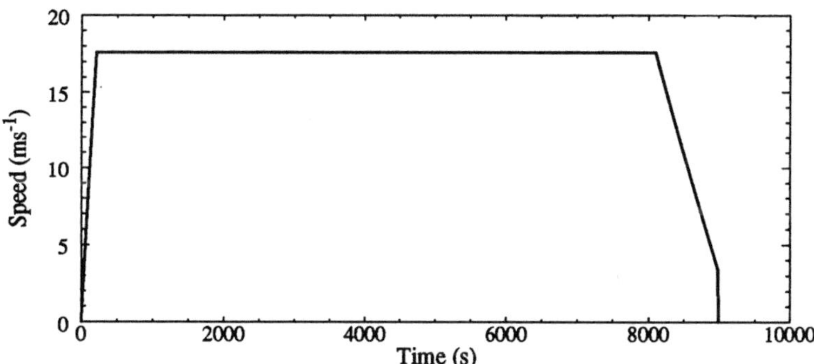

Figure 7-2: Speed profile for the optimal type strategy of Example 7.2

In the case of a strategy that contains no speed-hold phase the value of the parameter $\lambda$ is no longer specified by the maximum speed $\overline{V}$. For each such optimal strategy the value of the parameter $\lambda$ determines the speeds $V_1, V_2, ..., V_n$ from $\overline{V}$ and consequently determines the distance travelled. Thus we must now choose $\lambda$ so that

$x(\overline{V}, \lambda) = X$. Alternatively we can see that determination of any one value $V_i$ (other than the value $V_k = \overline{V}$) will determine $\lambda$ and hence determine all other values $V_j$. If the time constraint is satisfied this strategy will be feasible and can be costed.

### Example 7.3   An Optimal Strategy with No Speed-Holding

This example uses the same parameters as Example 7.1, but omits the speed-hold phase. The maximum speed is $\overline{V} = 24.223$. The control changes occur at speeds $V_1 = 0.778$, $V_2 = 5.401$, $V_3 = \overline{V} = 24.223$ and $V_4 = 2.745$. The Lagrange multiplier has the value $\lambda = 0.108537$. The cost of the strategy is $J = 13449$. The speed profile is shown in Figure 7-3.

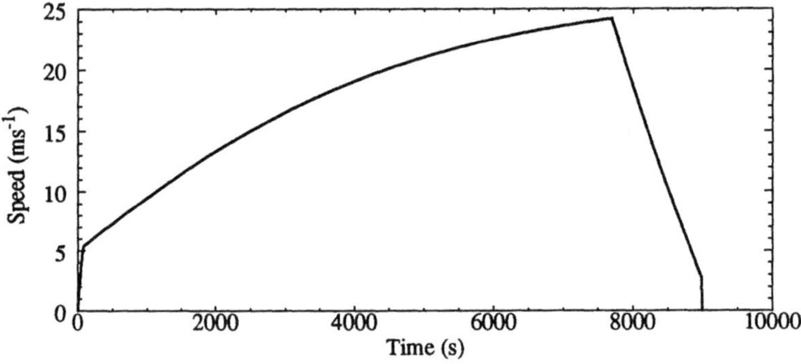

Figure 7-3: Speed profile for the optimal type strategy of Example 7.3

The parameter $\lambda$ is determined by an iterative procedure. Considerable precision is needed to give sufficient accuracy in journey distance and time.

## 7.5 Summary

The above analysis provides a clear answer to the type of strategy that must be adopted to achieve a minimum cost journey. The optimal strategy involves successive levels of constant applied acceleration with each subsequent level less than the preceding one and with the allowable levels restricted to the points at which the slope of the piecewise linear function $p(u)$ changes. If the maximum speed of the optimal strategy is given then a single real number parameter determines the complete journey. The appropriate value of this parameter can be determined by satisfying the distance constraint. By adjusting the maximum speed and the parameter the time constraint can be satisfied and the strategy will then be feasible. The parameters can be determined using an appropriate numerical procedure. For the more general problem where the function $p(u)$ is not piecewise linear it seems reasonable to assume that an approximate optimal strategy could be obtained by using a piecewise linear approximation to $p(u)$.

## 7.6  Limitations of the Model

We have assumed that a constant level of acceleration can be applied independently of the speed of the train. In practice, this is not so. In real trains the driver controls the level of power applied to the wheels, and maximum acceleration decreases as speed increases. To obtain realistic speed profiles with the model used in this chapter it is necessary to limit the acceleration to a maximum value consistent with a realistic limiting speed. In essence, we set $u_{max} = r(V_{max})$ in the above examples. Thus the initial acceleration is much less than it is in practice. To obtain realistic values for the initial acceleration it is necessary to have an unrealistically high limiting speed.

The cost function $p(u)$ used in this chapter is convex, and corresponds to a traction system that is more efficient in the mid-range of acceleration. In practice, it is likely that the traction system will be more efficient using maximum power. Many models assume constant efficiency, which corresponds to a cost function

$$p(u) = \begin{cases} 0 & u < 0 \\ u & u \geq 0 \end{cases}$$

Although the mechanical energy model was useful in developing *Metromiser*, a more realistic model, based on the locomotives used by Australian National, was developed by Benjamin et al [8] to investigate fuel efficiency on long-haul operations. This model is discussed in the next section.

# Analysis of the Fuel Consumption Model

# CHAPTER 8
# CRITICAL SPEEDS AND STRATEGIES OF OPTIMAL TYPE

When the fuel consumption model is used to formulate the train control problem on level track, an optimal strategy no longer exists. There is no feasible strategy that minimises fuel consumption. We will show that for any given sequence of control settings there exist optimal switching times that define a *strategy of optimal type*. This strategy gives the minimum fuel consumption for the given control sequence. Each strategy of optimal type is defined by two critical speeds.

This chapter was originally published as a paper in the IFAC journal *Automatica* [14].

## 8.1 Introduction

### 8.1.1 A Driver Perspective on the Problem

To understand the essence of the problem it is helpful to consider a simplified model.

Suppose the locomotive has only three discrete control settings; *power, coast* and *brake*. Consider the possible control options for the driver. The strategy must begin with a power phase and follow with alternate phases of coast and power. The strategy must end with a semi-final coast phase and a final brake phase although it is possible that one or other of these latter two phases could become degenerate. We assume that braking will be used only to stop the train.

The driver can decide the number of phases and the times when the control setting will be changed. These times are called the *switching times*.

The nature of the problem is not changed by allowing a greater range of discrete control settings. The driver must first decide the precise sequence of control settings and then determine the optimal switching times.

The driver must make these decisions in such a way that fuel consumption is minimised.

### 8.1.2  A Well-Posed Problem

Because the control strategy is restricted to a finite sequence of discrete settings the problem described above is not well-posed. In general there will be no admissible strategy for which the minimum fuel consumption is achieved, and hence no definitive conditions for optimality.

We overcome this difficulty in the following way. Consider the set of all possible control strategies. We divide the set into disjoint subsets and seek the best strategy from each subset. Each subset is defined by a fixed finite sequence of control settings. There are many different strategies in each subset; all strategies use the same sequence of control settings but each one is determined by different switching times.

To find a feasible strategy that minimises fuel consumption within the given control subset we need to find the optimal locations for the switching times subject to the distance and time constraints. A strategy satisfying these requirements will be called a strategy of optimal type. This problem is a standard finite dimensional constrained optimisation problem.

## 8.2  Formulation of the Train Control Problem

### 8.2.1  The Control Strategies

To describe the control mechanism we introduce a control variable $j$. Each non-negative value of the control variable determines a traction control and the single negative value determines a braking control. Let

$$C = \{-1, 0, 1, 2, \ldots, m\}$$

denote the set of all possible values for $j$. Let $f_j$ be the fuel supply rate corresponding to the control setting $j$ and assume that

$$0 = f_{-1} = f_0 < f_1 < f_2 < \ldots < f_m = 1.$$

Consider a fixed sequence $\{j(k+1)\}_{k=0, 1, \ldots, n}$ of control settings. We suppose that for $k < n$ the index $j(k+1)$ takes integer values between 0 and $m$ from the set $C$. The semi-final phase is a coast phase with $j(n) = 0$ and the final phase is a brake phase with $j(n+1) = -1$. This sequence defines a subset

$$S(\{j(k+1)\}_{k=0, 1, \ldots, n})$$

of control strategies each with $n + 1$ distinct control phases. Let us define

$$0 = t_0 \leq t_1 \leq \dots \leq t_{n+1}$$

where $t_0$ is the starting time, $\{t_k\}_{k=1,2,\dots,n}$ are the *switching times*, $t_{n+1}$ is the stopping time and $j(k+1)$ is the control setting on the interval $(t_k, t_{k+1})$. We use $\tau_{k+1}$ to denote the length of the interval $(t_k, t_{k+1})$. The control strategy from the given subset and with the above switching times is denoted by

$$S(\{\,[j(k+1);(t_k,t_{k+1})]\,\}_{k=0,1,\dots,n}).$$

We write $x_k = x(t_k)$ to denote the position of the train at time $t_k$, and use $\xi_{k+1} = x_{k+1} - x_k$ to denote the distance travelled during the interval $(t_k, t_{k+1})$. The points $\{x_k\}_{k+1,2,\dots,n}$ are called the *switching points*. The total fuel consumption for this strategy is given by

$$J = \sum_{k=0}^{n} f_{j(k+1)}\tau_{k+1}.$$

### 8.2.2  The Equations of Motion

When $j \geq 0$ the power developed by the locomotive is directly proportional to the rate of fuel supply. When $j < 0$ there is a constant negative acceleration applied to the train. If we write $K_j$ for the braking acceleration then $K_j = 0$ for $j \geq 0$ and $K_{-1} = -K$ is a negative constant. The equations of motion for a point mass train on level track are

$$\frac{dx}{dt} = v$$

and

$$\frac{dv}{dt} = \frac{Hf_j}{v} + K_j - r(v)$$

where $x$ is the position on the track, $v$ is the speed of the train, $-r(v)$ is the resistive acceleration due to friction, and $H$ is a constant. We assume that $r(0) > 0$, $r(v)$ is strictly increasing and that the graph $y = vr(v)$ is strictly convex.

### 8.2.3  A Precise Statement of the Problem

The problem can now be stated precisely. Let $X$ be the length of the journey and let $T$ be the time allowed for the journey. For a fixed control sequence $\{j(k+1)\}_{k=0,1,\ldots,n}$, we wish to choose the switching times $\{t_k\}_{k=1,2,\ldots,n}$ to define a control strategy

$$S(\{\,[j(k+1);(t_k,t_{k+1})\,]\,\}_{k=0,1,\ldots,n}) \in S(\{j(k+1)\}_{k=0,1,\ldots,n})$$

with $v(t_0) = 0$, $v(t_{n+1}) = 0$, $\Sigma_{k=0}^{n}\tau_{k+1} = T$ and $\Sigma_{k=0}^{n}\xi_{k+1} = X$ in such a way that

$$J = \sum_{k=0}^{n} f_{j(k+1)}\tau_{k+1}$$

is minimised. The corresponding strategy is called a *strategy of optimal type*.

In a subsequent chapter we will consider a systematic comparison of all different feasible strategies of optimal type to find a minimum cost strategy. The comparison procedure depends directly on the restricted optimality of the strategies discussed in this chapter.

## 8.3  The Nature of the Resistive Acceleration

The resistive acceleration due to friction depends only on the speed. In practice it is found that the Davis formula [16]

$$r(v) = a + bv + cv^2$$

where $a$, $b$ and $c$ are positive constants is a satisfactory approximation to the actual resistance. However in this paper we assume only that $r(0) > 0$, $r(v)$ is strictly increasing and that the graph

$$y = vr(v)$$

is strictly convex. These properties are certainly true if the Davis formula is used. In the examples we will use a Davis formula with realistic values for the constants.

## 8.4  The Fundamental Speed Profiles

Before we can proceed with the actual optimisation it is necessary to define speed functions that satisfy the equations of motion. The calculation of derivatives used in the optimisation procedure makes direct use of these definitions. To define the speed on the time interval $(t_k, t_{k+1})$ for each $k = 0, 1, ..., n-1$ we must first define $W_j$ as the solution to the equation

$$Hf_j - vr(v) = 0.$$

$W_j$ is the speed at which the frictional resistance equals the driving force. Secondly we use a direct integration of the equation of motion to define time as a function of speed. We set

$$t_k^*(v) = \begin{cases} \displaystyle\int_0^v \frac{w\,dw}{Hf_{j(k+1)} - wr(w)} & v \in [0, W_{j(k+1)}) \\[4mm] \displaystyle\int_v^{W_n} \frac{w\,dw}{wr(w) - Hf_{j(k+1)}} & v \in (W_{j(k+1)}, W_m] \end{cases}.$$

The value of $V_k$ determines which of the above definitions is used. Because the slope of the graph

$$y = vr(v)$$

is positive and bounded when $v \in [0, W_m]$ it is easy to show that $t_k^*(v) \to \infty$ as $v \to W_{j(k+1)}$. It now follows that if $V_k < W_{j(k+1)}$ then $V_{k+1} < W_{j(k+1)}$. Alternatively if $V_k > W_{j(k+1)}$ then $V_{k+1} > W_{j(k+1)}$. Whichever of the above definitions is used we denote the inverse function by $v_k^*$ and note that the actual speed on the interval $(t_k, t_{k+1})$ for $k = 0, 1, ..., n-1$ is given by

$$v(t) = v_k^*(t - t_k + t_k^*(V_k))$$

provided $V_k \neq W_{j(k+1)}$. The distance travelled in this time interval is

$$\xi_{k+1} = \int_0^{\tau_{k+1}} v_k^*(\tau + t_k^*(V_k))\,d\tau$$

and since

$$V_k = V_k(\tau_1, \tau_2, ..., \tau_k)$$

then

$$\xi_{k+1} = \xi_{k+1}(\tau_1, \tau_2, ..., \tau_{k+1}).$$

For the final interval we define

$$t_n^*(v) = \int_0^{W_m} \frac{dw}{K + r(w)}$$

for $v \in [0, W_m]$ and use $v_n^*(t)$ to denote the inverse function. The actual speed on the interval $(t_n, t_{n+1})$ is

$$v(t) = v_n^*(t - t_n + t_n^*(V_n))$$

and the distance travelled in this time interval is

$$\xi_{n+1} = \int_0^{\tau_{n+1}} v_n^*(\tau + t_n^*(V_n))d\tau.$$

Since

$$V_n = V_n(\tau_1, \tau_2, ..., \tau_n)$$

we can see that

$$\xi_{n+1} = \xi_{n+1}(\tau_1, \tau_2, ..., \tau_{n+1}).$$

It is important to realise that

$$\tau_{n+1} = t_n^*(0) - t_n^*(V_n)$$

and hence the dependence on $\tau_{n+1}$ can be eliminated. If we write $\tau = (\tau_1, \tau_2, ..., \tau_n) \in \mathfrak{R}^n$ then the key variables can be expressed as functions of $\tau$. In particular the cost of the strategy is

$$J(\tau) = \sum_{k=0}^{n-1} f_{j(k+1)}\tau_{k+1}$$

while the total distance travelled and the time taken are given respectively by

$$x(\tau) = \sum_{k=0}^{n-1} \int_0^{\tau_{k+1}} v_k^*(\tau + t_k^*(V_k))d\tau + \int_{t_n^*(V_n)}^{t_n^*(0)} v_n^*(\rho)d\rho$$

and

$$t(\tau) = \sum_{k=0}^{n-1} \tau_{k+1} + [t_n^*(0) - t_n^*(V_n)] .$$

## 8.5 Necessary Conditions for a Strategy of Optimal Type

We wish to minimise $J(\tau)$ subject to the equality constraints $x(\tau) = X$ and $t(\tau) = T$, and the inequality constraints $\tau_1 > 0, \tau_2 > 0, ..., \tau_n > 0$. For $\lambda, \mu \in \Re$ and $v = (v_1, v_2, ..., v_n) \in \Re^n$ we define a Lagrangean function

$$\mathcal{J}(\tau, \lambda, \mu, v) = HJ(\tau) + \lambda [X - x(\tau)] + \mu [t(\tau) - T] - \sum_{k=0}^{n-1} v_{k+1}\tau_{k+1}$$

and apply the Kuhn-Tucker conditions

$$\frac{\partial \mathcal{J}}{\partial \tau_{k+1}} = 0$$

for all $k$, and the complementary slackness conditions

$$\lambda [X - x(\tau)] = 0,$$

$$\mu [t(\tau) - T] = 0$$

and

$$v_{k+1}\tau_{k+1} = 0$$

for all $k$, where the Lagrange multipliers $v_{k+1}$ are guaranteed non-negative for all $k$. If we weaken the distance and time constraints to read $x(\tau) \geq X$ and $t(\tau) \leq T$ then we can also guarantee that $\lambda$ and $\mu$ are non-negative.

We will assume that $\tau$ lies in an open set such that $V_k(\tau_1, \tau_2, ..., \tau_n) \neq W_{j(k+1)}$ and begin by calculating some relevant partial derivatives. The details of these calculations are similar to calculations performed in Chapter 7, and the details are therefore omitted. Choose $h < n$. From

$$V_{h+1} = v_h^*(\tau_{h+1} + t_h^*(V_h))$$

it follows that

$$\frac{\partial V_{h+1}}{\partial \tau_{h+1}} = \frac{Hf_{j(h+1)} - V_{h+1}r(V_{h+1})}{V_{h+1}}$$

whereas when $k < h$ we find that

$$\frac{\partial V_{h+1}}{\partial \tau_{k+1}} = \frac{[Hf_{j(h+1)} - V_{h+1}r(V_{h+1})]}{[Hf_{jh+1} - V_h r(V_h)]} \frac{V_h}{V_{h+1}} \frac{\partial V_h}{\partial \tau_{h+1}}.$$

By recursive application of this formula it follows that

$$\frac{\partial V_{h+1}}{\partial \tau_{k+1}} = \frac{Hf_{j(h+1)} - V_{h+1}r(V_{h+1})}{V_{h+1}} \prod_{s=k+1}^{h} r_s$$

where we have used the notation

$$r_s = \frac{[Hf_{j(s)} - V_s r(V_s)]}{[Hf_{j(s+1)} - V_s r(V_s)]}$$

for $s < n$. Since

$$\xi_{h+1} = \int_0^{\tau_{h+1}} v_h^*(\tau + t_h^*(V_h))d\tau$$

we find that

$$\frac{\partial \xi_{h+1}}{\partial \tau_{h+1}} = V_{h+1}$$

while for $k < h$ we have

$$\frac{\partial \xi_{h+1}}{\partial \tau_{k+1}} = \int_0^{\tau_{h+1}} v_h^{*\prime}(\tau + t_h^*(V_h))t_h^{*\prime}(V_h) \frac{\partial V_h}{\partial \tau_{k+1}} \, d\tau$$

$$= (V_{h+1} - V_h) \prod_{s=k+1}^{h} r_s.$$

We also know that

$$\xi_{n+1} = \int_{t_n^*(V_n)}^{t_n^*(0)} v_n^*(\rho)\, d\rho$$

and therefore

$$\frac{\partial \xi_{n+1}}{\partial \tau_n} = (-1)\, V_n r_n$$

where

$$r_n = (-1)\, \frac{[Hf_{j(n)} - V_n r(V_n)]}{V_n\,[K + r(V_n)]}\,.$$

Now if we assume $\tau_{k+1} > 0$ then $v_{k+1} = 0$ for all $k$ and hence

$$\frac{\partial \mathcal{J}}{\partial \tau_n} = H\frac{\partial J}{\partial \tau_n} - \lambda\left[\frac{\partial \xi_n}{\partial \tau_n} + \frac{\partial \xi_{n+1}}{\partial \tau_n}\right] + \mu\left[1 - t_n^{*\prime}(V_n)\frac{\partial V_n}{\partial \tau_n}\right].$$
$$= Hf_{j(n)} - (\lambda V_n - \mu)\,(1 - r_n)\,.$$

If we assume that $f_{j(n)} = 0$ then the condition $\partial \mathcal{J}/\partial \tau_n = 0$ becomes

$$\lambda V_n - \mu = 0. \tag{8.1}$$

In general for $k + 1 < n$ it follows that

$$\frac{\partial \mathcal{J}}{\partial \tau_{k+1}} = H\frac{\partial J}{\partial \tau_{k+1}} - \lambda\left[\sum_{h=k}^{n}\frac{\partial \xi_{h+1}}{\partial \tau_{k+1}}\right] + \mu\left[1 - t_n^{*\prime}(V_n)\frac{\partial V_n}{\partial \tau_{k+1}}\right]$$
$$= Hf_{j(k+1)} - \lambda\left[V_{k+1} + \sum_{h=k+1}^{n}(V_{h+1} - V_h)\prod_{s=k+1}^{h} r_s\right.$$
$$\left. + \mu\left[1 - \prod_{s=k+1}^{n} r_s\right].\right.$$

Therefore the equation $\partial \mathcal{J}/\partial \tau_{k+1} = 0$ can be written in the form

$$Hf_{j(k+1)} - (\lambda V_n - \mu)\left(1 - \prod_{s=k+1}^{n} r_s\right)$$
$$- \lambda\left[V_{k+1} - V_n + \sum_{h=k+1}^{n-1}(V_{h+1} - V_h)\prod_{s=k+1}^{h} r_s\right] = 0.$$

By writing the equation $\partial \mathcal{J} / \partial \tau_k = 0$ in a similar form and by noting that

$$\frac{\partial \mathcal{J}}{\partial \tau_k} - r_k \frac{\partial \mathcal{J}}{\partial \tau_{k+1}} = 0$$

we obtain

$$H \left[ f_{j(k)} - r_k f_{j(k+1)} \right] - (\lambda V_k - \mu)(1 - r_k) = 0$$

and hence

$$\lambda V_k - \mu = V_k r(V_k). \tag{8.2}$$

Equations (8.1) and (8.2) define the critical speeds. There is only one solution $v = U$ to the equation

$$\lambda v - \mu = 0$$

hence (8.1) shows that $V_n = U$. Therefore $U$ is the speed at which braking begins. Since the graph $y = vr(v)$ is convex there are only two solutions to the equation

$$\lambda v - \mu = vr(v).$$

We let $v = V$ be the lesser and $v = W$ be the greater of the two solutions. For $k < n$ it follows that we must have either $V_k = V$ or $V_k = W$. The former speed is the speed at which a phase of negative net acceleration is changed to a phase of positive net acceleration, while the latter is the speed at which a phase of positive net acceleration is changed to a phase of negative net acceleration. The speeds $U$, $V$ and $W$ are the *critical speeds*. Because $\lambda$ and $\mu$ are positive it is clear that $0 < U < V < W$ and that

$$\lambda = \frac{Wr(W) - Vr(V)}{W - V}$$

and

$$\mu = \frac{VW \left[ r(W) - r(V) \right]}{W - V}.$$

For each prescribed sequence of fuel supply rates the parameters $(\lambda, \mu)$ determine a *strategy of optimal type*. The strategy of optimal type is also defined by the critical speeds $V$ and $W$. Nevertheless, this strategy will not be feasible if the distance and time constraints are not satisfied. It is necessary to adjust the values of the parameters in order to satisfy these constraints. From the formulae given above it is obvious that adjustments to the parameter values can be made by changing the values of the critical speeds $V$ and $W$. We have illustrated the relationship between the parameters and the critical speeds in Figure 8-1.

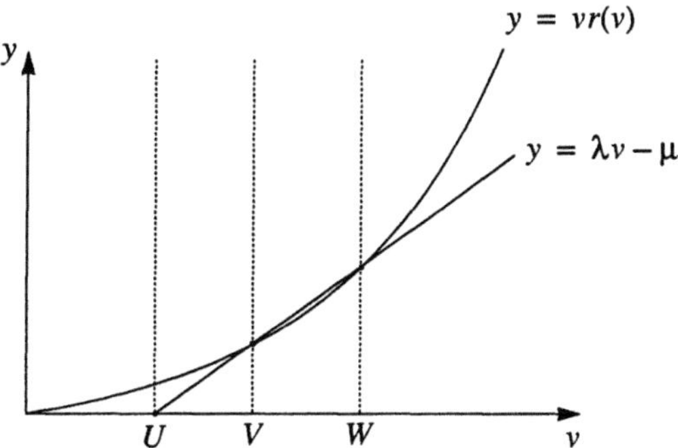


Figure 8-1: Relationship between the parameters and the critical speeds.

## 8.6 Calculating a Feasible Strategy of Optimal Type

In this section we discuss a systematic procedure for adjusting the critical speeds $V$ and $W$ so that we obtain a feasible strategy. For convenience we consider a strategy of optimal type with $n = 2q$, a control sequence given by

$$j(1) = j(3) = \ldots = j(2q-1) = m$$

and

$$j(2) = j(4) = \ldots = j(2q) = 0,$$

and with critical speeds

$$V_1 = V_3 = \ldots = V_{2q-1} = W,$$

$$V_2 = V_4 = \ldots = V_{2q-2} = V$$

and

$$V_{2q} = U(V, W)$$

where

$$U(V, W) = \frac{VW\,[r(W) - r(V)]}{Wr(W) - Vr(V)}.$$

We assume that

$$\frac{Wr(W)}{H} < f_m.$$

We calculate the time taken for the journey and the distance travelled by the formulae

$$t_q(V, W) = \int_0^W \frac{v\,dv}{Hf_m - vr(v)} + (q - 1) \int_V^W \frac{Hf_m\,dv}{[Hf_m - vr(v)]\,r(v)}$$

$$+ \int_{U(V, W)}^W \frac{dv}{r(v)} + \int_0^{U(V, W)} \frac{dv}{K + r(v)}$$

and

$$x_q(V, W) = \int_0^W \frac{v^2\,dv}{Hf_m - vr(v)} + (q - 1) \int_V^W \frac{Hf_m v\,dv}{[Hf_m - vr(v)]\,r(v)}$$

$$+ \int_{U(V, W)}^W \frac{v\,dv}{r(v)} + \int_0^{U(V, W)} \frac{v\,dv}{K + r(v)}.$$

We will show that if $W$ is fixed then $t_q(V, W)$ decreases as $V$ increases. Since

$$\frac{\partial t_q}{\partial V} = (-1)(q - 1) \frac{Hf_m}{[Hf_m - Vr(V)]\,r(V)} - \frac{K}{[K + r(U)]\,r(U)} \frac{\partial U}{\partial V}$$

and since $\partial U / \partial V > 0$ it is easy to see that

$$\frac{\partial t_q}{\partial V} < 0.$$

The maximum value of $t_q(V, W)$ occurs when $V = 0$ and the minimum value occurs in the limit as $V \uparrow W$. For convenience we will write

$$\lim_{V \uparrow W} t_q(V, W) = t_q(W, W)$$

and

$$\lim_{V \uparrow W} x_q(V, W) = x_q(W, W).$$

The range of values for $t_q(V, W)$ is illustrated in Figure 8-2 for the case $q = 3$.

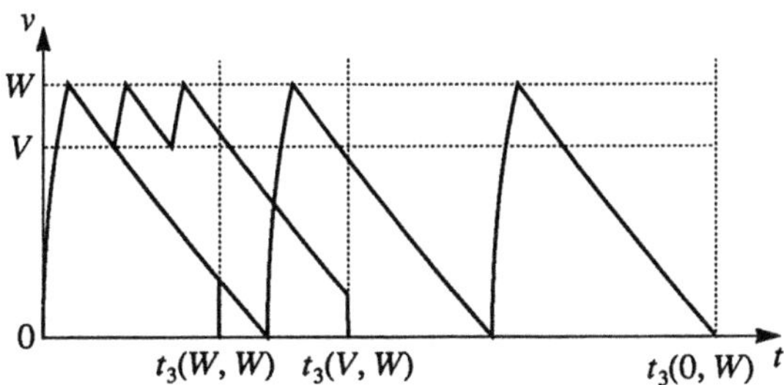


Figure 8-2: Range of values for $t_3(V, W)$

In general, provided

$$t_q(W, W) < T < t_q(0, W)$$

there must be a unique solution to the equation $t_q(V, W) = T$. If we now adjust $(V, W)$ in such a way that the relationship $t_q(V, W) = T$ is preserved then $x_q(V, W)$ increases as $W$ increases. If we regard $V = V(W)$ and note that the equation

$$\frac{dt_q}{dW} = 0$$

can be rewritten as

$$\frac{\partial t_q}{\partial V} \frac{dV}{dW} + \frac{\partial t_q}{\partial W} = 0$$

and hence

$$(-1)\,(q-1)\,\frac{Hf_m}{[Hf_m - Vr(V)]\,r(V)}\frac{dV}{dW} - \frac{H}{[H + r(U)]\,r(U)}\frac{dU}{dW}$$
$$+ q\,\frac{Hf_m}{[Hf_m - Wr(W)]\,r(W)} = 0. \qquad (8.3)$$

If $dV/dW$ is negative then (8.3) shows that

$$\frac{dU}{dW} > 0.$$

On the other hand we know that $\partial U/\partial V$ and $\partial V/\partial W$ are both positive, and hence if $dV/dW$ is positive we again see that

$$\frac{dU}{dW} = \frac{\partial U}{\partial V}\frac{dV}{dW} + \frac{\partial U}{\partial W} > 0.$$

We can now show that

$$\frac{dx_q}{dW} = (-1)(q-1)\frac{Hf_m V}{[Hf_m - Vr(V)]\,r(V)}\frac{dV}{dW}$$

$$- \frac{HU}{[H+r(U)]\,r(U)}\frac{dU}{dW} + q\frac{Hf_m W}{[Hf_m - Wr(W)]\,r(W)}$$

is also positive. In the case where $dV/dW$ is negative this is clear from the inequality

$$\frac{dx_q}{dW} > U\left\{(-1)(q-1)\frac{Hf_m}{[Hf_m - Vr(V)]\,r(V)}\frac{dV}{dW}\right.$$

$$\left. - \frac{H}{[H+r(U)]\,r(U)}\frac{dU}{dW} + q\frac{Hf_m}{[Hf_m - Wr(W)]\,r(W)}\right\}$$

which implies

$$\frac{dx_q}{dW} > U\frac{dt_q}{dW} = 0.$$

When $dV/dW$ is positive a similar argument shows that

$$\frac{dx_q}{dW} > V\frac{dt_q}{dW} = 0.$$

Therefore $x_q(V(W), W)$ increases as $W$ increases. As long as we can find $\overline{W}$ with

$$t_q(\overline{W}, \overline{W}) = T$$

and

$$(x_q(\overline{W}, \overline{W})) > X$$

it now follows that we can find a unique pair $(X_q, Y_q)$ with $0 < X_q < Y_q < \overline{W}$ and such that $t_q(X_q, Y_q) = T$ and $x_q(X_q, Y_q) = X$. The cost of the strategy is given by

$$J = \int_0^{Y_q} \frac{v\,dv}{Hf_m - vr(v)} + (q-1) \int_{X_q}^{Y_q} \frac{v\,dv}{Hf_m - vr(v)}.$$

## 8.7  Numerical Examples

The following examples are based on data obtained from train models used by the Scheduling and Control Group. Length is measured in metres and time is measured in seconds. All calculations were originally performed on an IBM compatible PC using the software package MATLAB.

We consider a journey with $X = 18000$ and $T = 1500$. We assume that $r(v) = a + bv + cv^2$ where $a = 1.5 \times 10^{-2}$, $b = 3 \times 10^{-5}$ and $c = 6 \times 10^{-6}$. We take $H = 1.5$ and $K = 1$, and assume $C = \{-1, 0, 1\}$ with $f_{-1} = f_0 = 0$ and $f_1 = 1$.

### Example 8.1  A Minimum Time Strategy

To travel the distance in the least possible time we consider a strategy with only an acceleration phase and a braking phase. We take $n = 1$ and set $j(0) = 1$ and $j(2) = -1$. We set $V_1 = W$ and note that the distance travelled is given by

$$x(W) = \int_0^W \left[ \frac{v^2}{H - vr(v)} + \frac{v}{K + r(v)} \right] dv$$

By adjusting $W$ using a Newton iteration we can find an approximate solution to the equation $x(W) = X$. The time taken for the journey is now calculated by

$$t(W) = \int_0^W \left[ \frac{v}{H - vr(v)} + \frac{1}{K + r(v)} \right] dv$$

and the cost of the strategy is

$$J = \int_0^W \frac{v\,dv}{H - vr(v)}.$$

The results of the calculation were $W = 35.98$, $t(W) = 708.93$ and $J = 673.59$. Since $t(W) \leq 1500$ the strategy is feasible. There is no optimisation available with this strategy.

$\square$

### Example 8.2  A Strategy of Optimal Type

If we use a strategy with $n = 2$, $j(1) = 1$, $j(2) = 0$ and $j(3) = -1$ then a minimum cost strategy is obtained as follows. We set $V_1 = W$ and $V_2 = U$, and according to the optimisation procedure we require $\lambda$ and $\mu$ such that $\lambda W - \mu = Wr(W)$ and $\lambda U - \mu = 0$. If we assume that the journey is completed ahead of schedule (i.e. $t_3 < T$) then the Kuhn-Tucker conditions require that $\mu = 0$ and hence $U = 0$. In this case the braking phase is effectively eliminated and the distance travelled by the train is given by

$$x(W) = \int_0^W \left[ \frac{v^2}{H - vr(v)} + \frac{v}{r(v)} \right] dv.$$

By adjusting $W$ using a Newton iteration we find an approximate solution to the equation $x(W) = X$. The time taken for the journey is now calculated by

$$t(W) = \int_0^W \left[ \frac{v}{H - vr(v)} + \frac{1}{r(v)} \right] dv.$$

The results of the calculation show that $W = 22.41$ and $t(W) = 1581.26$. Since $t(W) > T$ this strategy is not feasible and hence the assumption that the journey is completed ahead of schedule is false. We must therefore have $t_3 = T$ and hence $\mu > 0$ and $U > 0$ are allowed. The time taken for the journey is given by

$$t(U, W) = \int_0^W \frac{v\,dv}{H - vr(v)} + \int_U^W \frac{dv}{r(v)} + \int_0^U \frac{dv}{K + r(v)}$$

and the distance travelled is

$$x(U, W) = \int_0^W \frac{v^2\,dv}{H - vr(v)} + \int_U^W \frac{v\,dv}{r(v)} + \int_0^U \frac{v\,dv}{K + r(v)}.$$

The cost of the strategy is

$$J = \int_0^W \frac{v\,dv}{H - vr(v)}.$$

We set an initial value for $W$ and solve the equation $t(U, W) = T$ to find $U$. We now calculate $x(U, W)$. If the distance is too large $W$ is decreased and the process is repeated. If the distance is too small $W$ is increased. The results of the calculation were $W = 22.45$, $U = 1.28$ and $J = 204.39$. The dramatic cost saving is due to the use of coasting.

□

The speed profiles for Examples 8.1 and 8.2 are shown in Figures 8-3 and 8-4.

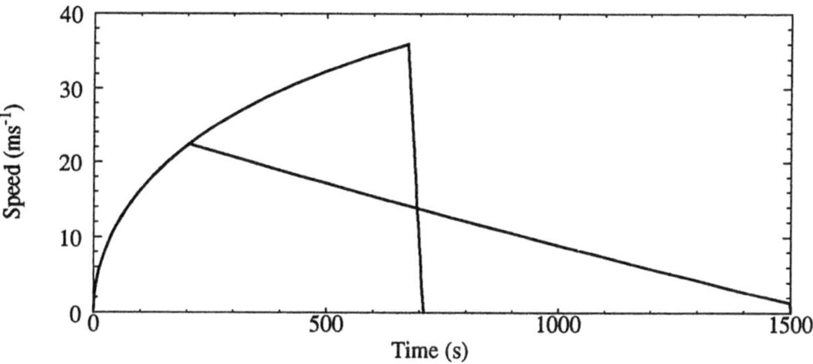

Figure 8-3: Speed $v(t)$ for Examples 8.1 and 8.2

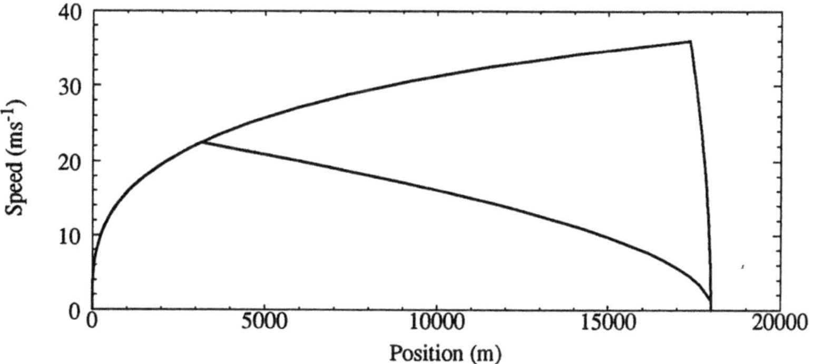

Figure 8-4: Speed $v(x)$ for Examples 8.1 and 8.2

### *Example 8.3  An Improved Strategy of Optimal Type*

If we use a driving strategy with $n = 4$, $j(1) = j(3) = 1$, $j(2) = j(4) = 0$ and $j(5) = -1$ then a minimum cost strategy is obtained as follows. We select critical speeds $V$ and $W$ with $V < W$ and determine the third critical speed $U = U(V, W)$. We set $V_1 = V_3 = W$, $V_2 = V$ and $V_4 = U$, and adjust $V$ and $W$ to satisfy the time and distance constraints. The results of the calculation were $W = 18.59$, $V = 13.37$, $U = 2.68$ and $J = 202.15$.

$\square$

### *Example 8.4  A Further Improved Strategy of Optimal Type*

If we use a driving strategy with $n = 8$, $j(1) = j(3) = j(5) = j(7) = 1$, $j(2) = j(4) = j(6) = j(8) = 0$ and $j(9) = -1$ then a minimum cost strategy is obtained as follows. We select critical speeds $V$ and $W$ with $V < W$ and determine the third critical speed $U = U(V, W)$. We set $V_1 = V_3 = V_5 = V_7 = W$, $V_2 = V_4 = V_6 = V$ and $V_8 = U$, and adjust $V$ and $W$ to satisfy the time and distance constraints. The results of the calculation were $W = 17.25$, $V = 15.03$, $U = 2.81$ and $J = 201.93$.

$\square$

### *Example 8.5  Nine Coast-Power Pairs*

A similar strategy with $n = 20$ gives $W = 16.58$, $V = 15.76$, $U = 2.83$ and $J = 201.89$.

$\square$

The speed profiles for Examples 8.3–8.5 are illustrated in Figures 8-5 and 8-6.

We have graphed the speed profiles against time and against position. When speed is graphed against time, the distance travelled is represented by the area under the curve. When speed is graphed against position, however, there is no direct way of visualising the time taken for the journey.

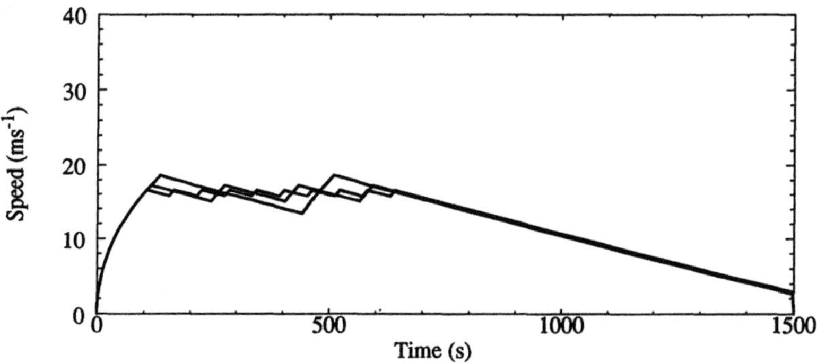

Figure 8-5: Speed $v(t)$ for Examples 8.3–8.5

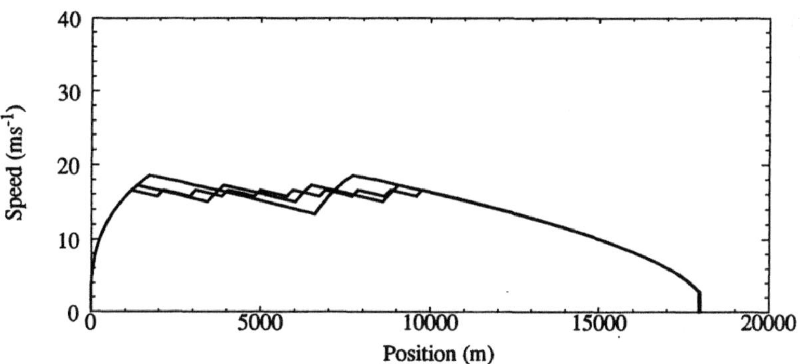

Figure 8-6: Speed $v(x)$ for Examples 8.3–8.5

## 8.8  Conclusions and Future Developments

For a prescribed sequence of fuel supply rates we have shown that a strategy of optimal type depends on two parameters which determine three critical values of the speed. By adjusting the values of the critical speeds we can ensure that the strategy is feasible. We have found a minimum cost strategy given that a prescribed sequence of fuel supply rates must be used. In the next chapter we will show that the wider question of an overall minimum cost strategy can now be solved by comparing all possible strategies of optimal type. We have in effect simplified the overall problem by selecting one optimal representative from each class of strategies. From the examples presented in this chapter it is indeed possible to conjecture that a strategy with an idealised speed-hold phase will provide the ultimate solution. We will confirm this conjecture precisely in a subsequent chapter.

It is pertinent to note that the problem can be formulated using position rather than time as the independent variable. The results obtained in this chapter can also be obtained in a similar way using the alternative formulation. The alternative formulation is important because it will be used to solve the problem for a track with non-constant gradient.

# CHAPTER 9
# MINIMISATION OF FUEL CONSUMPTION

In the previous chapter we showed that for each given sequence of control settings, fuel consumption is minimised if the settings are changed only when the speed reaches a critical value. The duration of each phase is determined by these critical speeds. A strategy in which the control setting is changed at each of the corresponding switching times is called a strategy of optimal type. Critical speeds were discussed in the previous chapter, and it was shown how these speeds could be adjusted to produce a feasible strategy.

In this chapter we show how the strategies of optimal type can be used to find a minimum cost strategy. We show that fuel consumption is decreased as the number of phases is increased. This idea will be extended to show that an idealised strategy of optimal type with a phase of maximum power followed by phases of speed-hold, coast and brake will provide the desired minimum fuel consumption. The speed-hold segment is constructed using infinitesimal *coast–power* control pairs. Although the idealised strategy cannot be realised in practice, we will show that it can be approximated to any required accuracy by a strategy of optimal type.

This chapter was originally published as a paper in the *IEEE Transactions on Automatic Control* [15].

## 9.1 Additional Notation

We can specify a strategy or indeed any segment of a strategy by specifying the sequence of controls and the corresponding sequence of switching speeds. Thus the general strategy of this section can be specified by the sequence

$$S = \{ [j(k+1); (V_k, V_{k+1})] \}_{k = 0, 1, \dots, n}$$

where $V_0 = V_{n+1} = 0$.

A strategy of optimal type $\mathcal{T}$ contains alternate phases of positive and negative net acceleration and is best described by reference to the so called *critical speeds* $U$, $V$ and $W$ where $0 \le V < W$ and $U = U(V, W)$. In general we have

$$\mathcal{T} = \{ \ [j(1);(0,W)] \ , \ [j(2);(W,V)] \ , \dots$$
$$[j(n-1);(V,W)] \ , \ [0;(W,U)] \ , \ [-1;(U,0)] \ \} \ .$$

We have shown that there are non-negative parameters $\lambda$ and $\mu$ such that $V$ and $W$ are given by the two solutions to the equation

$$\lambda v - \mu = vr(v)$$

and such that $U$ is given by the unique solution to the equation

$$\lambda v - \mu = 0.$$

We have also shown how to adjust the values of the parameters to produce a feasible strategy of optimal type.


## 9.2 Approximating the Minimum-Cost Strategy

We define a special strategy of optimal type. Let $q$ be a non-negative integer and let $\mathcal{T}_q$ be a strategy of optimal type with $2q+3$ phases. These phases are an initial phase of maximum power, $q$ pairs of alternate coast and maximum power, a semi-final coast phase, and a final brake phase. Thus $\mathcal{T}_q = \mathcal{T}_q(V, W)$ is given by

$$\mathcal{T}_q = \{ \ [m;(0,W)] \ , \ [0;(W,V)] \ , \dots$$
$$[m;(V,W)] \ , \ [0;(W,U)] \ , \ [-1;(U,0)] \ \}$$

where the critical speed at which braking begins is defined by

$$U(V,W) = \frac{VW\,[r(W)-r(V)]}{Wr(W)-Vr(V)}$$

$$= W - \frac{Wr(W)}{\left[\dfrac{Wr(W)-Vr(V)}{W-V}\right]} \, .$$

We can calculate the time $t_q(V, W)$ taken for the journey, the distance $x_q(V, W)$ travelled by the train and the cost $J_q(V, W)$ of the strategy. We note that

$$t_q(V, W) = \int_0^W \frac{v\,dv}{Hf_m - vr(v)} + q\int_V^W \frac{Hf_m\,dv}{[Hf_m - vr(v)]\,r(v)}$$

$$+ \int_{U(V, W)}^W \frac{dv}{r(v)} + \int_0^{U(V, W)} \frac{dv}{K + r(v)},$$

$$x_q(V, W) = \int_0^W \frac{v^2\,dv}{Hf_m - vr(v)} + q\int_V^W \frac{Hf_m v\,dv}{[Hf_m - vr(v)]\,r(v)}$$

$$+ \int_{U(V, W)}^W \frac{v\,dv}{r(v)} + \int_0^{U(V, W)} \frac{v\,dv}{K + r(v)}$$

and

$$J_q(V, W) = \int_0^W \frac{f_m v\,dv}{Hf_m - vr(v)} + q\int_V^W \frac{f_m v\,dv}{Hf_m - vr(v)}. \qquad (9.1)$$

For a feasible strategy we require a pair $(X_q, Y_q)$ of critical speeds such that $t_q(X_q, Y_q) = T$ and $x_q(X_q, Y_q) = X$.

We also define an *idealised strategy of optimal type* $\mathcal{T}$ with a *power–hold–coast–brake* structure. This strategy will be obtained as the limit of a sequence of feasible strategies of the form $\mathcal{T}_q$ where the critical speeds $X_q$ and $Y_q$ converge to a common value $Z$ as $q \to \infty$. Initially we consider an idealised strategy $\mathcal{T} = \mathcal{T}(Y)$ in which speed-holding occurs at some arbitrary critical speed $Y > 0$. To hold at this speed would require a fuel supply rate given by

$$f = \frac{Yr(Y)}{H}$$

Unfortunately this rate will not in general be an allowable rate of fuel supply. To overcome this difficulty we assume that speed-holding is approximated by short periods of coast alternating with short periods of maximum power. It will be shown in the next section that the effective rate of fuel supply for the approximate strategy approaches the required rate of fuel supply for the idealised strategy as the magnitude of the speed oscillations is decreased.

For the strategy $\mathcal{T}$ we define the critical speed $U(Y)$ at which braking begins by the formula

$$U(Y) = \lim_{V, W \to Y} U(V, W) = Y - \frac{Yr(Y)}{\dfrac{d}{dv}[vr(v)]\Big|_{v=Y}}.$$

We also define

$$\tau(Y) = \lim_{V, W \to Y} t_q(V, W)$$

and note that if $\tau(Y) < T$ then the time allowed for the speed-hold phase is given by $T - \tau(Y)$.

We define

$$\xi(Y) = \lim_{V, W \to Y} x_q(V, W)$$

and calculate the total distance travelled as

$$x(Y) = \xi(Y) + Y[T - \tau(Y)].$$

The cost of the strategy is given by

$$J(Y) = f_m \int_0^Y \frac{v\,dv}{Hf_m - vr(v)} + \frac{Yr(Y)}{H}[T - \tau(Y)]. \tag{9.2}$$

For a feasible strategy we need to find $Z$ such that $\tau(Z) < T$ and $x(Z) = X$.

We will assume that apart from the final brake phase a strategy of optimal type should be constructed using only phases of maximum power and coast. This assumption is justified by showing that a strategy using intermediate levels of power can be replaced by a more effective strategy using only the extreme throttle positions. We can now state our two main results.


### Theorem 9.1

*If there is a positive number $\bar{Y}$ with $\tau(\bar{Y}) = T$ and $\xi(\bar{Y}) > X$ then for each $q = 0, 1, \ldots$ we can find a unique pair of critical speeds $(X_q, Y_q)$ with $0 < X_q < Y_q < \bar{Y}$ such that the strategy $\mathcal{T}_q = \mathcal{T}_q(X_q, Y_q)$ is feasible. There is a unique number $Z$ such that the idealised strategy $\mathcal{T} = \mathcal{T}(Z)$ is feasible. We note that $X_q < Z < Y_q$ and $J(Z) < J_q(X_q, Y_q)$ for all $q = 0, 1, \ldots$ and also that $X_q \to Z$, $Y_q \to Z$ and $J_q(X_q, Y_q) \to J(Z)$ as $q \to \infty$.*

$\square$

***Theorem 9.2***

*Let $Z$ be the unique positive number with $\tau(Z) < T$ and $x(Z) = X$. The strategy $\mathcal{T} = \mathcal{T}(Z)$ is feasible and is the minimum cost strategy.*

$\square$

## 9.3 Approximate Speed-Holding Strategies

To show that a speed-hold phase can be approximated to any desired accuracy using alternate phases of maximum power and coast we require the following results.

***Lemma 9.3***

*Let $Y > 0$ and $\Delta T > 0$, and suppose that*

$$\int_0^Y \frac{Hf_m dv}{[Hf_m - vr(v)]\, r(v)} \geq \Delta T. \tag{9.3}$$

*For each $q = 1, 2, \ldots$ we can find $R_q$ and $S_q$ with $0 < R_q < Y < S_q$ and $[R_q, S_q] \subset (R_{q-1}, S_{q-1})$, and such that*

$$q \int_{R_q}^{S_q} \frac{Hf_m dv}{[Hf_m - vr(v)]\, r(v)} = \Delta T$$

*and*

$$q \int_{R_q}^{S_q} \frac{Hf_m v dv}{[Hf_m - vr(v)]\, r(v)} = Y\Delta T.$$

***Proof***

Let $\bar{R}_1 = \underline{S}_1 = Y$. Choose $\bar{S}_1 > Y$ and $\underline{R}_1 < Y$ so that

$$\int_{\bar{R}_1}^{\bar{S}_1} \frac{Hf_m dv}{[Hf_m - vr(v)]\, r(v)} = \int_{\underline{R}_1}^{\underline{S}_1} \frac{Hf_m dv}{[Hf_m - vr(v)]\, r(v)} = \Delta T.$$

There exists a continuous, strictly increasing mapping $\gamma_1 : [\underline{R}_1, \bar{R}_1] \to [\underline{S}_1, \bar{S}_1]$ such that for each point $R \in [\underline{R}_1, \bar{R}_1]$ we can find a corresponding point $\gamma_1(R) \in [\underline{S}_1, \bar{S}_1]$ with

$$\int_{R}^{\gamma_1(R)} \frac{Hf_m dv}{[Hf_m - vr(v)]\, r(v)} = \Delta T.$$

From Figure 9-1 it can be seen that

$$\int_{\underline{R}_1}^{\gamma_1(\underline{R}_1)} \frac{Hf_m v dv}{[Hf_m - vr(v)]\, r(v)} < V\Delta T < \int_{\bar{R}_1}^{\gamma_1(\bar{R}_1)} \frac{Hf_m v dv}{[Hf_m - vr(v)]\, r(v)}$$

and hence we can find a point $R_1 \in (\underline{R}_1, \bar{R}_1)$ and a corresponding point $S_1 = \gamma_1(R_1) \in (\underline{S}_1, \bar{S}_1)$ with

$$\int_{R_1}^{S_1} \frac{Hf_m v dv}{[Hf_m - vr(v)]\, r(v)} = Y\Delta T.$$

Note that $R_1 < \bar{R}_1 = Y = \underline{S}_1 < S_1$.

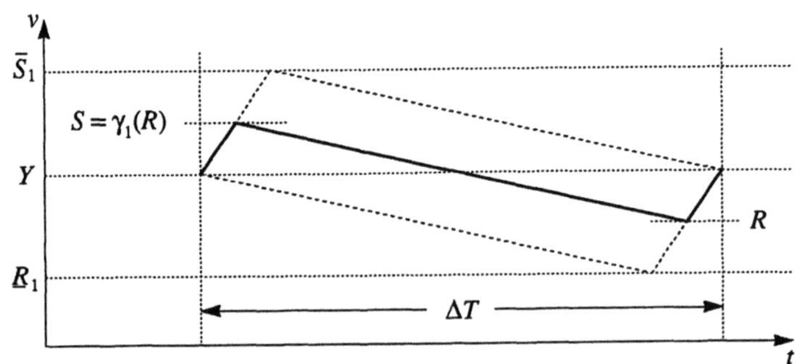


Figure 9-1: Choosing $R_1$ to satisfy the distance constraint

In general we suppose that there exists an interval $[R_p, S_p]$ with $R_p < Y < S_p$ such that

$$p \int_{R_p}^{S_p} \frac{Hf_m dv}{[Hf_m - vr(v)] \, r(v)} = \Delta T$$

and

$$p \int_{R_p}^{S_p} \frac{Hf_m v dv}{[Hf_m - vr(v)] \, r(v)} = Y \Delta T$$

for each $p = 1, 2, \ldots, q-1$. We set $\underline{R}_q = R_{q-1}$ and $\bar{S}_q = S_{q-1}$, and define $\bar{R}_q, \underline{S}_q \in (R_{q-1}, S_{q-1})$ such that

$$q \int_{\bar{R}_q}^{\bar{S}_q} \frac{Hf_m dv}{[Hf_m - vr(v)] \, r(v)} = q \int_{\underline{R}_q}^{\underline{S}_q} \frac{Hf_m dv}{[Hf_m - vr(v)] \, r(v)} = \Delta T. \tag{9.4}$$

There exists a continuous strictly increasing mapping $\gamma_q : [\underline{R}_q, \bar{R}_q] \to [\underline{S}_q, \bar{S}_q]$ such that for each point $R \in [\underline{R}_q, \bar{R}_q]$ there is a corresponding point $\gamma_q(R) \in [\underline{S}_q, \bar{S}_q]$ with

$$q \int_{R}^{\gamma_q(R)} \frac{Hf_m dv}{[Hf_m - vr(v)] \, r(v)} = \Delta T.$$

Since $[\bar{R}_q, \bar{S}_q] \subseteq [R_{q-1}, S_{q-1}]$ we can use the inductive hypothesis and (9.4) to show that

$$q \int_{\bar{R}_q}^{\bar{S}_q} \frac{Hf_m v dv}{[Hf_m - vr(v)] \, r(v)} - (q-1) \int_{R_{q-1}}^{S_{q-1}} \frac{Hf_m v dv}{[Hf_m - vr(v)] \, r(v)}$$

$$= q \int_{\bar{R}_q}^{S_{q-1}} \frac{Hf_m v dv}{[Hf_m - vr(v)] \, r(v)} - (q-1) \int_{R_{q-1}}^{S_{q-1}} \frac{Hf_m v dv}{[Hf_m - vr(v)] \, r(v)}$$

$$= \int_{\bar{R}_q}^{S_{q-1}} \frac{Hf_m v dv}{[Hf_m - vr(v)]\, r(v)} - (q-1) \int_{R_{q-1}}^{\bar{R}_q} \frac{Hf_m v dv}{[Hf_m - vr(v)]\, r(v)}$$

$$> \bar{R}_q \left[ \int_{\bar{R}_q}^{S_{q-1}} \frac{Hf_m dv}{[Hf_m - vr(v)]\, r(v)} - (q-1) \int_{R_{q-1}}^{\bar{R}_q} \frac{Hf_m dv}{[Hf_m - vr(v)]\, r(v)} \right]$$

$$= \bar{R}_q \left[ q \int_{\bar{R}_q}^{S_{q-1}} \frac{Hf_m dv}{[Hf_m - vr(v)]\, r(v)} - (q-1) \int_{R_{q-1}}^{S_{q-1}} \frac{Hf_m dv}{[Hf_m - vr(v)]\, r(v)} \right]$$

$$= \bar{R}_q \left[ q \int_{\bar{R}_q}^{\bar{S}_q} \frac{Hf_m dv}{[Hf_m - vr(v)]\, r(v)} - (q-1) \int_{R_{q-1}}^{S_{q-1}} \frac{Hf_m dv}{[Hf_m - vr(v)]\, r(v)} \right]$$

$$= 0.$$

A similar argument shows that

$$q \int_{\underline{R}_q}^{S_q} \frac{Hf_m v dv}{[Hf_m - vr(v)]\, r(v)} - (q-1) \int_{R_{q-1}}^{S_{q-1}} \frac{Hf_m v dv}{[Hf_m - vr(v)]\, r(v)} < 0$$

and hence

$$q \int_{\underline{R}_q}^{S_q} \frac{Hf_m v dv}{[Hf_m - vr(v)]\, r(v)} < Y\Delta T < q \int_{\bar{R}_q}^{\bar{S}_q} \frac{Hf_m v dv}{[Hf_m - vr(v)]\, r(v)}.$$

It follows that we can find a point $R_q \in (\underline{R}_q, \bar{R}_q)$ and a corresponding point $S_q = \gamma_q(R_q) \in [\underline{S}_q, \bar{S}_q]$ with

$$q \int_{R_q}^{S_q} \frac{Hf_m v dv}{[Hf_m - vr(v)]\, r(v)} = Y\Delta T.$$

Clearly $[R_q, S_q] \subset (R_{q-1}, S_{q-1})$. Since

$$q \int_{R_q}^{S_q} \frac{Hf_m v dv}{[Hf_m - vr(v)]\, r(v)} > R_q \left[ q \int_{R_q}^{S_q} \frac{Hf_m dv}{[Hf_m - vr(v)]\, r(v)} \right] = R_q \Delta T$$

and

$$q \int_{R_q}^{S_q} \frac{Hf_m v\, dv}{[Hf_m - vr(v)]\, r(v)} \; < \; S_q \left[ q \int_{R_q}^{S_q} \frac{Hf_m\, dv}{[Hf_m - vr(v)]\, r(v)} \right] = S_q \Delta T$$

it follows that $R_q < Y < S_q$. The general result now follows by induction.

$\square$

### Corollary 9.4

*Let $Y > 0$ and $\Delta T > 0$, and suppose that $q_0$ is a natural number with*

$$q_0 \int_0^Y \frac{Hf_m\, dv}{[Hf_m - vr(v)]\, r(v)} \geq \Delta T.$$

*For each natural number $q > q_0$ we can find $R_q$ and $S_q$ with $0 < R_q < Y < S_q$ and $[R_q, S_q] \subset (R_{q-1}, S_{q-1})$, and such that*

$$q \int_{R_q}^{S_q} \frac{Hf_m\, dv}{[Hf_m - vr(v)]\, r(v)} = \Delta T$$

*and*

$$q \int_{R_q}^{S_q} \frac{Hf_m v\, dv}{[Hf_m - vr(v)]\, r(v)} = Y \Delta T.$$

$\square$

### Example 9.1  A Modified Strategy

Consider the strategy of optimal type in Example 8.3. Denote the critical speeds by $R_1 = V = 13.37$ and $S_1 = W = 18.59$, and denote the strategy by $\mathcal{T}_1$. Now choose $(R_2, S_2)$ such that

$$2\int_{R_2}^{S_2} \frac{dv}{[H - vr(v)]\, r(v)} = \int_{R_1}^{S_1} \frac{dv}{[H - vr(v)]\, r(v)}$$

and

$$2\int_{R_2}^{S_2} \frac{vdv}{[H - vr(v)]\, r(v)} = \int_{R_1}^{S_1} \frac{vdv}{[H - vr(v)]\, r(v)}.$$

The results of this calculation were $R_2 = 14.68$ and $S_2 = 17.29$. We now replace $\mathcal{T}_1$ by a strategy $S_2$ with $n = 6$ and $j(1) = j(3) = j(5) = 1$, $j(2) = j(4) = j(6) = 0$ and $j(7) = -1$. The switching speeds are $V_1 = V_3 = S_2$, $V_2 = V_4 = R_2$, $V_5 = S_1$ and $V_6 = U(R_1, S_1)$. The cost of this strategy is given by $J(S_2) = 202.02$, which is less than $J_1(R_1, S_1) = 202.15$. Note that $S_2$ remains feasible but is not a strategy of optimal type. The speed profiles for the two strategies are shown in Figure 9-2.

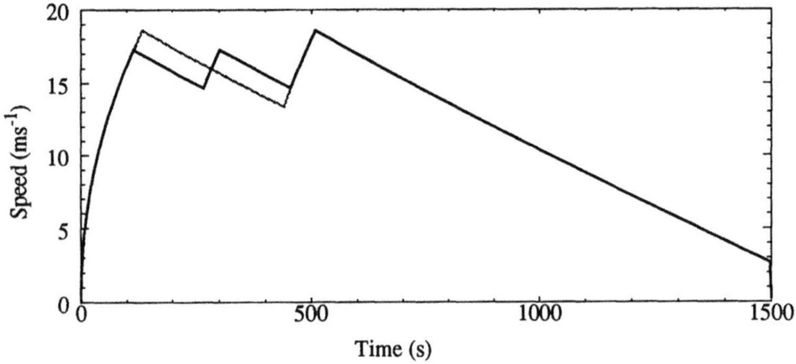

Figure 9-2: Speed $v(t)$ for the strategies in Example 9.1

☐

***Example 9.2  An Improved Strategy of Optimal Type***

We replace $S_2$ by a strategy $\mathcal{T}_2$ of optimal type from the same class of strategies. Thus we have $n = 6$ with $j(1) = j(3) = j(5) = 1$, $j(2) = j(4) = j(6) = 0$ and $j(7) = -1$. We select critical speeds $V = X_2$ and $W = Y_2$ where $0 \le X_2 < Y_2$ and determine the critical speed $U$ from the formula $U = U(V, W)$. The switching speeds are given by $V_1 = V_3 = V_5 = W$, $V_2 = V_4 = V$ and $V_6 = U$. The results of this calculation were $X_2 = 14.55$, $Y_2 = 17.67$ and hence $U = 2.78$. The cost of this strategy is given by $J_2 = 201.98$. The speed profile for the improved strategy is shown in Figure 9-3.

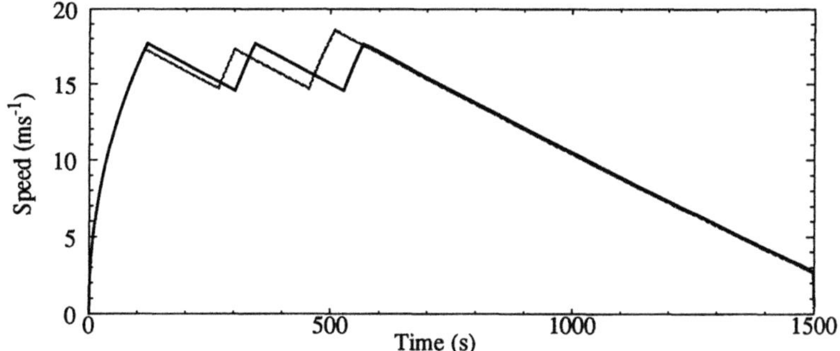

Figure 9-3: Speed $v(t)$ for the strategies in Example 9.2

☐

These examples are not intended to imply that the modification process is a useful practical procedure but rather to illustrate the theoretical results. In practice it is relevant to observe that often it is sufficient to construct a strategy of optimal type with relatively few phases. The percentage reduction in cost that can be achieved by a strategy of optimal type with more phases is likely to be minimal.

### *Lemma 9.5*

*Suppose that $Y > 0$ and $\Delta T > 0$, and let $q_0$ be a natural number such that inequality (9.3) is satisfied. Let $\bar{Y} > Y$ be chosen such that*

$$q_0 \int_{Y}^{\bar{Y}} \frac{Hf_m dv}{[Hf_m - vr(v)]\, r(v)} = \Delta T.$$

*For each natural number $q \geq q_0$ we have*

$$S_q - R_q \leq \frac{r(Y)\Delta T}{q}.$$

**Proof**

Since

$$q \int_{R_q}^{S_q} \frac{H f_m \, dv}{[H f_m - v r(v)] \, r(v)} = \Delta T$$

for each natural number $q \geq q_0$ it follows that

$$R_q < Y < S_q < \bar{Y}$$

and since

$$\frac{H f_m}{[H f_m - v r(v)] \, r(v)} \geq \frac{1}{r(\bar{\bar{Y}})}$$

for all $v \in [R_q, S_q]$ we can also deduce that

$$\frac{q \, (S_q - R_q)}{r(\bar{\bar{Y}})} \leq \Delta T.$$

$\square$

We now replace a speed-holding segment $\Delta S$ at speed $Y$ and of duration $\Delta T$ by an approximate speed-holding segment $\Delta S_q$ with $2q$ phases given by

$$\Delta S_q = \{ \, [m; (Y, S_q)] \, , \, [0; (S_q, R_q)] \, , \, [m; (R_q, S_q)] \, , \, \ldots$$
$$[0; (S_q, R_p)] \, , \, [m; (R_q, Y)] \, \}$$

where $R_q$ and $S_q$ are chosen according to Corollary 9.4, which shows that the time taken and the distance travelled are the same for each segment of strategy. The cost of the approximate speed-holding strategy is given by

$$\Delta J(R_q, S_q) = q \left[ f_m \int_{R_p}^{S_q} \frac{v \, dv}{H f_m - v r(v)} \right]$$

and since $v r(v)$ increases as $v$ increases it can be seen from Corollary 9.4 that

$$\left[ \frac{R_q r(R_q)}{H} \right] \Delta T \leq \Delta J(R_q, S_q) \leq \left[ \frac{S_q r(S_q)}{H} \right] \Delta T. \tag{9.5}$$

Using Lemma 9.5 and the inequality (9.5), and by taking the limit as $q \to \infty$, the cost of an *idealised speed-hold* phase of duration $\Delta T$ at speed $Y$ is given by

$$\Delta J(Y) = \left[\frac{Yr(Y)}{H}\right]\Delta T.$$

Since the idealised speed-hold phase can be realised to any desired accuracy it will be used where necessary without further comment.

If we replace the approximate speed-holding segment $\Delta S_{q-1}$ by the approximate speed-holding segment $\Delta S_q$ we can see that

$$\Delta J\left(R_{q-1}, S_{q-1}\right) - \Delta J(R_q, S_q)$$

$$= (q-1)\left\{\int_{R_{q-1}}^{R_q} \frac{f_m v\,dv}{Hf_m - vr(v)} + \int_{S_q}^{S_{q-1}} \frac{f_m v\,dv}{Hf_m - vr(v)}\right\} - \int_{R_q}^{S_q} \frac{f_m v\,dv}{Hf_m - vr(v)}.$$

Because the graph $y = vr(v)$ is convex we can find $\lambda$ and $\mu$ such that $\lambda v - \mu < vr(v)$ when $v \in (R_q, S_q)$ and $\lambda v - \mu > vr(v)$ when $v \notin [R_q, S_q]$. It follows that

$$\Delta J\left(R_{q-1}, S_{q-1}\right) - \Delta J(R_q, S_q)$$

$$> (q-1)\left\{\int_{R_{q-1}}^{R_q} \frac{f_m(\lambda v - \mu)\,dv}{[Hf_m - vr(v)]\,r(v)} + \int_{S_q}^{S_{q-1}} \frac{f_m(\lambda v - \mu)\,dv}{[Hf_m - vr(v)]\,r(v)}\right\}$$

$$- \int_{R_q}^{S_q} \frac{f_m(\lambda v - \mu)\,dv}{[Hf_m - vr(v)]\,r(v)}$$

$$= 0$$

by Corollary 9.4. Therefore

$$\Delta J(R_{q-1}, S_{q-1}) - \Delta J(R_q, S_q) > 0.$$

Thus the cost of $\Delta S_q$ decreases as $q$ increases.

## 9.4 The Structure of an Optimal Strategy

The following results can be used to show that an overall minimum cost strategy must contain an initial phase of maximum power followed by phases of speed-hold, coast and brake. These arguments are based on a modification procedure that preserves the feasibility of the strategy but decreases the cost.

### 9.4.1 The Power Phase

Consider the two strategy segments shown in Figure 9-4. The curve $v = v_1(t)$ corresponds to a net acceleration phase with fuel supply rate $f$ such that

$$\frac{Wr(W)}{H} < f < f_m,$$

whereas the curve $v = v_2(t)$ corresponds to the sequence of *maximum power–hold– maximum power*.

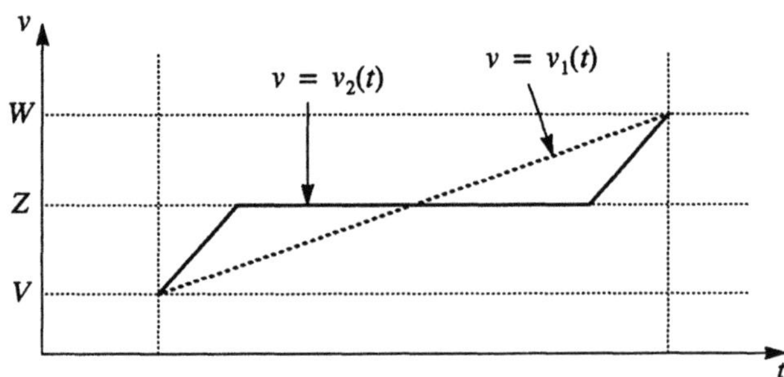


Figure 9-4: The power phase

The time $\tau$ spent speed-holding is such that both strategy segments have the same time, and is given by

$$\tau = \int_V^W \frac{v\,dv}{Hf - vr(v)} - \int_V^W \frac{v\,dv}{Hf_m - vr(v)}.$$

The holding speed $Z$ is chosen so that the distance travelled using each strategy is the same. That is

$$\int_V^W \frac{v^2 dv}{Hf_m - vr(v)} + Z \left\{ \int_V^W \frac{v dv}{Hf - vr(v)} - \int_V^W \frac{v dv}{Hf_m - vr(v)} \right\} = \int_V^W \frac{v^2 dv}{Hf - vr(v)}.$$

Rearranging gives

$$\int_V^W \frac{v(v - Z) dv}{Hf_m - vr(v)} = \int_V^W \frac{v(v - Z) dv}{Hf - vr(v)}. \tag{9.6}$$

The costs of the two strategy segments are

$$\Delta J_1 = f \int_V^W \frac{v dv}{Hf - vr(v)}$$

and

$$\Delta J_2 = f_m \int_V^W \frac{v dv}{Hf_m - vr(v)} + \frac{Zr(Z)}{H} \left[ \int_V^W \frac{v dv}{Hf - vr(v)} - \int_V^W \frac{v dv}{Hf_m - vr(v)} \right].$$

We can now use the following lemma to show that the second strategy segment uses less fuel than the first.

**Lemma 9.6**

*Let $0 < V < W$ and let*

$$0 < \frac{Wr(W)}{H} < f < f_m.$$

*If $Z \in (V, W)$ is chosen so that*

$$\int_V^W \frac{v(v - Z) dv}{Hf_m - vr(v)} = \int_V^W \frac{v(v - Z) dv}{Hf - vr(v)}$$

*then*

$$f_m \int_V^W \frac{v\,dv}{Hf_m - vr(v)} + \frac{Zr(Z)}{H}\left[\int_V^W \frac{v\,dv}{Hf - vr(v)} - \int_V^W \frac{v\,dv}{Hf_m - vr(v)}\right]$$

$$< f \int_V^W \frac{v\,dv}{Hf - vr(v)}.$$

**Proof**

Since we assume that the graph $y = vr(v)$ is strictly convex the tangent line $y = \lambda v - \mu$ at the point $v = Z$ is below the graph for all $v \neq Z$. It follows that

$$f_m \int_V^W \frac{v\,dv}{Hf_m - vr(v)} + \frac{Zr(Z)}{H}\left[\int_V^W \frac{v\,dv}{Hf - vr(v)} - \int_V^W \frac{v\,dv}{Hf_m - vr(v)}\right] - f\int_V^W \frac{v\,dv}{Hf - vr(v)}$$

$$= (f_m - f)\int_V^W \frac{[Zr(Z) - vr(v)]\,v\,dv}{[Hf_m - vr(v)]\,[Hf - vr(v)]}$$

$$< (f_m - f)\int_V^W \frac{\lambda(Z - v)\,v\,dv}{[Hf_m - vr(v)]\,[Hf - vr(v)]}$$

$$= 0.$$

$\square$

### 9.4.2 The Transition from Power to Speed-Hold

Now consider the two strategy segments illustrated in Figure 9-5. The curve $v = v_1(t)$ corresponds to a *hold–power–hold–coast* sequence with hold speeds $V$ and $W$ respectively, and the curve $v = v_2(t)$ corresponds to a *power–hold* sequence with hold speed $Z$.

Let $\tau_V$, $\tau_W$ and $\tau_Z$ be the times spent speed-holding at speeds $V$, $W$ and $Z$ respectively. For the segment times to be equal we require

$$\int_V^Z \frac{v\,dv}{Hf_m - vr(v)} + \tau_Z = \tau_V + \int_V^W \frac{v\,dv}{Hf_m - vr(v)} + \tau_W + \int_Z^W \frac{dv}{r(v)}. \tag{9.7}$$

For the segment distances to be equal we require

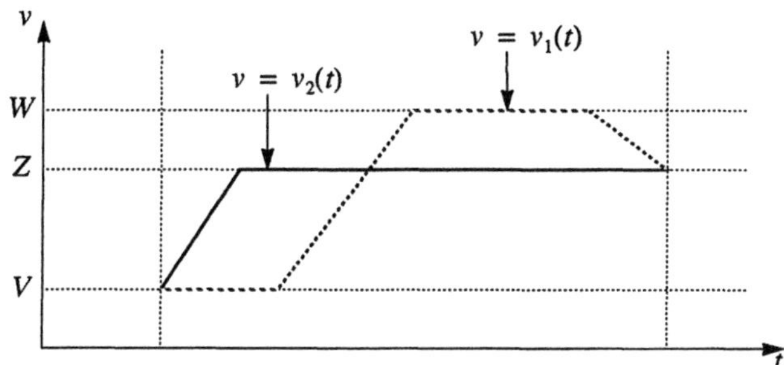

Figure 9-5: The transition from power to speed-hold

$$\int_V^Z \frac{v^2 dv}{Hf_m - vr(v)} + Z\tau_Z = V\tau_V + \int_V^W \frac{v^2 dv}{Hf_m - vr(v)} + W\tau_W + \int_Z^W \frac{v\,dv}{r(v)}. \qquad (9.8)$$

The costs of the two segments are

$$\Delta J_1 = \frac{Vr(V)}{H}\tau_V + f_m \int_V^W \frac{v\,dv}{Hf_m - vr(v)} + \frac{Wr(W)}{H}\tau_W$$

and

$$\Delta J_2 = f_m \int_V^Z \frac{v\,dv}{Hf_m - vr(v)} + \frac{Zr(Z)}{H}\tau_Z.$$

We can now show that the second strategy segment uses less fuel than the first.

### Lemma 9.7

*Let $0 < V < W$ and let $\tau_V$ and $\tau_W$ be given. If $Z \in (V, W)$ and $\tau_Z$ are chosen so that*

$$\int_V^Z \frac{v\,dv}{Hf_m - vr(v)} + \tau_Z = \tau_V + \int_V^W \frac{v\,dv}{Hf_m - vr(v)} + \tau_W + \int_Z^W \frac{dv}{r(v)}$$

*and*

$$\int_V^Z \frac{v^2 dv}{Hf_m - vr(v)} + Z\tau_Z = V\tau_V + \int_V^W \frac{v^2 dv}{Hf_m - vr(v)} + W\tau_W + \int_Z^W \frac{vdv}{r(v)}$$

*then*

$$f_m\int_V^Z \frac{vdv}{Hf_m - vr(v)} + \frac{Zr(Z)}{H}\tau_Z < \frac{Vr(V)}{H}\tau_V + f_m\int_V^W \frac{vdv}{Hf_m - vr(v)} + \frac{Wr(W)}{H}\tau_W.$$

### Proof

Since we assume that the graph $y = vr(v)$ is strictly convex the tangent line $y = \lambda v - \mu$ at the point $v = Z$ is below the graph for all $v \neq Z$. Then

$$f_m\int_V^Z \frac{vdv}{Hf_m - vr(v)} + \frac{Zr(v)}{H}\tau_Z - \left\{ \frac{Vr(V)}{H}\tau_V + f_m\int_V^W \frac{vdv}{Hf_m - vr(v)} + \frac{Wr(W)}{H}\tau_W \right\}$$

$$= -f_m\int_Z^W \frac{vr(v)dv}{[Hf_m - vr(v)]\,r(v)} + \frac{Zr(Z)}{H}\tau_Z - \frac{Vr(V)}{H}\tau_V - \frac{Wr(W)}{H}\tau_W$$

$$< -f_m\int_Z^W \frac{(\lambda v - \mu)\,dv}{[Hf_m - vr(v)]\,r(v)} + \frac{\lambda Z - \mu}{H}\tau_Z - \frac{\lambda V - \mu}{H}\tau_V - \frac{\lambda W - \mu}{H}\tau_W$$

$$= 0$$

*since*

$$\int_Z^W \frac{Hf_m dv}{Hf_m - vr(v)r(v)} - \tau_Z + \tau_V + \tau_W = 0$$

*and*

$$\int_Z^W \frac{Hf_m vdv}{[Hf_m - vr(v)]\,r(v)} - Z\tau_Z + V\tau_V + W\tau_W = 0.$$

$\square$

### 9.4.3  The Transition from Speed-Hold to Coast

Now consider the two strategy segments illustrated in Figure 9-6. The curve

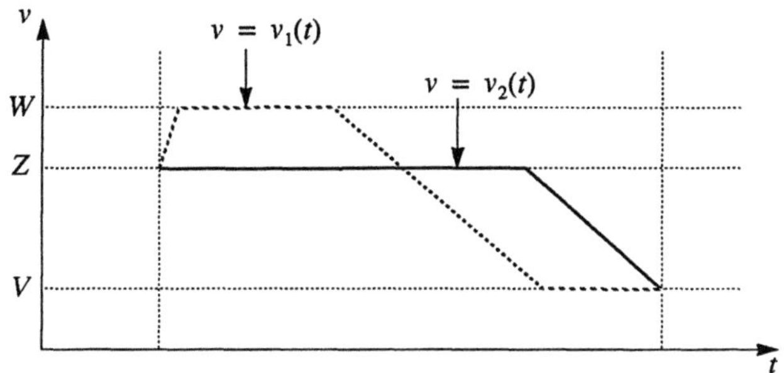


Figure 9-6: The transition from speed-hold to coast

$v = v_1(t)$ corresponds to a *power–hold–coast–hold* sequence with hold speeds $W$ and $V$ respectively, and the curve $v = v_2(t)$ corresponds to a *hold–coast* sequence with hold speed $Z$.

Let $\tau_V$, $\tau_W$ and $\tau_Z$ be the times spent speed-holding at speeds $V$, $W$ and $Z$ respectively. For the segment times to be equal we require

$$\tau_Z + \int_V^Z \frac{dv}{r(v)} = \int_Z^W \frac{vdv}{Hf_m - vr(v)} + \tau_W + \int_Z^W \frac{dv}{r(v)} + \tau_V. \tag{9.9}$$

For the segment distances to be equal we require

$$Z\tau_Z + \int_V^Z \frac{vdv}{r(v)} = \int_Z^W \frac{v^2 dv}{Hf_m - vr(v)} + W\tau_W + \int_Z^W \frac{vdv}{r(v)} + V\tau_V. \tag{9.10}$$

The costs of the two segments are

$$\Delta J_1 = f_m \int_Z^W \frac{vdv}{Hf_m - vr(v)} + \frac{Wr(W)}{H}\tau_W + \frac{Vr(V)}{H}\tau_V$$

and

$$\Delta J_2 = \frac{Zr(Z)}{H}\tau_Z.$$

Equations (9.9) and (9.10) are equivalent to (9.7) and (9.8). Since

$$\Delta J_2 - \Delta J_1 = \frac{Zr(Z)}{H}\tau_Z - \left\{ f_m \int_Z^W \frac{v\,dv}{Hf_m - vr(v)} + \frac{Wr(W)}{H}\tau_W + \frac{Vr(V)}{H}\tau_V \right\}$$

$$= f_m \int_V^Z \frac{v\,dv}{Hf_m - vr(v)} + \frac{Zr(v)}{H}\tau_Z$$

$$- \left\{ \frac{Vr(V)}{H}\tau_V + f_m \int_V^W \frac{v\,dv}{Hf_m - vr(v)} + \frac{Wr(W)}{H}\tau_W \right\}$$

we can now use Lemma 9.7 to show that the second strategy segment uses less fuel than the first.

### 9.4.4 The Coast Phase

Finally, consider the two strategy segments shown in Figure 9-7. The curve $v = v_1(t)$ corresponds to a negative net acceleration phase with fuel supply rate $g$ such that

$$0 < g < \frac{Vr(V)}{H},$$

whereas the curve $v = v_2(t)$ corresponds to a *coast–hold–coast* sequence.

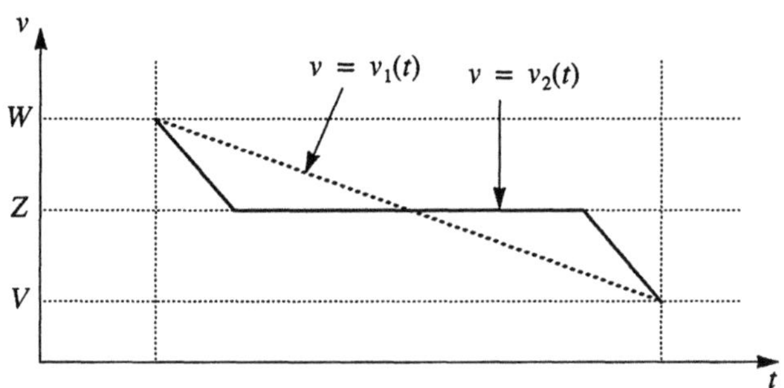


Figure 9-7: The coast phase

The time $\tau$ spent speed-holding is such that both strategy segments have the same time, and is given by

$$\tau = \int_V^W \frac{vdv}{vr(v) - Hg} - \int_V^W \frac{dv}{r(v)}.$$

The holding speed $Z$ is chosen so that the distance travelled using each strategy is the same. That is

$$\int_V^W \frac{vdv}{r(v)} + Z\left\{\int_V^W \frac{vdv}{vr(v) - Hg} - \int_V^W \frac{dv}{r(v)}\right\} = \int_V^W \frac{v^2 dv}{vr(v) - Hg}.$$

Rearranging gives

$$\int_V^W \frac{(v - Z)\, dv}{r(v)} = \int_V^W \frac{v\,(v - Z)\, dv}{vr(v) - Hg}. \tag{9.11}$$

The costs of the two strategy segments are

$$\Delta J_1 = g \int_V^W \frac{vdv}{vr(v) - Hg}$$

and

$$\Delta J_2 = \frac{Zr(Z)}{H}\left[\int_V^W \frac{vdv}{vr(v) - Hg} - \int_V^W \frac{dv}{r(v)}\right].$$

We can now use the following lemma to show that the second strategy segment uses less fuel than the first.

**Lemma 9.8**

*Let $0 < V < W$ and let*

$$0 < g < \frac{Vr(V)}{H}.$$

*If $Z \in (V, W)$ is chosen so that*

$$\int_V^W \frac{(v-Z)\,dv}{r(v)} = \int_V^W \frac{v\,(v-Z)\,dv}{vr(v)-Hg}$$

*then*

$$\frac{Zr(Z)}{H}\left[\int_V^W \frac{v\,dv}{vr(v)-Hg} - \int_V^W \frac{dv}{r(v)}\right] < g\int_V^W \frac{v\,dv}{vr(v)-Hg}.$$

**Proof**

The proof is similar to the proof of Lemma 9.6.

$\square$

## 9.5  A Speed-Holding Strategy of Optimal Type

Consider the strategies $T_q(V, W)$ and $T(Y)$ of optimal type described in Section 9.2. We have shown in Chapter 8 that under certain reasonable conditions we can find a feasible strategy $T_q(X_q, Y_q)$ with this general structure. We can also find a feasible strategy $T(Z)$. We have

$$t(Z) = t_q(X_q, Y_q) = T \qquad\qquad (9.12)$$

and

$$x(Z) = x_q(X_q, Y_q) = X. \qquad\qquad (9.13)$$

From (9.12) we have the time spent speed-holding given by

$$\tau_Z = T - \tau(Z)$$

$$= \int_Z^{Y_q} \frac{Hf_m\,dv}{[Hf_m - vr(v)]\,r(v)} + q\int_{X_q}^{Y_q} \frac{Hf_m\,dv}{[Hf_m - vr(v)]\,r(v)}$$

$$+ \int_{U(X_q,Y_q)}^{U(Z)} \frac{K\,dv}{[K + r(v)]\,r(v)} \qquad\qquad (9.14)$$

and from (9.13) we have the distance travelled during speed-holding given by

$$Z\tau_Z = Z(T - \tau_Z)$$

$$= \int_Z^{Y_q} \frac{Hf_m v\, dv}{[Hf_m - vr(v)]\, r(v)} + q\int_{X_q}^{Y_q} \frac{Hf_m v\, dv}{[Hf_m - vr(v)]\, r(v)}$$

$$+ \int_{U(X_q, Y_q)}^{U(Z)} \frac{Kv\, dv}{[K + r(v)]\, r(v)} \tag{9.15}$$

It is now possible to show that $X_q < Z < Y_q$. We have the following result.

### *Lemma 9.9*

*Let* $0 < X_q < Z < Y_q$. *Then* $J(Z) < J_q(X_q, Y_q)$.

### *Proof*

Because both strategies are feasible we have $t_q(X_q, Y_q) = T$ and $x_q(X_q, Y_q) = x(Z) = X$. By multiplying (9.14) by $Z$ and subtracting from (9.15) we obtain the equation

$$Hf_m \left\{ \int_Z^{Y_q} \frac{(v - Z)\, dv}{[Hf_m - vr(v)]\, r(v)} + q\int_{X_q}^{Y_q} \frac{(v - Z)\, dv}{[Hf_m - vr(v)]\, r(v)} \right\}$$

$$+ K \int_{U(X_q, Y_q)}^{U(Z)} \frac{(v - Z)\, dv}{[K + r(v)]\, r(v)} = 0. \tag{9.16}$$

From (9.1) and (9.2) the cost difference is given by

$$J(Z) - J_q(X_q, Y_q)$$

$$= (-f_m) \left\{ \int_Z^{Y_q} \frac{vr(v)\, dv}{[Hf_m - vr(v)]\, r(v)} + q\int_{X_a}^{Y_q} \frac{vr(v)\, dv}{[Hf_m - vr(v)]\, r(v)} \right\}$$

$$+ \frac{Zr(Z)}{H} \left\{ \int_Z^{Y_q} \frac{Hf_m\, dv}{[Hf_m - vr(v)]\, r(v)} + q\int_{X_q}^{Y_q} \frac{Hf_m\, dv}{[Hf_m - vr(v)]\, r(v)} \right.$$

$$+ \int_{U(X_q, Y_q)}^{U(Z)} \frac{K\,dv}{[K + r(v)]\, r(v)} \bigg\}.$$

Since we assume that the graph $y = vr(v)$ is strictly convex the tangent line $y = \lambda v - \mu$ at the point $v = Z$ is below the graph for all $v \neq Z$. Thus

$$J(Z) - J_q(X_q, Y_q)$$

$$< (-f_m) \left\{ \int_Z^{Y_q} \frac{\lambda\,(v - Z)\,dv}{[Hf_m - vr(v)]\, r(v)} + q \int_{X_q}^{Y_q} \frac{\lambda\,(v - Z)\,dv}{[Hf_m - vr(v)]\, r(v)} \right\}$$

$$+ \frac{K}{H} \int_{U(X_q, Y_q)}^{U(Z)} \frac{Zr(Z)dv}{[K + r(v)]\, r(v)}.$$

From (9.16) it follows that

$$J(Z) - J_q(X_q, Y_q) < \frac{K}{H} \int_{U(X_q, Y_q)}^{U(Z)} \frac{(\lambda v - \mu)\,dv}{[K + r(v)]\, r(v)}.$$

Since $v = U(Z)$ is the solution to the equation $\lambda v - \mu = 0$ it can be seen that $\lambda v - \mu > 0$ when $v > U(Z)$ and that $\lambda v - \mu < 0$ when $v < U(Z)$. It now follows that

$$J(Z) - J_q(X_q, Y_q) < 0.$$

We complete this section with an example that uses the same basic data considered in previous examples.

### Example 9.3  A Minimum Cost Strategy

Take an idealised strategy $T = T(Z)$ as described in Section 9.2 and choose $Z$ such that $t(Z) < T$ and $x(Z) = X$. The calculations show that $Z = 16.18$ and hence that $U(Z) = 2.84$. The cost of the strategy is $J = 201.89$. The speed profile for this strategy is shown in Figure 9-8.

□

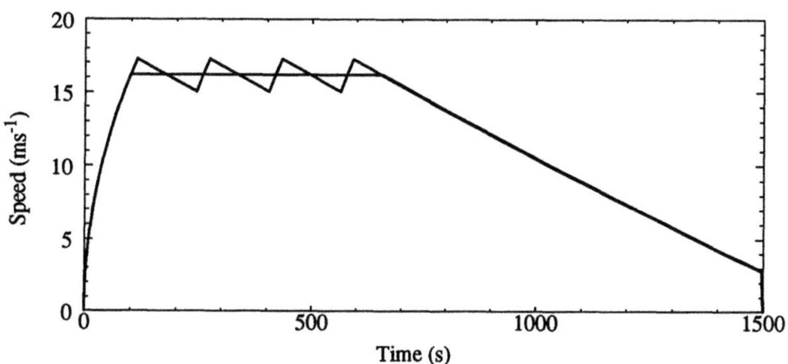

Figure 9-8: Speed $v(t)$ for Examples 8.4 and 9.3

# CHAPTER 10
# A MORE GENERAL MODEL

So far we have assumed that power is constant for each fuel supply rate, and hence that the applied acceleration is inversely proportional to the speed of the train. In this chapter we use a more general model of applied acceleration, and obtain results similar to those of the previous two chapters. The more general model can be used to describe tractive effort and dynamic braking curves such as those shown in Figure 2-1 and Figure 2-3.

These results were described briefly in *IEEE Transactions on Automatic Control* [15].

## 10.1  The Equations of Motion

We begin by discussing what happens when power is applied. Let $f$ denote the rate of fuel supply. In the previous two chapters we assume that the power developed by the locomotive is directly proportional to the rate of fuel supply. In the absence of friction the assumption can be expressed in the form

$$\frac{d}{dt}\left[\int v\,dv\right] = Hf$$

That is,

$$\frac{dv}{dt} = \frac{Hf}{v}.$$

Although the assumption has a realistic basis it represents an ideal situation and is too restrictive to allow general conclusions. In this chapter we use a more flexible condition defined by

$$\frac{d}{dt}\left[\int \frac{dv}{\theta(v)}\right] = p(f).$$

That is,

$$\frac{dv}{dt} = p(f)\theta(v)$$

where $\theta(v)$ is positive for $v > 0$ and $\theta(v)$ decreases as $v$ increases. We assume that $p(0) = 0$ and that $p(f)$ increases as $f$ increases. We also assume that the graph $y = p(f)$ is convex. This assumption is based on the performance characteristics of a typical diesel-electric locomotive. The new conditions do not demand ideal performance of the locomotive and hence can be more widely applied. The resistive acceleration due to friction depends on the speed of the train and when this is taken into account the equation of motion becomes

$$\frac{dv}{dt} = p(f)\theta(v) - r(v)$$

where $r(v)$ is positive and increases as $v$ increases. This equation will also be used when the train is coasting, in which case the rate of fuel supply is zero. When the brakes are applied a similar assumption is made. If friction is ignored we write

$$\frac{d}{dt}\left[ \int \frac{dv}{\psi(v)} \right] = -1.$$

That is,

$$\frac{dv}{dt} = -\psi(v)$$

where $\psi(v)$ is positive and $\psi(v)$ is bounded on $[0, \infty)$ . When friction is taken into account the equation of motion becomes

$$\frac{dv}{dt} = -\psi(v) - r(v)$$

When the brakes are applied the rate of fuel supply is zero.


## 10.2  Statement of the Train Control Problem

We use the same basic formulation of the train control problem as we used in the previous two chapters. Thus we let

$$C = \{-1, 0, 1, ..., m\}$$

denote the set of all possible values for the control variable $j$ and let $f_j$ be the fuel supply rate corresponding to the control setting $j$. We write

$$p_j = p(f_j)$$

for the power generated with control setting $j$.

For a prescribed sequence of fuel supply rates we wish to determine optimal switching times such that all constraints are satisfied and the cost of the strategy is minimised. The resulting strategy is called a strategy of optimal type. We will then consider a systematic comparison of all different feasible strategies of optimal type to find a minimum cost strategy.


## 10.3  The Nature of the Model

For studies of long-haul freight trains, the Scheduling and Control Group used a model described in Benjamin et al [4] by the formulae

$$p(f) = Hf,$$

$$\theta(v) = \begin{cases} \dfrac{1}{v_0} & 0 < v \le v_0 \\[2ex] \dfrac{1}{v} & v > v_0 \end{cases}$$

and

$$\psi(v) = -K$$

where $H$ and $K$ are positive constants. Define

$$\varphi(v) = \frac{r(v)}{\theta(v)}$$

for $v > 0$. In this chapter we assume only that $r(0) > 0$, $r(v)$ is strictly increasing and that the graph $y = \varphi(v)$ has positive slope and is strictly convex in the region $v > 0$. These assumptions are valid for the above model if the Davis formula for resistance is used.

## 10.4 The Main Results

As before, we suppose that a prescribed control sequence $\{j(k+1)\}_{k=0, 1, ..., n}$ is given and that we wish to select optimal values $t_k$ for the switching times so that all constraints are satisfied and such that the cost of the strategy is minimised. We will show that in principle this determination depends on two positive parameters $\lambda$ and $\mu$ which fix critical values of the speed in the following way. If, for each $j$, we let $v = W_j$ denote the unique solution to the equation

$$\varphi(v) = p_j$$

then $W_j$ is the speed at which the frictional resistance equals the driving force. For each $i < j$ we write

$$\varphi_{ij}(v) = \frac{f_j - f_i}{p_j - p_i}\,[\varphi(v) - p_j] + f_j.$$

The critical speeds are $U$, $X_{ij}$ and $Y_{ij}$, where $v = U$ is the solution to the equation

$$\lambda v - \mu = 0,$$

$v = X_{ij}$ is the lesser and $v = Y_{ij}$ is the greater of the two solutions to the equation

$$\lambda v - \mu = f_{ij}(v).$$

$U$ is the speed at which braking begins, while $X_{ij}$ and $Y_{ij}$ are the speeds at which the fuel supply rate is changed either from $f = f_i$ to $f = f_j$ or else from $f = f_j$ to $f = f_i$. For a strategy that contains alternate phases of positive net acceleration with $f = f_j$ and negative net acceleration with $f = f_i$ it is necessary that $W_i < X_{ij} < Y_{ij} < W_j$. Each allowable pair $(\lambda, \mu)$ defines a strategy of optimal type using the prescribed sequence of fuel supply rates. Nevertheless the strategy may not be feasible because the time and distance constraints are not necessarily satisfied. If the strategy is not feasible it may be possible to adjust the parameters $\lambda$ and $\mu$ to produce a strategy of optimal type that is feasible.

We will again show that an idealised strategy with a phase of maximum power followed by phases of speed-hold, coast and brake will provide the desired minimum fuel consumption.

## 10.5  The Fundamental Speed Profiles

To define the speed on the interval $(t_k, t_{k+1})$ for each $k = 0, 1, ..., n-1$ we begin by using a direct integration of the equation of motion to define time as a function of speed by setting

$$
t_k^*(v) = \begin{cases} \displaystyle\int_0^v \frac{dw}{p_{j(k+1)}\theta(w) - r(w)} & v \in [0, W_{j(k+1)}) \\[4ex] \displaystyle\int_v^{W_m} \frac{dw}{r(w) - p_{j(k+1)}\theta(w)} & v \in (W_{j(k+1)}, W_m] \end{cases}
$$

The value of $V_k$ determines which of the above definitions is used. Because the slope of the graph

$$
y = \varphi(v)
$$

is positive and bounded when $v \in [0, W_m]$ it is easy to show that $t_k^*(v) \to \infty$ as $v \to W_{j(k+1)}$. It now follows that if $V_k < W_{j(k+1)}$ then $V_{k+1} < W_{j(k+1)}$. Alternatively if $V_k > W_{j(k+1)}$ then $V_{k+1} > W_{j(k+1)}$. Whichever of the above definitions is used we denote the inverse function by $v_k^*(t)$ and note that the actual speed on the interval $(t_k, t_{k+1})$ for $k = 0, 1, ..., n-1$ is

$$
v(t) = v_k^*(t - t_k + t_k^*(V_k))
$$

provided $V_k \neq W_{j(k+1)}$. A similar approach is used for the final interval where we define

$$
t_n^*(v) = \int_0^{W_m} \frac{dw}{\psi(w) + r(w)}
$$

for $v \in [0, W_m]$ and use $v_n^*(t)$ to denote the inverse function. The actual speed on the interval $(t_n, t_{n+1})$ is given by

$$
v(t) = v_n^*(t - t_n + t_n^*(V_n)).
$$

## 10.6  Necessary Conditions for a Strategy of Optimal Type

We wish to minimise

$$J(\tau) \; = \; \sum_{k=0}^{n} f_{j(k+1)} \tau_{k+1}$$

subject to the equality constraints $x(\tau) = X$ and $t(\tau) = T$, and the inequality constraints $\tau_1 > 0, \tau_2 > 0, ..., \tau_n > 0$. For $\lambda, \mu \in \Re$ and $v = (v_1, v_2, ..., v_n) \in \Re^n$ we define a Lagrangean function

$$\mathcal{H}(\tau, \lambda, \mu, v) \; = \; J(\tau) + \lambda \, [X - x(\tau)] + \mu \, [t(\tau) - T] - \sum_{k=0}^{n-1} v_{k+1} \tau_{k+1}$$

and apply the Kuhn-Tucker conditions

$$\frac{\partial \mathcal{J}}{\partial \tau_{k+1}} \; = \; 0,$$

for all $k$ and the complementary slackness conditions

$$\lambda \, [X - x(\tau)] \; = \; 0,$$

$$\mu \, [t(\tau) - T] \; = \; 0$$

and

$$v_{k+1} \tau_{k+1} \; = \; 0$$

for all $k$, where the Lagrange multipliers $v_{k+1}$ are guaranteed non-negative. If we weaken the time and distance constraints to read $t(\tau) \leq T$ and $x(\tau) \geq X$ then we can also guarantee that $\lambda$ and $\mu$ are non-negative. We will assume that $\tau$ lies in an open set such that $V_k(\tau_1, \tau_2, ..., \tau_n) \neq W_{j(k+1)}$. Applying the Kuhn-Tucker conditions gives a set of complicated equations which can be simplified using the same procedures applied in the previous two chapters. The detailed calculations are therefore omitted. When $h < n$ we have

$$\frac{\partial V_{h+1}}{\partial \tau_{h+1}} \; = \; p_{j(h+1)} \theta(V_{h+1}) - r(V_{h+1}).$$

When $k < h$ we find that

$$\frac{\partial V_{h+1}}{\partial \tau_{k+1}} = [p_{j(h+1)}\theta(V_{h+1}) - r(V_{h+1})] \prod_{s=k+1}^{h} r_s$$

where

$$r_s = \frac{[p_{j(s)}\theta(V_s) - r(V_s)]}{[p_{j(s+1)}\theta(V_s) - r(V_s)]}$$

for $s < n$. We also have

$$\frac{\partial \xi_{h+1}}{\partial \tau_{h+1}} = V_{h+1},$$

$$\frac{\partial \xi_{h+1}}{\partial \tau_{k+1}} = (V_{h+1} - V_h) \prod_{s=k+1}^{h} r_s$$

when $k < h$, and

$$\frac{\partial \xi_{n+1}}{\partial \tau_n} = (-1) V_n r_n$$

where

$$r_n = (-1) \frac{[p_{j(n)}\theta(V_n) - r(V_n)]}{[\psi(V_n) + r(V_n)]}.$$

If we assume $\tau_{k+1} > 0$ then $v_{k+1} = 0$ and hence

$$\frac{\partial \mathcal{J}}{\partial \tau_n} = f_{j(n)} - (\lambda V_n - \mu)(1 - r_n).$$

If we assume that $f_{j(n)} = 0$ then the condition $\partial \mathcal{J}/\partial \tau_n = 0$ becomes

$$\lambda V_n - \mu = 0. \tag{10.1}$$

In general for $k + 1 < n$ it follows that

$$\frac{\partial \mathcal{J}}{\partial \tau_{k+1}} = f_{j(k+1)} - \lambda \left[ V_{k+1} + \sum_{h=k+1}^{n} (V_{h+1} - V_h) \prod_{s=k+1}^{h} r_s \right]$$

$$+ \mu \left[ 1 - \prod_{s=k+1}^{n} r_s \right].$$

By noting that

$$\frac{\partial \mathcal{J}}{\partial \tau_k} - r_k \frac{\partial \mathcal{J}}{\partial \tau_{k+1}} = 0$$

we obtain

$$f_{j(k)} - r_k f_{j(k+1)} - (\lambda V_k - \mu)(1 - r_k) = 0$$

and hence

$$\lambda V_k - \mu = \varphi_{j(k)j(k+1)}(V_k). \tag{10.2}$$

From (10.1) and (10.2) it can be seen that $V_k = v(t_k)$ is one of the critical speeds described in Section 10.4.


## 10.7 The Critical Speeds

For each $i < j$ the graph $y = \varphi_{ij}(v)$ is strictly convex and hence the equation

$$\lambda v - \mu = \varphi_{ij}(v)$$

will generally have two solutions $v = X_{ij}$ and $v = Y_{ij}$ where $X_{ij} < Y_{ij}$. In the case where $h < i < j < k$ and $p(f)$ is strictly convex we can use the convexity to deduce that

$$\frac{f_j - f_i}{p_j - p_i} - \frac{f_j - f_h}{p_j - p_h} < 0$$

and

$$\frac{f_j - f_i}{p_j - p_i} - \frac{f_k - f_i}{p_k - p_i} < 0.$$

Therefore $\varphi_{ij}(v) > \varphi_{hj}(v)$ when $v < W_j$ and $\varphi_{ij}(v) > \varphi_{ik}(v)$ when $v > W_i$. A similar argument shows that $\varphi_{hj}(v) > \varphi_{hk}(v) = \varphi$ when $v > W_h$ and $\varphi_{ik}(v) > \varphi_{hk}(v)$ when $v < W_k$. By combining these inequalities it follows that $\varphi_{ij}(v) > \varphi_{hk}(v)$ for all $v$. It is now obvious that $X_{hk} < X_{ij} < Y_{ij} < Y_{hk}$. We have illustrated these observations in Figure 10-1.

We will not consider a comprehensive analysis of the various possible relationships of the critical speeds. We do note however that when $f = f_j$ the limiting speed is

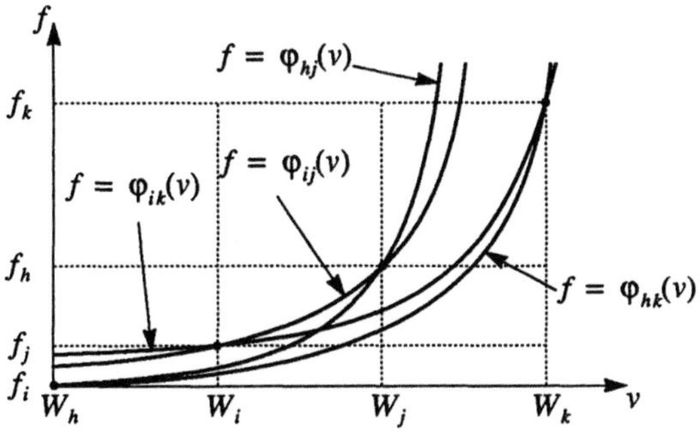


Figure 10-1: The graphs $f = \varphi_{hj}(v)$, $\varphi_{hk}(v)$, $\varphi_{ij}(v)$ and $\varphi_{ik}(v)$.

$v = W_j$. In fact if $f_{j(k+1)} = f_j$ and $V_k < W_j$ then $v(t) < V_{k+1} < W_j$ for all $t \in [t_k, t_{k+1})$. On the other hand, if $V_k > W_j$ then $v(t) > V_{k+1} > W_j$ for all $t \in [t_k, t_{k+1})$. Consequently a strategy that contains alternate phases of positive net acceleration with $f = f_j$ and negative net acceleration with $f = f_i$ must have $W_i < X_{ij} < Y_{ij} < W_j$. In this case $X_{ij}$ is the critical speed at which negative net acceleration with $f = f_i$ changes to positive net acceleration with $f = f_j$, and $Y_{ij}$ is the critical speed at which positive net acceleration with $f = f_j$ changes to negative net acceleration with $f = f_i$. Other configurations are possible but we will show that it is not necessary to investigate all possibilities to determine the minimum cost strategy. The determination of the critical speeds is illustrated in Figure 10-2.

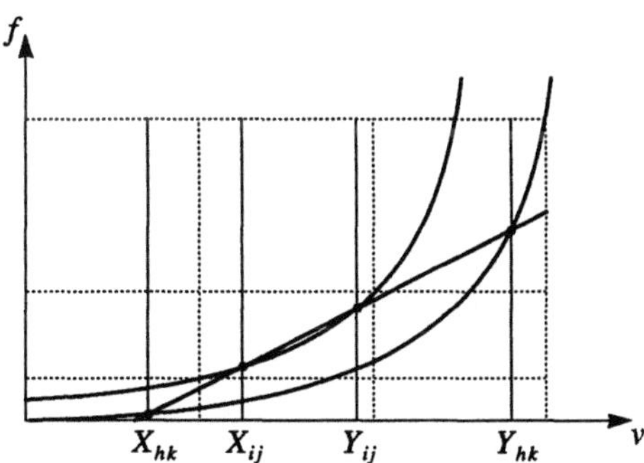


Figure 10-2: Setting the critical speeds $X_{hk} < X_{ij} < Y_{ij} < Y_{hk}$.

## 10.8 Examples

The following examples are based on data obtained from train models used by the Scheduling and Control Group at the South Australian Institute of Technology. Length is measured in metres and time is measured in seconds. All calculations were originally performed on an IBM compatible PC using the software package MATLAB. We assume that

$$\theta(v) = \frac{1}{v},$$

$$\psi(v) = -K$$

and

$$r(v) = a + bv + cv^2.$$

where $H = 1.5$, $K = 1$, $a = 1.5 \times 10^{-2}$, $b = 3 \times 10^{-5}$ and $c = 6 \times 10^{-6}$. We will assume that there are five available throttle positions with power and fuel supply rates defined by the following table

**Table 10-1: Fuel supply rates and power for examples**

| $j$ | 0 | 1 | 2 | 3 | 4 |
|-----|---|---|---|---|---|
| $f_j$ | 0 | 0.344 | 0.616 | 0.776 | 1 |
| $p_j$ | 0 | 0.419 | 0.737 | 0.957 | 1.5 |

In each example we will use $X = 100000$ and $T = 4800$. For the purposes of numerical calculation we use the convenient formulae

$$\tau_{k+1} = \int_{V_k}^{V_{k+1}} \frac{dv}{p_{j(k+1)}\theta(v) - r(v)}$$

and

$$\xi_{k+1} = \int_{V_k}^{V_{k+1}} \frac{v\,dv}{p_{j(k+1)}\theta(v) - r(v)}.$$

Calculation procedures are similar to those in Chapter 8.

### *Example 10.1  A Feasible Strategy*

We use a strategy with $\{j(k+1)\}_{k=0, 1, ...,4} = \{2, 1, 3, 0, -1\}$. We set $V_1 = V_3 = Y$, $V_2 = X$ and $V_4 = 7$, and adjust the parameters $X$ and $Y$ to satisfy the time and distance constraints. The results of this calculation were $X = 22.54$, $Y = 25.44$ and $J = 1557.50$. This strategy is not a strategy of optimal type. The speed profile is shown in Figure 10-3.

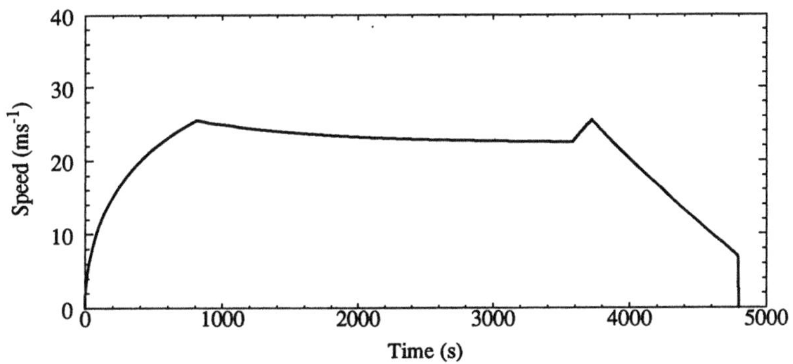


Figure 10-3: Speed $v(t)$ for Example 10.1

☐

### *Example 10.2  A Strategy of Optimal Type*

We use the same strategy as the strategy of Example 10.1. In this example, however, we set $V_1 = Y_{12}$, $V_2 = X_{13}$, $V_3 = Y_{03}$ and $V_4 = U$ where the critical speeds are determined from $\lambda$ and $\mu$ as described in Section 10.4. The parameters $\lambda$ and $\mu$ are adjusted to satisfy the time and distance constraints. The results of this calculation were $X_{13} = 22.47$, $Y_{03} = 28.74$, $Y_{12} = 23.66$, $U = 6.97$ and $J = 1555.01$. This strategy is a strategy of optimal type. The speed profile is shown in Figure 10-4.

☐

### *Example 10.3  A Strategy of Optimal Type*

We use a strategy with $\{j(k+1)\}_{k=0, 1, ...,4} = \{3, 0, 3, 0, -1\}$. We set $V_1 = V_3 = Y_{03}$, $V_2 = X_{03}$ and $V_4 = U$ where the critical speeds are determined

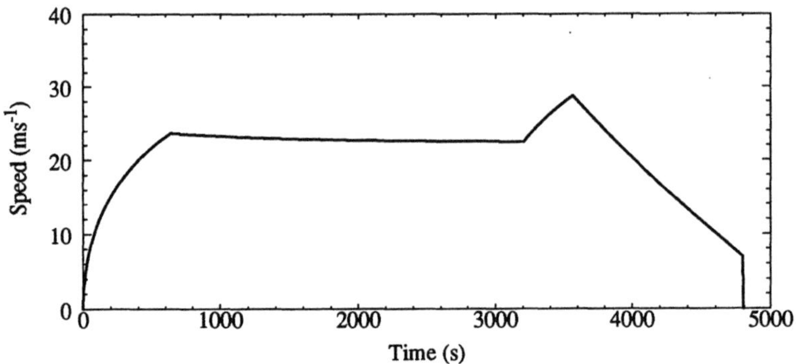

Figure 10-4: Speed $v(t)$ for Example 10.2

from $\lambda$ and $\mu$ as described in Section 10.4. The parameters $\lambda$ and $\mu$ are adjusted to satisfy the time and distance constraints. The results of this calculation were $X_{03} = 10.29$, $Y_{03} = 31.85$, $U = 3.72$ and $J = 1571.27$. This strategy is a strategy of optimal type. The speed profile is shown in Figure 10-5.

□

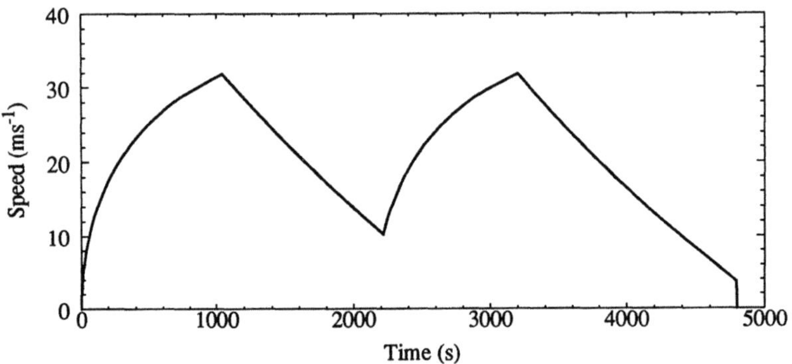

Figure 10-5: Speed $v(t)$ for Examples 10.1–10.3

The apparent speed-holding in Examples 10.1 and 10.2 uses an inefficient control setting, and these strategies have a higher fuel consumption than might be expected from the speed profile.

### *Example 10.4  An Improved Strategy of Optimal Type*

We use a strategy with $\{j(k+1)\}_{k=0,1,\ldots,6} = \{3, 0, 3, 0, 3, 0, -1\}$. We set $V_1 = V_3 = V_5 = Y_{03}$, $V_2 = V_4 = X_{03}$ and $V_6 = U$ where the critical speeds are determined from $\lambda$ and $\mu$ as described in Section 10.4. The parameters $\lambda$ and $\mu$ are adjusted to satisfy the time and distance constraints. The results of this calculation were $X_{03} = 15.69$, $Y_{03} = 29.15$, $U = 5.33$ and $J = 1542.97$. The speed profile is shown in Figure 10-6

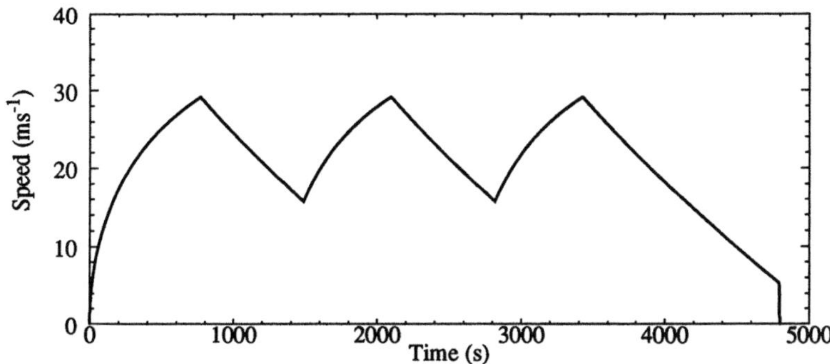


Figure 10-6: Speed $v(t)$ for Example 10.4

☐

### *Example 10.5  A Further Improved Strategy of Optimal Type*

We use a strategy with $\{j(k+1)\}_{k=0,1,\ldots,6} = \{4, 0, 4, 0, 4, 0, -1\}$. We set $V_1 = V_3 = V_5 = Y_{04}$, $V_2 = V_4 = X_{04}$ and $V_6 = U$ where the critical speeds are determined from $\lambda$ and $\mu$ as described in Section 10.4. The parameters $\lambda$ and $\mu$ are adjusted to satisfy the time and distance constraints. The results of this calculation were $X_{03} = 12.96$, $Y_{03} = 31.59$, $U = 4.72$ and $J = 1281.09$. The speed profile is shown in Figure 10-7.

☐

In Example 10.5 the fuel consumption is decreased dramatically because much more time is spent coasting. This is a consequence of the more efficient power phases.

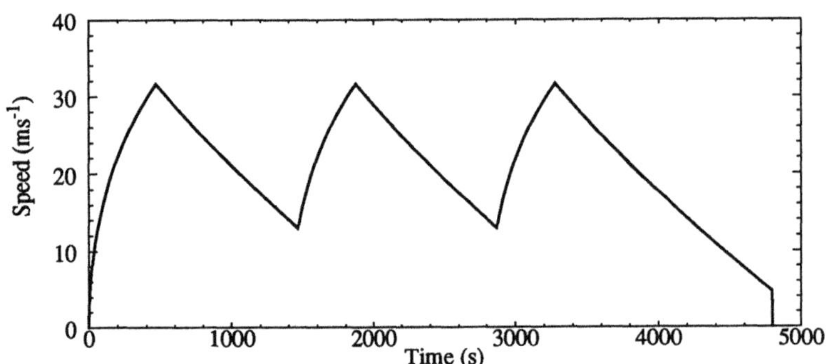

Figure 10-7: Speed $v(t)$ for Examples 10.4 and 10.5

## 10.9 Approximating the Minimum-Cost Strategy

We define a special strategy of optimal type $\mathcal{T}_q$ with $2q+3$ phases using only maximum power, coast and brake, as described in Section 9.2. The critical speed at which braking begins is defined by

$$
\begin{aligned}
U(V, W) &= \frac{V\varphi(W) - W\varphi(V)}{\varphi(W) - \varphi(V)} \\
&= W - \frac{\varphi(W)}{\left[\dfrac{\varphi(W) - \varphi(V)}{W - V}\right]}.
\end{aligned}
$$

We can calculate the time $t_q(V, W)$ taken for the journey, the distance $x_q(V, W)$ travelled by the train and the cost

$$
J_q(V, W) = f_m \left\{ \int_0^W \frac{dv}{p_m\theta(v) - r(v)} + q\int_V^W \frac{dv}{p_m\theta(v) - r(v)} \right\} \tag{10.3}
$$

of the strategy as in Section 9.2. For a feasible strategy we require a pair $(X_q, Y_q)$ of critical speeds such that $t_q(X_q, Y_q) = T$ and $x_q(X_q, Y_q) = X$.

We also define an idealised strategy $\mathcal{T}$ of optimal type with an initial phase of maximum power followed by phases of speed-hold, coast and brake, as in Section 9.2. This strategy will be obtained as the limit of a sequence of feasible strategies of the form $\mathcal{T}_q$ where the critical speeds $X_q$ and $Y_q$ converge to a common value $Z$ as $q \to \infty$. Initially we consider an idealised strategy $\mathcal{T} = \mathcal{T}(Y)$ in which speed-holding occurs at some arbitrary critical speed $Y > 0$. To hold at this speed would require a fuel supply rate given by

$$f = p^{-1} [\varphi(Y)] \tag{10.4}$$

Unfortunately this rate will not in general be an allowable rate of fuel supply. To overcome this difficulty we assume that speed-holding is approximated by short periods of coast alternating with short periods of maximum power. It will be shown in the next section that as the magnitude of the speed oscillations is decreased the effective rate of fuel supply for the approximate strategy approaches a limit that is less than or equal to the required rate of fuel supply for the idealised strategy.

For the strategy $\mathcal{T}$ the critical speed $U(Y)$ at which braking begins is given by the formula

$$U(Y) = \lim_{V, W \to Y} U(V, W) = Y - \frac{\varphi(Y)}{\dfrac{d}{dv}[\varphi(V)]\Big|_{v = Y}} .$$

Let

$$\tau(Y) = \lim_{V, W \to Y} t_q(V, W)$$

and not that if $\tau(Y) < T$ then the time allowed for the speed-hold phase is given by $T - \tau(Y)$. We define

$$\xi(Y) = \lim_{V, W \to Y} x_q(V, W)$$

and calculate the total distance travelled as

$$x(Y) = \xi(Y) + Y [T - \tau(Y)] .$$

The cost of the strategy is given by

$$J(Y) = f_m \int_0^Y \frac{dv}{p_m \theta(v) - r(v)} + p^{-1} [\varphi(Y)] [T - \tau(Y)] . \tag{10.5}$$

For a feasible strategy we need to find $Z$ such that $\tau(Z) < T$ and $x(Z) = X$.

Once again we will assume that apart from the final brake phase a strategy of optimal type should be constructed using only maximum power and coast. We will show that Theorems 9.1 and 9.2 remain valid.

## 10.10  Approximate Speed-Holding Strategies

We can prove the following lemmas.

### *Lemma 10.1*

*Let $Y > 0$ and $\Delta T > 0$, and suppose that*

$$\int_0^Y \frac{p_m \theta(v)\,dv}{[p_m \theta(v) - r(v)]\, r(v)} \geq \Delta T. \tag{10.6}$$

*For each $q = 1, 2, \ldots$ we can find $R_q$ and $S_q$ with $0 < R_q < Y < S_q$ and $[R_q, S_q] \subset (R_{q-1}, S_{q-1})$, and such that*

$$q \int_{R_q}^{S_q} \frac{p_m \theta(v)\,dv}{[p_m \theta(v) - r(v)]\, r(v)} = \Delta T$$

*and*

$$q \int_{R_q}^{S_q} \frac{p_m \theta(v)\,v\,dv}{[p_m \theta(v) - r(v)]\, r(v)} = Y\Delta T.$$

### *Proof*

Let $\bar{R}_1 = \underline{S}_1 = Y$. Choose $\bar{S}_1 > Y$ and $\underline{R}_1 < Y$ so that

$$\int_{\bar{R}_1}^{\bar{S}_1} \frac{p_m \theta(v)\,dv}{[p_m \theta(v) - r(v)]\, r(v)} = \int_{\underline{R}_1}^{\underline{S}_1} \frac{p_m \theta(v)\,dv}{[p_m \theta(v) - r(v)]\, r(v)} = \Delta T.$$

There exists a continuous, strictly increasing mapping $\gamma_1 : [\underline{R}_1, \bar{R}_1] \to [\underline{S}_1, \bar{S}_1]$ such that for each point $R \in [\underline{R}_1, \bar{R}_1]$ we can find a corresponding point $\gamma_1(R) \in [\underline{S}_1, \bar{S}_1]$ with

$$\int_R^{\gamma_1(R)} \frac{p_m \theta(v)\,dv}{[p_m \theta(v) - r(v)]\, r(v)} = \Delta T.$$

It follows that

$$\int_{\underline{R}_1}^{\gamma_1(\underline{R}_1)} \frac{p_m\theta(v)v\,dv}{[p_m\theta(v) - r(v)]\,r(v)} < Y\Delta T < \int_{\bar{R}_1}^{\gamma_1(\bar{R}_1)} \frac{p_m\theta(v)v\,dv}{[p_m\theta(v) - r(v)]\,r(v)}$$

and hence we can find a point $R_1 \in (\underline{R}_1, \bar{R}_1)$ and a corresponding point $S_1 = \gamma_1(R_1) \in (\underline{S}_1, \bar{S}_1)$ with

$$\int_{R_1}^{S_1} \frac{p_m\theta(v)v\,dv}{[p_m\theta(v) - r(v)]\,r(v)} = Y\Delta T.$$

Note that $R_1 < \bar{R}_1 = Y = \underline{S}_1 < S_1$. In general we suppose that there exists an interval $[R_p, S_p]$ with $R_p < Y < S_p$ such that

$$p\int_{R_p}^{S_p} \frac{p_m\theta(v)\,dv}{[p_m\theta(v) - r(v)]\,r(v)} = \Delta T$$

and

$$p\int_{R_p}^{S_p} \frac{p_m\theta(v)v\,dv}{[p_m\theta(v) - r(v)]\,r(v)} = Y\Delta T$$

for each $p = 1, 2, ..., q-1$. We set $\underline{R}_q = R_{q-1}$ and $\bar{S}_q = S_{q-1}$, and define $\bar{R}_q, \underline{S}_q \in (R_{q-1}, S_{q-1})$ such that

$$q\int_{\bar{R}_q}^{\bar{S}_q} \frac{p_m\theta(v)\,dv}{[p_m\theta(v) - r(v)]\,r(v)} = q\int_{\underline{R}_q}^{\underline{S}_q} \frac{p_m\theta(v)\,dv}{[p_m\theta(v) - r(v)]\,r(v)} = \Delta T.$$

There exists a continuous strictly increasing mapping $\gamma_q : [\underline{R}_q, \bar{R}_q] \to [\underline{S}_q, \bar{S}_q]$ such that for each point $R \in [\underline{R}_q, \bar{R}_q]$ there is a corresponding point $\gamma_q(R) \in [\underline{S}_q, \bar{S}_q]$ with

$$q\int_{R}^{\gamma_q(R)} \frac{p_m\theta(v)\,dv}{[p_m\theta(v) - r(v)]\,r(v)} = \Delta T.$$

Using an argument similar to the argument in Lemma 9.3 it can now be shown that

$$q \int_{\underline{R}_q}^{\underline{S}_q} \frac{p_m \theta(v) v \, dv}{[p_m \theta(v) - r(v)] \, r(v)} < Y\Delta T < q \int_{\overline{R}_q}^{\overline{S}_q} \frac{p_m \theta(v) v \, dv}{[p_m \theta(v) - r(v)] \, r(v)}.$$

It follows that we can find a point $R_q \in (\underline{R}_q, \overline{R}_q)$ and a corresponding point $S_q = \gamma_q(R_q) \in [\underline{S}_q, \overline{S}_q]$ with

$$q \int_{R_q}^{S_q} \frac{p_m \theta(v) v \, dv}{[p_m \theta(v) - r(v)] \, r(v)} = Y\Delta T.$$

Clearly $[R_q, S_q] \subset (R_{q-1}, S_{q-1})$. It is easily shown that $R_q < Y < S_q$. The general result now follows by induction.

### Corollary 10.2

*Let $Y > 0$ and $\Delta T > 0$, and suppose that $q_0$ is a natural number with*

$$q_0 \int_0^Y \frac{p_m \theta(v) \, dv}{[p_m \theta(v) - r(v)] \, r(v)} \geq \Delta T.$$

*For each natural number $q > q_0$ we can find $R_q$ and $S_q$ with $0 < R_q < Y < S_q$ and $[R_q, S_q] \subset (R_{q-1}, S_{q-1})$, and such that*

$$q \int_{R_q}^{S_q} \frac{p_m \theta(v) \, dv}{[p_m \theta(v) - r(v)] \, r(v)} = \Delta T$$

*and*

$$q \int_{R_q}^{S_q} \frac{p_m \theta(v) v \, dv}{[p_m \theta(v) - r(v)] \, r(v)} = Y\Delta T.$$

$\square$

### Lemma 10.3

*Suppose that $Y > 0$ and $\Delta T > 0$, and let $q_0$ be a natural number such that inequality (10.6) is satisfied. Let $\bar{Y} > Y$ be chosen such that*

$$q_0 \int_Y^{\bar{Y}} \frac{p_m \theta(v)\,dv}{[p_m \theta(v) - r(v)]\,r(v)} = \Delta T.$$

*For each natural number $q \geq q_0$ we have*

$$S_q - R_q \leq \frac{r(Y)\Delta T}{q}.$$

### Proof

Since

$$q \int_{R_q}^{S_q} \frac{p_m \theta(v)\,dv}{[p_m \theta(v) - r(v)]\,r(v)} = \Delta T$$

for each natural number $q \geq q_0$ it follows that

$$R_q < Y < S_q < \bar{Y}$$

and since

$$\frac{p_m \theta(v)}{[p_m \theta(v) - r(v)]\,r(v)} \geq \frac{1}{r(\bar{Y})}$$

for all $v \in [R_q, S_q]$ we can also deduce that

$$\frac{q\,(S_q - R_q)}{r(\bar{Y})} \leq \Delta T.$$

$\square$

We now replace a speed-holding segment $\Delta S$ at speed $Y$ and of duration $\Delta T$ by an approximate speed-holding segment $\Delta S_q$ with $2q$ phases given by

$$\Delta S_q = \{ [m;(Y,S_q)] , [0;(S_q,R_q)], [m;(R_q,S_q)], \dots$$
$$[0;(S_q,R_p)], [m;(R_q,Y)] \}$$

where $R_q$ and $S_q$ are chosen according to Corollary 10.2, which shows that the time taken and the distance travelled are the same for each segment of strategy. The cost of the approximate speed-holding strategy is given by

$$\Delta J(R_q,S_q) = f_m \left[ q \int_{R_p}^{S_q} \frac{dv}{p_m\theta(v)-r(v)} \right]$$

and since $\varphi(v)$ increases as $v$ increases it can be seen from Corollary 10.2 that

$$\frac{f_m}{p_m}\varphi(R_q)\Delta T = \frac{f_m}{p_m}\varphi(R_q) \left[ q \int_{R_p}^{S_q} \frac{p_m dv}{[p_m\theta(v)-r(v)]\,\varphi(v)} \right]$$

$$< \frac{f_m}{p_m} \left[ q \int_{R_p}^{S_q} \frac{p_m dv}{[p_m\theta(v)-r(v)]} \right]$$

$$= \Delta J(R_q,S_q)$$

$$= \frac{f_m}{p_m} \left[ q \int_{R_p}^{S_q} \frac{p_m dv}{[p_m\theta(v)-r(v)]} \right]$$

$$< \frac{f_m}{p_m}\varphi(S_q) \left[ q \int_{R_p}^{S_q} \frac{p_m dv}{[p_m\theta(v)-r(v)]\,\varphi(v)} \right]$$

$$= \frac{f_m}{p_m}\varphi(S_q)\Delta T . \tag{10.7}$$

Using Lemma 10.3 and the inequality (10.7), and by taking the limit as $q \to \infty$, the cost of an idealised speed-hold phase of duration $\Delta T$ at speed $Y$ is given by

$$\Delta J(Y) = \left[ \frac{f_m}{p_m}\varphi(Y) \right] \Delta T. \tag{10.8}$$

Since the idealised speed-hold phase can be realised to any desired accuracy it will be used where necessary without further comment. If we replace the approximate speed-hold segment $\Delta S_{q-1}$ by the approximate speed-holding segment $\Delta S_q$ we can see that

$$\Delta J\left(R_{q-1}, S_{q-1}\right) - \Delta J(R_q, S_q)$$

$$= (q-1)\left\{ \int_{R_{q-1}}^{R_q} \frac{f_m dv}{p_m \theta(v) - r(v)} + \int_{S_q}^{S_{q-1}} \frac{f_m dv}{p_m \theta(v) - r(v)} \right\}$$

$$- \int_{R_q}^{S_q} \frac{f_m dv}{p_m \theta(v) - r(v)}.$$

Because the graph $y = \varphi(v)$ is convex we can find $\lambda$ and $\mu$ such that $\lambda v - \mu < \varphi(v)$ when $v \in (R_q, S_q)$ and $\lambda v - \mu > \varphi(v)$ when $v \notin [R_q, S_q]$. It follows that

$$\Delta J\left(R_{q-1}, S_{q-1}\right) - \Delta J(R_q, S_q)$$

$$> (q-1)\left\{ \int_{R_{q-1}}^{R_q} \frac{f_m (\lambda v - \mu)\, dv}{[p_m \theta(v) - r(v)]\, \varphi(v)} + \int_{S_q}^{S_{q-1}} \frac{f_m (\lambda v - \mu)\, dv}{[p_m \theta(v) - r(v)]\, \varphi(v)} \right\}$$

$$- \int_{R_q}^{S_q} \frac{f_m (\lambda v - \mu)\, dv}{[p_m \theta(v) - r(v)]\, \varphi(v)}$$

$$= 0$$

by Corollary 10.2. Therefore

$$\Delta J(R_{q-1}, S_{q-1}) - \Delta J(R_q, S_q) > 0.$$

Thus the cost of the strategy $\Delta S_q$ decreases as $q$ increases.


### The Cost of Speed-Holding

Note that as the magnitude of the speed oscillations decreases, (10.8) shows that the average rate of fuel supply for an approximate speed-hold phase at speed $Y$ approaches a limiting value given by

$$g = \frac{f_m}{p_m} \varphi(Y).$$

On the other hand, (10.4) shows that the fuel supply rate required for an idealised speed-hold phase is given by

$$f = p^{-1}[\varphi(Y)] \; .$$

Using the convexity of $p$ it can be seen that

$$p(f) \le \frac{f}{f_m} p(f_m)$$

and hence

$$g = \frac{f_m}{p_m}\varphi(Y) = \frac{f_m}{p_m}p(f) \le f.$$

This shows that approximate speed-holding is at least as efficient as true speed-holding.

## 10.11  The Structure of an Optimal Strategy

The following results can be used to show that an overall minimum cost strategy must contain an initial phase of maximum power followed by phases of speed-hold, coast and brake. These arguments are based on a modification procedure which preserves the feasibility of the strategy but decreases the cost. The results of this section are completely analogous to the results of Section 9.4.

### 10.11.1  The Power Phase

Consider the two strategy segments shown in Figure 9-4. The curve $v = v_1(t)$ corresponds to a net acceleration phase with a fuel supply rate $f$ such that

$$p^{-1}[\varphi(W)] \; < f < f_m,$$

whereas the curve $v = v_2(t)$ corresponds to a *maximum power–hold–maximum power* sequence. The time $\tau$ spent speed-holding is given by

$$\tau = \int_V^W \frac{dv}{p(f)\theta(v) - r(v)} - \int_V^W \frac{dv}{p_m\theta(v) - r(v)}.$$

The holding speed $Z$ is chosen so that

$$\int_V^W \frac{(v-Z)\,dv}{p_m\theta(v)-r(v)} = \int_V^W \frac{(v-Z)\,dv}{p(f)\theta(v)-r(v)}. \tag{10.9}$$

The costs of the two strategy segments are

$$\Delta J_1 = f \int_V^W \frac{dv}{p(f)\theta(v)-r(v)}$$

and

$$\Delta J_2 = f_m \int_V^W \frac{dv}{p_m\theta(v)-r(v)}$$

$$+ \frac{f_m}{p_m}\varphi(Z)\left[\int_V^W \frac{dv}{p(f)\theta(v)-r(v)} - \int_V^W \frac{dv}{p_m\theta(v)-r(v)}\right].$$

The following lemma shows that the second strategy segment uses less fuel than the first.

### Lemma 10.4

*Let $0 < V < W$ and let*

$$0 < p^{-1}[\varphi(W)] < f < f_m.$$

*If $Z \in (V, W)$ is chosen so that*

$$\int_V^W \frac{(v-Z)\,dv}{p_m\theta(v)-r(v)} = \int_V^W \frac{(v-Z)\,dv}{p(f)\theta(v)-r(v)}$$

*then*

$$f_m \int_V^W \frac{dv}{p_m\theta(v)-r(v)} + \frac{f_m}{p_m}\varphi(Z)\left[\int_V^W \frac{dv}{p(f)\theta(v)-r(v)} - \int_V^W \frac{dv}{p_m\theta(v)-r(v)}\right]$$

$$< f \int_V^W \frac{dv}{p(f)\theta(v)-r(v)}.$$

***Proof***

Since we assume that the graph $y = \varphi(v)$ is strictly convex the tangent line $y = \lambda v - \mu$ at the point $v = Z$ is below the graph for all $v \neq Z$. Since

$$p(f) < \frac{p_m f}{f_m}$$

and

$$\varphi(Z) < p_m$$

it follows that

$$f_m \int_V^W \frac{dv}{p_m \theta(v) - r(v)} + \frac{f_m}{p_m} \varphi(Z) \left[ \int_V^W \frac{dv}{p(f)\theta(v) - r(v)} - \int_V^W \frac{dv}{p_m \theta(v) - r(v)} \right]$$

$$- f \int_V^W \frac{dv}{p(f)\theta(v) - r(v)}$$

$$= \int_V^W \frac{\{ (f_m p(f) - f p_m) - (f_m - f)\varphi(v) + \dfrac{f_m \varphi(Z)}{p_m}[p_m - p(f)] \} \theta(v)dv}{[p_m \theta(v) - r(v)]\,[p(f)\theta(v) - r(v)]}$$

$$= \int_V^W \frac{\{ (f_m - f)\lambda(Z - v) + [f_m p(f) - f p_m]\left[1 - \dfrac{\varphi(Z)}{p_m}\right] \} \theta(v)dv}{[p_m \theta(v) - r(v)]\,[p(f)\theta(v) - r(v)]}$$

$$< (f_m - f) \int_V^W \frac{\lambda(Z - v)\theta(v)dv}{[p_m \theta(v) - r(v)]\,[p(f)\theta(v) - r(v)]}$$

$$= 0 \,.$$

$\square$

### 10.11.2 The Transition from Power to Speed-Hold

Now consider the two strategy segments illustrated in Figure 9-5. The curve $v = v_1(t)$ corresponds to a *hold–power–hold–coast* sequence with hold speeds $V$ and $W$ respectively, and the curve $v = v_2(t)$ corresponds to a *power–hold* sequence with hold speed $Z$. Let $\tau_V$, $\tau_W$ and $\tau_Z$ be the times spent speed-holding at speeds $V$, $W$ and $Z$ respectively. For the segment times to be equal we require

$$\int_V^Z \frac{dv}{p_m\theta(v)-r(v)} + \tau_Z = \tau_V + \int_V^W \frac{dv}{p_m\theta(v)-r(v)} + \tau_W + \int_Z^W \frac{dv}{r(v)}. \qquad (10.10)$$

For the segment distances to be equal we require

$$\int_V^Z \frac{v\,dv}{p_m\theta(v)-r(v)} + Z\tau_Z = V\tau_V + \int_V^W \frac{v\,dv}{p_m\theta(v)-r(v)} + W\tau_W + \int_Z^W \frac{v\,dv}{r(v)}. \qquad (10.11)$$

The costs of the two segments are

$$\Delta J_1 = \frac{f_m}{p_m}\varphi(V)\tau_V + f_m\int_V^W \frac{dv}{p_m\theta(v)-r(v)} + \frac{f_m}{p_m}\varphi(W)\tau_W$$

and

$$\Delta J_2 = f_m\int_V^Z \frac{dv}{p_m\theta(v)-r(v)} + \frac{f_m}{p_m}\varphi(Z)\tau_Z.$$

We can now show that the second strategy segment uses less fuel than the first.

### Lemma 10.5

*Let $0 < V < W$ and let $\tau_V$ and $\tau_W$ be given. If $Z \in (V, W)$ and $\tau_Z$ are chosen so that*

$$\int_V^Z \frac{dv}{p_m\theta(v)-r(v)} + \tau_Z = \tau_V + \int_V^W \frac{dv}{p_m\theta(v)-r(v)} + \tau_W + \int_Z^W \frac{dv}{r(v)}$$

*and*

$$\int_V^Z \frac{v\,dv}{p_m\theta(v)-r(v)} + Z\tau_Z = V\tau_V + \int_V^W \frac{v\,dv}{p_m\theta(v)-r(v)} + W\tau_W + \int_Z^W \frac{v\,dv}{r(v)}$$

*then*

$$f_m \int_V^Z \frac{dv}{p_m \theta(v) - r(v)} + \frac{f_m}{p_m} \varphi(Z) \tau_Z$$

$$< \frac{f_m}{p_m} \varphi(V) \tau_V + f_m \int_V^W \frac{dv}{p_m \theta(v) - r(v)} + \frac{f_m}{p_m} \varphi(W) \tau_W.$$

### Proof

Since we assume that the graph $y = \varphi(v)$ is strictly convex the tangent line $y = \lambda v - \mu$ at the point $v = Z$ is below the graph for all $v \neq Z$. Then

$$f_m \int_V^Z \frac{dv}{p_m \theta(v) - r(v)} + \frac{f_m}{p_m} \varphi(Z) \tau_Z$$

$$- \left\{ \frac{f_m}{p_m} \varphi(V) \tau_V + f_m \int_V^W \frac{dv}{p_m \theta(v) - r(v)} + \frac{f_m}{p_m} \varphi(W) \tau_W \right\}$$

$$< -f_m \int_Z^W \frac{(\lambda v - \mu)\, dv}{[p_m \theta(v) - r(v)]\, \varphi(v)}$$

$$+ \frac{f_m}{p_m} \left\{ [\lambda Z - \mu]\, \tau_Z - [\lambda V - \mu]\, \tau_V - [\lambda W - \mu]\, \tau_W \right\}$$

$$= 0$$

since

$$\int_Z^W \frac{p_m dv}{[p_m \theta(v) - r(v)]\, \varphi(v)} - \tau_Z + \tau_V + \tau_W = 0$$

and

$$\int_Z^W \frac{p_m v dv}{[p_m \theta(v) - r(v)]\, \varphi(v)} - Z \tau_Z + V \tau_V + W \tau_W = 0.$$

$\square$

### 0.11.3 The Transition from Speed-Hold to Coast

Now consider the two strategy segments illustrated in Figure 9-6. The curve $v = v_1(t)$ corresponds to a *power–hold–coast–hold* sequence with hold speeds $W$ and $V$ respectively, and the curve $v = v_2(t)$ corresponds to a *hold–coast* sequence with hold speed $Z$. Let $\tau_V$, $\tau_W$ and $\tau_Z$ be the times spent speed-holding at speeds $V$, $W$ and $Z$ respectively. For the segment times to be equal we require

$$\tau_Z + \int_V^Z \frac{dv}{r(v)} = \int_Z^W \frac{dv}{p_m\theta(v) - r(v)} + \tau_W + \int_Z^W \frac{dv}{r(v)} + \tau_V. \qquad (10.12)$$

For the segment distances to be equal we require

$$Z\tau_Z + \int_V^Z \frac{v\,dv}{r(v)} = \int_Z^W \frac{v\,dv}{p_m\theta(v) - r(v)} + W\tau_W + \int_Z^W \frac{v\,dv}{r(v)} + V\tau_V. \qquad (10.13)$$

The costs of the two segments are

$$\Delta J_1 = f_m \int_Z^W \frac{dv}{p_m\theta(v) - r(v)} + \frac{f_m}{p_m}\varphi(W)\tau_W + \frac{f_m}{p_m}\varphi(V)\tau_V$$

and

$$\Delta J_2 = \frac{f_m}{p_m}\varphi(Z)\tau_Z.$$

Since (10.12) and (10.13) are equivalent to (10.10) and (10.11), Lemma 10.5 can be used to show that the second strategy segment uses less fuel than the first.


### 0.11.4 The Coast Phase

Finally, consider the two strategy segments shown in Figure 9-7. The curve $v = v_1(t)$ corresponds to a negative net acceleration phase with fuel supply rate $g$ such that

$$0 < g < p^{-1}[\varphi(V)],$$

whereas the curve $v = v_2(t)$ corresponds to a *coast–hold–coast* sequence. The time spent speed-holding is given by

$$\tau = \int_V^W \frac{dv}{r(v) - p(g)\theta(v)} - \int_V^W \frac{dv}{r(v)}.$$

The holding speed $Z$ is chosen so that the distance travelled using each strategy is the same. That is

$$\int_V^W \frac{(v-Z)\,dv}{r(v)} = \int_V^W \frac{(v-Z)\,dv}{r(v) - p(g)\theta(v)}. \tag{10.14}$$

The costs of the two strategy segments are

$$\Delta J_1 = g \int_V^W \frac{dv}{r(v) - p(g)\theta(v)}$$

and

$$\Delta J_2 = \frac{f_m}{p_m}\varphi(Z)\left[\int_V^W \frac{dv}{r(v) - p(g)\theta(v)} - \int_V^W \frac{dv}{r(v)}\right].$$

We can now use the following lemma to show that the second strategy segment uses less fuel than the first.

### *Lemma 10.6*

*Let $0 < V < W$ and let*

$$0 < g < p^{-1}\left[\varphi(V)\right].$$

*If $Z \in (V, W)$ is chosen so that*

$$\int_V^W \frac{(v-Z)\,dv}{r(v)} = \int_V^W \frac{(v-Z)\,dv}{r(v) - p(g)\theta(v)}$$

*then*

$$\frac{f_m}{p_m}\varphi(Z)\left[\int_V^W \frac{dv}{r(v) - p(g)\theta(v)} - \int_V^W \frac{dv}{r(v)}\right] < g \int_V^W \frac{dv}{r(v) - p(g)\theta(v)}.$$

## *Proof*

The proof is similar to the proof of Lemma 10.4,

□

## 10.12  A Speed-Holding Strategy of Optimal Type

Consider the strategies $T_q(V, W)$ and $T(Y)$ of optimal type described in Section 10.9. Under certain reasonable conditions we can find a feasible strategy $T_q(X_q, Y_q)$. We will show that if $W$ is fixed then $t_q(V, W)$ decreases as $V$ increases. Since $\partial U/\partial V > 0$ and since

$$\frac{\partial t_q}{\partial V} = (-1)\,q\frac{p_m\theta(x)}{[p_m\theta(v) - r(v)]\,r(v)} - \frac{\psi(U)}{[\psi(U) + r(U)]\,r(U)}\frac{\partial U}{\partial V}$$

it follows that $\partial t_q/\partial V < 0$. Provided $\tau(W) < T < t_q(0, W)$ we can therefore solve the equation $t_q(V, W) = T$. If we now adjust $(V, W)$ in such a way that the relationship $t_q(V, W) = T$ is preserved then $x_q(V, W)$ increases as $W$ increases. If we regard $V = V(W)$ and note that

$$\frac{dt_q}{dW} = \frac{\partial t_q}{\partial V}\frac{dV}{dW} + \frac{\partial t_q}{\partial W} = 0$$

it follows that

$$\frac{p_m\theta(W)}{[p_m\theta(W) - r(W)]\,r(W)} - \frac{\psi(U)}{[\psi(U) + r(U)]\,r(U)}\frac{dU}{dW}$$
$$-\,q\frac{p_m\theta(V)}{[p_m\theta(V) - r(v)]\,r(V)}\frac{dV}{dW} + q\frac{p_m\theta(W)}{[p_m\theta(W) - r(W)]\,r(W)} = 0.$$

Now if $dV/dW < 0$ then this equation shows that $dU/dW > 0$. On the other hand, if $dV/dW > 0$ then since $\partial U/\partial V > 0$ and $\partial U/\partial W > 0$ we also know that

$$\frac{dU}{dW} = \frac{\partial U}{\partial V}\frac{dV}{dW} + \frac{\partial U}{\partial W} > 0.$$

Since

$$\frac{dx_q}{dY} = \frac{p_m W\theta(W)}{[p_m\theta(W) - r(W)]\,r(W)} - \frac{U\psi(U)}{[\psi(U) + r(U)]\,r(U)}\frac{dU}{dW}$$

$$-q\frac{p_m V\theta(V)}{[p_m\theta(V) - r(V)]\, r(V)}\frac{dV}{dW} + q\frac{p_m W\theta(W)}{[p_m\theta(W) - r(W)]\, r(W)}$$

an argument similar to the argument of Section 8.6 shows that

$$\frac{dx_q}{dW} > U\frac{dt_q}{dW} = 0$$

when $dV/dW < 0$ and

$$\frac{dx_q}{dW} > V\frac{dt_q}{dW} = 0$$

when $dV/dW > 0$. Therefore $x_q(V(W), W)$ increases as $W$ increases. As long as we can find $\overline{W}$ with $\tau(\overline{W}) = T$ and $\xi(\overline{W}) > X$ it now follows that we can find a unique pair $(X_q, Y_q)$ with $0 < X_q < Y_q < \overline{W}$ such that $t_q(X_q, Y_q) = T$ and $x_q(X_q, Y_q) = X$. We can also find a feasible strategy $\mathcal{T}(Z)$. We have

$$t(Z) = t_q(X_q, Y_q) = T \tag{10.15}$$

and

$$x(Z) = x_q(X_q, Y_q) = X. \tag{10.16}$$

From (10.15) we have the time spent speed-holding given by

$$\begin{aligned}
\tau_Z &= T - \tau(Z) \\
&= \int_Z^{Y_q} \frac{p_m\theta(v)\,dv}{[p_m\theta(v) - r(v)]\, r(v)} + q\int_{X_q}^{Y_q} \frac{p_m\theta(v)\,dv}{[p_m\theta(v) - vr(v)]\, r(v)} \\
&\quad + \int_{U(X_q, Y_q)}^{U(Z)} \frac{\psi(v)\,dv}{[\psi(v) + r(v)]\, r(v)}
\end{aligned} \tag{10.17}$$

and from (10.16) we have the distance travelled during speed-holding given by

$$\begin{aligned}
Z\tau_Z &= Z(T - \tau_Z) \\
&= \int_Z^{Y_q} \frac{p_m\theta(v)v\,dv}{[p_m\theta(v) - r(v)]\, r(v)} + q\int_{X_q}^{Y_q} \frac{p_m\theta(v)v\,dv}{[p_m\theta(v) - r(v)]\, r(v)}
\end{aligned}$$

$$+ \int_{U(X_q, Y_q)}^{U(Z)} \frac{\psi(v)\, v\, dv}{[\psi(v) + r(v)]\, r(v)} \tag{10.18}$$

It is now possible to show that $X_q < Z < Y_q$. We have the following result.

### Lemma 10.7

Let $0 < X_q < Z < Y_q$. Then $J(Z) < J_q(X_q, Y_q)$.

### Proof

Because both strategies are feasible we have $t_q(X_q, Y_q) = T$ and $x_q(X_q, Y_q) = x(Z) = X$. By multiplying (10.17) by $Z$ and subtracting from (10.18) we obtain the equation

$$\int_{Z}^{Y_q} \frac{p_m\,(v - Z)\, dv}{[p_m\theta(v) - r(v)]\, \varphi(v)} + q\int_{X_q}^{Y_q} \frac{p_m\,(v - Z)\, dv}{[p_m\theta(v) - r(v)]\, \varphi(v)}$$

$$+ \int_{U(X_q, Y_q)}^{U(Z)} \frac{\psi(v)\,(v - Z)\, dv}{[\psi(v) + r(v)]\, r(v)} = 0. \tag{10.19}$$

From (10.3) and (10.5) the cost difference is given by

$$J(Z) - J_q(X_q, Y_q)$$

$$= (-f_m)\left\{ \int_{Z}^{Y_q} \frac{dv}{[p_m\theta(v) - r(v)]} + q\int_{X_q}^{Y_q} \frac{dv}{[p_m\theta(v) - vr(v)]} \right\}$$

$$+ \frac{f_m}{p_m}\varphi(Z)\left\{ \int_{Z}^{Y_q} \frac{p_m dv}{[p_m\theta(v) - r(v)]\, \varphi(v)} + q\int_{X_q}^{Y_q} \frac{p_m dv}{[p_m\theta(v) - vr(v)]\, \varphi(v)} \right.$$

$$\left. + \int_{U(X_q, Y_q)}^{U(Z)} \frac{\psi(v)\, dv}{[\psi(v) + r(v)]\, r(v)} \right\}.$$

Since we assume that the graph $y = \varphi(v)$ is strictly convex the tangent line

$$y = \lambda v - \mu = \varphi'(Z)\,(v - Z) + \varphi(Z)$$

at the point $v = Z$ is below the graph for all $v \neq Z$. Thus

$$J(Z) - J_q(X_q, Y_q)$$

$$< (-f_m)\left\{ \int_Z^{Y_q} \frac{\varphi'(Z)\,(v - Z)\,dv}{[p_m\theta(v) - r(v)]\,\varphi(v)} + q\int_{X_q}^{Y_q} \frac{\varphi'(Z)\,(v - Z)\,dv}{[p_m\theta(v) - r(v)]\,\varphi(v)} \right\}$$

$$+ \frac{f_m}{p_m}\varphi(Z) \int_{U(X_q,Y_q)}^{U(Z)} \frac{\psi(v)\,dv}{[\psi(v) + r(v)]\,r(v)}.$$

From (10.19) it follows that

$$J(Z) - J_q(X_q, Y_q) < \frac{f_m}{p_m} \int_{U(X_q,Y_q)}^{U(Z)} \frac{(\lambda v - \mu)\,\psi(v)\,dv}{[\psi(v) + r(v)]\,r(v)}.$$

Since $v = U(Z)$ is the solution to the equation $\lambda v - \mu = 0$ it can be seen that $\lambda v - \mu > 0$ when $v > U(Z)$ and that $\lambda v - \mu < 0$ when $v < U(Z)$. It now follows that

$$J(Z) - J_q(X_q, Y_q) < 0.$$

## CHAPTER 11

# SPEED LIMITS

In this chapter we again use a more general model to formulate and solve the train control problem for a level track with varying speed limits. We find key equations that determine critical speeds for a strategy of optimal type. In the idealised case, there is a unique hold speed for the journey. On intervals of track where the speed limit is below the hold speed, the speed must be held at the limit. If braking is necessary on an interval, the speed at which braking commences is determined in part by the hold speed for the interval.

The material in this chapter was first published in the *Journal of the Australian Mathematical Society, Series B* [11].


## 11.1  Vehicle Model

In this chapter we assume that the vehicle has discrete control settings denoted by

$$j \in C = \{-q, \ldots, -1, 0, 1, \ldots, Q\}$$

where $j = -q$ represents maximum braking effort, $j = 0$ represents coasting, and $j = Q$ represents maximum tractive effort. On level track the acceleration of the vehicle is given by

$$v\frac{dv}{dx} = u_j(v) - r(v)$$

where $v$ is the speed of the vehicle, $x$ is the position, $u_j(v)$ is the tractive or braking acceleration developed by the vehicle for control setting $j$ at speed $v$, and $-r(v)$ is the acceleration of the vehicle due to resistance at speed $v$.

We assume a constant cost rate $f_j$ for each control setting, and that $f_j = 0$ for $j \le 0$. We also assume that $u_j(v)$ is decreasing and that $r(v)$ is increasing, and that each of these functions satisfies a Lipschitz condition on each bounded interval.

## 11.2 Journey Model

Consider a strategy with $n + 1$ distinct phases. The control setting is specified for each phase. The positions at which the control must be changed are to be determined. These positions are denoted by $0 = x_0 < x_1 < \ldots < x_{n+1} = X$. The journey starts at position $x = 0$ and finishes at position $x = X$.

The track speed limit changes at positions $0 = X_0 < X_1 < \ldots < X_p = X$. The speed limit for the interval $(X_i, X_{i+1})$ is $M_{i+1}$.

It is necessary to determine the locations of the switching points $\{x_k\}$ for $k = 1, 2, \ldots, n$ relative to the fixed points $\{X_i\}$ for $i = 0, 1, \ldots, p$. We use the notation $\{q(i)\}$ to denote the sequence with $q(0) = 0$ and $q(p) = n + 1$, and with

$$x_{q(i)} \leq X_i < x_{q(i)+1}$$

for $i = 1, 2, \ldots, p - 1$. Let $\xi_{k+1} = x_{k+1} - x_k$ denote the length of the control interval $[x_k, x_{k+1}]$. The control setting in this interval is $j(k+1)$. If the time taken to traverse the interval $[x_k, x_{k+1}]$ is $\tau_{k+1}$ then the cost of the journey is

$$J = \sum_{k=0}^{n} f_{j(k+1)} \tau_{k+1}.$$

If the time allowed for the journey is $T$, we require

$$\sum_{k=0}^{n} \tau_{k+1} \leq T.$$

The initial speed $V_0$ and final speed $V_X$ are given. The switching speeds $V_k$ at the locations $x_k$ must be determined for $k = 1, 2, \ldots, n$. Because the control sequence is specified, the switching speed $V_k$ depends only on $\xi_1, \ldots, \xi_k$.

## 11.3 Fundamental Speed Profiles

The limiting speed for a control setting $j$ with $u_j(0) \geq r(0)$ is the unique speed $W_j$ satisfying

$$u_j(W_j) = r(W_j).$$

For control settings $j$ with $u_j(0) < r(0)$, the limiting speed is taken to be $W_j = 0$. We can define position as a function of speed by setting

$$x_k^*(v) = \begin{cases} \displaystyle\int_0^v \frac{w\,dw}{u_{j(k+1)}(w) - r(w)} & v < W_{j(k+1)} \\[4mm] \displaystyle\int_v^{W_q} \frac{w\,dw}{r(w) - u_{j(k+1)}(w)} & v \geq W_{j(k+1)} \end{cases}$$

Since $[u_j(v) - r(v)]$ is Lipschitz, we can show that $x_k^*(v) \to \infty$ as $v \to W_{j(k+1)}$. Therefore

$$V_k < W_{j(k+1)} \quad \Rightarrow \quad V_{k+1} < W_{j(k+1)}$$

and

$$V_k > W_{j(k+1)} \quad \Rightarrow \quad V_{k+1} > W_{j(k+1)}.$$

The monotonic nature of the speed on the control interval $[x_k, x_{k+1}]$ means that speed limits can be imposed on the entire interval simply by constraining $V_k$ and $V_{k+1}$. The function $v_k^*$ is the inverse of $x_k^*$. The actual speed on the control interval $[x_k, x_{k+1}]$ is given by

$$v(x) = v_k^*(x - x_k + x_k^*(V_k))$$

if $V_k \neq W_{j(k+1)}$. The speed $U_i$ at the speed limit change point $X_i$ is given by

$$U_i = v_{q(i)}^*(X_i - x_{q(i)} + x_{q(i)}^*(V_{q(i)}))$$

for $i = 1, 2, ..., p - 1$. The time required to traverse the control interval $[x_k, x_{k+1}]$ is given by

$$\tau_{k+1} = \int_0^{\xi_{k+1}} \frac{1}{v_k^*(\xi + x_k^*(V_k))}\,d\xi.$$

## 11.4 Strategies of Optimal Type

Since

$$V_k = V_k(\xi_1, \xi_2, ..., \xi_k)$$

we have

$$\tau_k = \tau_k(\xi_1, \xi_2, ..., \xi_k).$$

If we write $\xi = (\xi_1, \xi_2, ..., \xi_{n+1}) \in \mathfrak{R}^{n+1}$ then the key variables can be written as functions of $\xi$. Thus the cost of the journey is

$$J(\xi) = \sum_{k=0}^{n} f_{j(k+1)}\tau_{k+1},$$

the total distance travelled is

$$x(\xi) = \sum_{k=0}^{n} \xi_{k+1}$$

and the time taken for the journey is

$$t(\xi) = \sum_{k=0}^{n} \tau_{k+1}.$$

We wish to minimise the cost of the journey $J(\xi)$. We require that the total distance travelled and the total time taken satisfy the constraints

$$x(\xi) = X$$

and

$$t(\xi) \leq T.$$

Note that the constraint $x(\xi) = X$ can be weakened to read $x(\xi) \geq X$ with no change to the solution. In addition to the distance and time constraints, we must also satisfy the speed constraints

$$V_{k+1} \leq M_i$$

for $q(i-1) < k+1 \leq q(i)$ and $k = 0, 1, ..., n-1$, and

$$U_i \leq N_i = \min \{M_i, M_{i+1}\}$$

for $i = 1, 2, ..., p$. The first of these constraints ensures that the switching speeds do not exceed the speed limit on any interval $(X_{i-1}, X_i)$. This is sufficient to ensure that the vehicle speed will not exceed the speed limit, since the speed of the vehicle will always lie between the previous and next switching speeds. The second constraint ensures that speed limits are not exceeded at any point $X_i$ where the speed limit changes.

We ensure that the journey finishes at the correct speed by adding the constraint

$$V_{n+1} = V_X.$$

We now define a Lagrangean function

$$\mathcal{J}(\xi) = J(\xi) + \lambda\,[X - x(\xi)] + \mu\,[t(\xi) - T]$$

$$+ \sum_{i=1}^{p-1} \sum_{k=q(i-1)}^{q(i)-1} \rho_{k+1}\,[V_{k+1} - M_i] + \sum_{k=q(p-1)}^{q(p)-2} \rho_{k+1}\,[V_{k+1} - M_p]$$

$$+ \sigma\,[V_{n+1} - V_x] + \sum_{i=1}^{p-1} \eta_i\,[U_i - N_j]$$

where $\lambda, \mu \in \mathfrak{R}$, $\rho \in \mathfrak{R}^n$, $\sigma \in \mathfrak{R}$ and $\eta \in \mathfrak{R}^{p-1}$ are Lagrange multipliers. We then apply the Kuhn-Tucker conditions

$$\frac{\partial \mathcal{J}}{\partial \xi_{k+1}} = 0$$

for $k = 0, 1, \ldots, n$, and the complementary slackness conditions

$$\lambda\,[X - x(\xi)] = 0,$$

$$\mu\,[t(\xi) - T] = 0,$$

$$\rho_{k+1}\,[V_{k+1} - M_i] = 0,$$

for all $q(i-1) < k+1 \leq q(i)$ and $i = 1, 2, \ldots, p-1$,

$$\sigma\,[V_{n+1} - V_X] = 0,$$

and

$$\eta_i\,[U_i - N_i] = 0$$

for all $i = 1, 2, \ldots, p-1$. The Lagrange multipliers $\lambda$, $\mu$, $\rho_{k+1}$ and $\eta_i$ are all non-negative. We also assume that $\xi$ lies in an open set such that $V_k(\xi) \neq W_{j(k)}$ for all $k$. Thus for $k = 0, 1, \ldots, n-1$ and $i$ such that $q(i-1) < k+1 \leq q(i)$ we have

$$\frac{\partial \mathcal{J}}{\partial \xi_{k+1}} = \sum_{h=k}^{n} f_{j(h+1)}\frac{\partial \tau_{h+1}}{\partial \xi_{k+1}} - \lambda + \mu \sum_{h=k}^{n} \frac{\partial \tau_{h+1}}{\partial \xi_{k+1}}$$

$$+ \sum_{h=k}^{n-1} \rho_{h+1}\frac{\partial V_{h+1}}{\partial \xi_{k+1}} + \sigma\frac{\partial V_{n+1}}{\partial \xi_{k+1}} + \sum_{s=i}^{p-1} \eta_s\frac{\partial U_s}{\partial \xi_{k+1}}$$

The calculation of the partial derivatives is similar to the calculation of the corresponding derivatives in the previous three chapters. At the end of the journey, the equation

$$\frac{\partial \mathcal{J}}{\partial \xi_{n+1}} = 0$$

gives

$$\lambda V_{n+1} - \mu = f_{j(n+1)} - \sigma\left[u_{j(n+1)}(V_{n+1}) - r(V_{n+1})\right]$$

Since $V_{n+1} = V_X$ is known, this equation simply relates the three Lagrange multipliers $\lambda$, $\mu$ and $\sigma$. For $k = 1, 2, \ldots, n$, the equation

$$\frac{\partial \mathcal{J}}{\partial \xi_k} - r_k\frac{\partial \mathcal{J}}{\partial \xi_{k+1}} = 0$$

gives

$$\left(\sum_{s=i}^{p-1} \eta_s\frac{u_{j(q(s)+1)}(U_s) - r(U_s)}{U_s} + \lambda\right)V_k - \mu$$

$$= \frac{f_{j(k)} - r_k f_{j(k+1)}}{1 - r_k} + \rho_k\frac{u_{j(k)}(V_k) - r(V_k)}{1 - r_k}$$

for $i$ such that $q(i-1) < k < k+1 \leq q(i)$, where

$$r_k = \frac{u_{j(k)}(V_k) - r(V_k)}{u_{j(k+1)}(V_k) - r(V_k)}.$$

If we let

$$\lambda_i = \begin{cases} \lambda & i = p \\ \displaystyle\sum_{s=i}^{p-1} \eta_s\frac{u_{j(q(s)+1)}(U_s) - r(U_s)}{U_s} + \lambda & i < p \end{cases}$$

then we have $V_k = M_i$ or $\rho_k = 0$. If $\rho_k = 0$ we have

$$\lambda_i V_k - \mu = \frac{f_{j(k)} - r_k f_{j(k+1)}}{1 - r_k}.$$

If $\eta_i \neq 0$, the complementary slackness conditions require that $U_i = N_i$. If, on the other hand, $\eta_i = 0$ and $U_i < N_i$, then $\lambda_i = \lambda_{i+1}$. If $U_i < N_i$ for all $i$ then $\lambda_i = \lambda$ and the conditions are identical to the conditions obtained in Chapter 8 for the case where there are no speed limits.


## 11.5 Critical Speeds

We now consider a vehicle with three control settings, power, coast and brake, and with tractive acceleration given by

$$u_j(v) = \begin{cases} u(v) & j = 1 & \textit{(power)} \\ 0 & j = 0 & \textit{(coast)} \\ -K & u = -1 & \textit{(brake)} \end{cases}$$

where $u$ is the acceleration corresponding to full tractive effort and $-K$ is the acceleration corresponding to full braking effort. We assume that the cost rate $f_1 = 1$ for power, and that $f_j$ is zero for coast and brake.

By alternating between coast and power, and by braking when necessary, the three control modes can be used to follow any possible speed profile as closely as we please. We will further assume that the control is never changed between power and brake without an intermediate coast phase.

When changing between coast and power, the conditions for an optimal strategy require $V_k = M_i$ or $\rho_k = 0$. If $\rho_k = 0$ we get

$$\lambda_i V_k - \mu = \varphi(V_k) \tag{11.1}$$

where

$$\varphi(v) = \frac{r(v)}{u(v)}$$

for $v \geq 0$. The function $\varphi(v)$ is often strictly convex with $\varphi(0) \geq 0$. This is certainly true for suburban and long-haul trains, which typically have $u(v) = H/v$ where $H$ is the tractive power per unit mass, and $r(v) = a + bv + cv^2$. In such cases, (11.1) will generally have two solutions, $V_k = V^{(i)}$ and $V_k = W^{(i)}$, where $0 \leq V^{(i)} \leq W^{(i)}$.

Furthermore, for $\lambda_i > \lambda_h$ we have $0 \le V^{(i)} < V^{(h)} \le W^{(h)} < W^{(i)}$. In fact, if we rewrite (11.1) in the form

$$E_\mu(V_k) = \lambda_i \tag{11.2}$$

where

$$E_\mu(v) = \frac{\mu + \varphi(v)}{v}$$

is the *energy density* function for the strategy then we can justify these assertions with the following elementary result.

### Lemma 11.1

*Let $\varphi : \mathfrak{R}_+ \to \mathfrak{R}$ where $\varphi(v)$ is strictly convex and $\varphi(v)/v \to \infty$ as $v \to \infty$. Let $\mu \in \mathfrak{R}$ with $\mu + \varphi(0) > 0$. Define $E_\mu : \mathfrak{R}_+ \to \mathfrak{R}$ by the formula*

$$E_\mu(v) = \frac{\mu + \varphi(v)}{v}.$$

*There is a unique point $w_\mu > 0$ with $E_\mu{}'(w_\mu) = 0$ and such that $E_\mu{}'(v) < 0$ for $v < w_\mu$ and $E_\mu{}'(v) > 0$ for $v > w_\mu$.*

### Proof

Define $\psi_\mu : \mathfrak{R}_+ \to \mathfrak{R}$ by the formula

$$\begin{aligned}
\psi_\mu(v) &= v^2 E_\mu{}'(v) \\
&= v\varphi'(v) - [\mu + \varphi(v)].
\end{aligned}$$

We show that the equation $\psi_\mu(v) = 0$ has a unique solution $v = w_\mu$ in the region $v > 0$. First, note that

$$\psi_\mu{}'(v) = v\varphi''(v) > 0$$

and hence $\psi_\mu(v)$ is strictly monotone increasing. Since $\varphi(v)/v \to \infty$ and $v \to \infty$ it follows that $E_\mu(v) \to \infty$ as $v \to \infty$. Therefore we can find some point $w$ with $E_\mu{}'(w) > 0$. Thus $\psi_\mu(w) > 0$. Since $\psi_\mu(0) = -[\mu + \varphi(0)] < 0$ there is precisely one point $w_\mu \in (0, w)$ with $\psi_\mu(w_\mu) = 0$. For $v \in (0, w_\mu)$ we have $\psi_\mu(v) < 0$ and for

$v \in (w_\mu, \infty)$ we have $\psi_\mu(v) > 0$. Since $\psi_\mu(v) = v^2 E_\mu{}'(v)$, the result is now established.

$\square$

From this lemma it follows that the function $E_\mu(v)$ has a unique minimum turning point $v = w_\mu$ in the region $v > 0$. For values of $\lambda$ above the minimum there are precisely two solutions $V$ and $W$ with $V < w_\mu < W$ to the equation $\lambda = E_\mu(v)$ in the region $v > 0$.

When changing between coast and brake, the conditions for an optimal strategy require $V_k = M_i$ or $\rho_k = 0$. If $\rho_k = 0$ we get

$$\lambda_i V_k - \mu = 0. \tag{11.3}$$

Equation (11.3) will have one solution, $V_k = U^{(i)}$, where $0 < U^{(i)} < V^{(i)} \leq W^{(i)}$. The speeds $U^{(i)}$, $V^{(i)}$ and $W^{(i)}$ are *critical speeds* for the interval $(X_{i-1}, X_i)$.

In an interval $(X_{i-1}, X_i)$ with speed limit $M_i$, the control may be changed from coast to power only when $V_k = V^{(i)}$, and requires $V^{(i)} < M_i$. Similarly, the control may be changed from power to coast only when $V_k = \min \{W^{(i)}, M_i\}$. A change from coast to brake may occur when $V_k = U^{(i)}$. Once braking has commenced, the speed of the vehicle will remain below the critical speeds for the interval, and so the control cannot be changed again within the interval.

At a boundary $X_i$ between two intervals, a change from power to coast may occur if $V_k = W^{(i)} = W^{(i+1)} \leq N_i$ or if $V_k = N_i \leq \min \{W^{(i)}, W^{(i+1)}\}$. A change from coast to power can occur at this point only if $V_k = V^{(i)} = V^{(i+1)} < N_i$. A change from brake to coast can occur if $N_i < U^{(i)}$. These changes allow us to maintain a feasible strategy.


### Construction of an Optimal Strategy

From the above analysis it follows that the feasible control changes are:

- *power* to *coast* if $V_k = \min \{W^{(i)}, M_i\}$

- *coast* to *power* if $V_k = V^{(i)}$

- *coast* to *brake* if $V_k = U^{(i)}$

for $x_k \in (X_{i-1}, X_i)$, and

- *power* to *coast* if $V_k = W^{(i)} = W^{(i+1)} \leq N_i$
  $$\text{or } V_k = N_i \leq \min \{W^{(i)}, W^{(i+1)}\}$$

- *coast* to *power* if $V_k = V^{(i)} = V^{(i+1)} < N_i$

- *brake* to *coast* if $V_k = N_i \le U^{(i)}$

for $x_k = X_i$.

The construction of an optimal strategy is illustrated in Figure 11-1. The Lagrange multipliers $\mu$ and $\lambda_i$ determine critical speeds $U^{(i)}$, $V^{(i)}$ and $W^{(i)}$ for each track interval $(X_i, X_{i+1})$. We can construct an approximate speed-hold phase on a track interval using *coast–power* pairs, with the speed of the vehicle oscillating between $V^{(i)} < M_i$ and $V = \min\{W^{(i)}, M_i\}$. If the number of *coast–power* pairs is given for each track interval, then for each $\mu$ we can adjust the Lagrange multiplier $\lambda_i = \lambda_i(\mu)$ so that the distance and speed constraints are satisfied for each interval. We can then adjust $\mu$ to satisfy the time constraint for the journey.

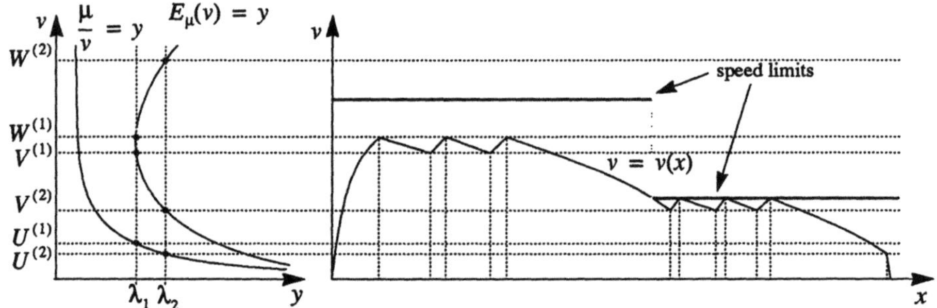


Figure 11-1: Construction of an optimal strategy

As we increase the number of *coast–power* pairs, either $V^{(i)}, W^{(i)} \to w_\mu$, or $V^{(i)} \to M_i$. Thus in the idealised case (11.2) becomes either

$$E_\mu(w_\mu) = \lambda_i$$

or else

$$E_\mu(M_i) = \lambda_i$$

and hence $\lambda_i$ can be easily determined from $\mu$. We can regard $w_\mu$ as the nominal hold speed for the strategy. On track intervals where the hold speed is above the speed limit, the speed of the vehicle must be held at the limit.

## 11.6 Examples

In the following examples, the vehicle acceleration is given by

$$
u_j(v) = \begin{cases} 1.5/v & j = 1 \\ 0 & j = 0 \\ -1 & j = -1 \end{cases}
$$

and resistance is given by

$$
r(v) = 0.015 + 0.00003v + 0.000006v^2.
$$

In all examples we use *coast–power* pairs to approximate speed-holding. The number of pairs in each speed limit interval is specified.

### *Example 11.1*

The track has length $X = 18000$ and the required journey time is $T = 1500$. The speed limit for the track is

$$
M(x) = \begin{cases} 25 & x < 7000 \\ 15 & x \geq 7000 \end{cases}
$$

We use one *coast–power* pair in each speed interval. The optimal strategy has Lagrange multipliers $\mu = 0.097400$, $\lambda_1 = 0.022861$ and $\lambda_2 = 0.025623$. The cost of the journey is $J = 203.63$. The speed profile is shown in Figure 11-2. The shaded region represents the allowable speed range.

$\square$

### *Example 11.2*

Consider the same track and speed limits, with no *coast–power* pairs in the first speed interval and two in the second. The optimal strategy has Lagrange multipliers $\mu = 0.065715$, $\lambda_1 = 0.0212728$ and $\lambda_2 = 0.0214888$. The cost of the journey is $J = 202.52$. The speed profile is shown in Figure 11-3.

$\square$

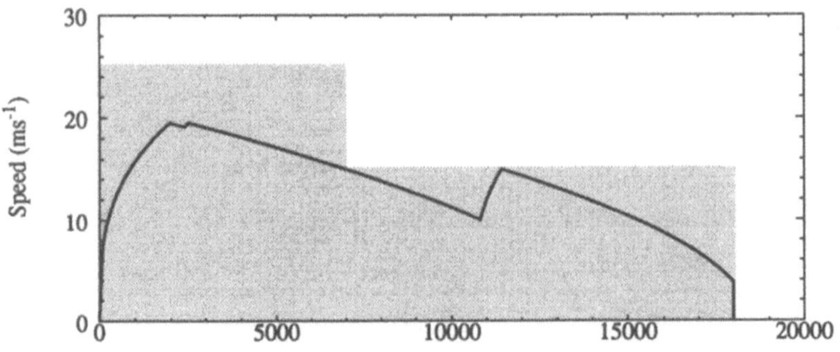

Figure 11-2: Speed $v(x)$ for Example 11.1

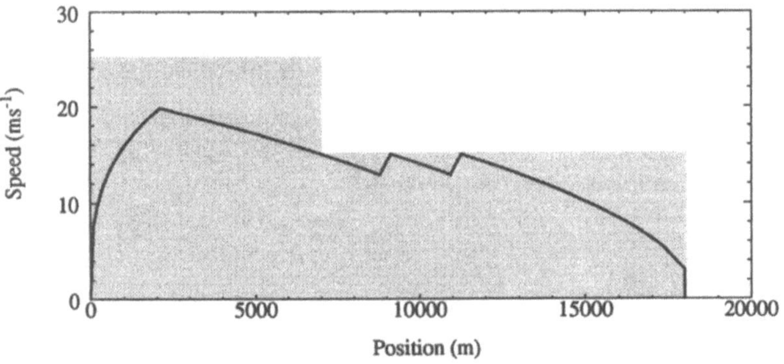

Figure 11-3: Speed $v(x)$ for Example 11.2

The second example has the same number of phases as the first, but is less expensive. Without speed limits, the minimum cost for a strategy with seven phases is $J = 201.98$.

### Example 11.3

We now use four *coast–power* pairs in each track interval. The optimal strategy has Lagrange multipliers $\mu = 0.067873$, $\lambda_1 = 0.021239$ and $\lambda_2 = 0.021449$. The cost of the journey is $J = 202.16$. Without speed limits, the journey cost is $J = 201.89$. The speed profile is shown in Figure 11-4.

$\square$

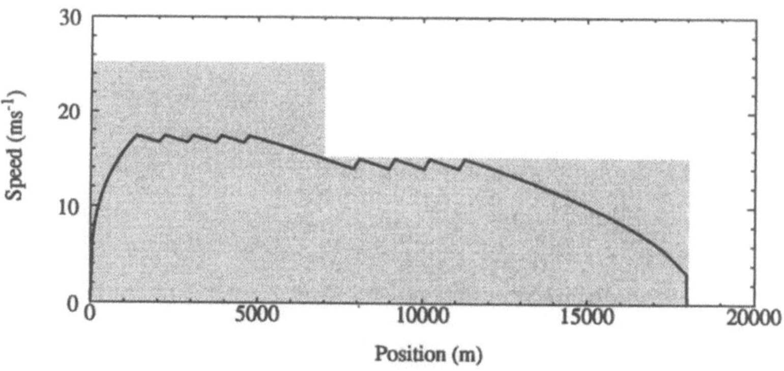

Figure 11-4: Speed $v(x)$ for Example 11.3

In the next two examples, journey time is increased to $T = 1620$, and the speed imit for the track is

$$M(x) = \begin{cases} 20 & x < 9000 \\ 10 & 9000 \leq x \leq 10000 \\ 15 & x > 10000 \end{cases}.$$

### *Example 11.4*

We use one *coast–power* pair in the first interval, two in the second, and one in the hird. The optimal strategy has Lagrange multipliers $\mu = 0.060353$, $\lambda_1 = 0.020798$, $\lambda_2 = 0.022309$ and $\lambda_3 = 0.020866$. The cost of the journey is $J = 199.14$. Without speed limits, the cost of the journey is $J = 198.01$. The speed profile is shown in Figure 11-5.
⅂

### *Example 11.5*

We use four *coast–power* pairs in the first interval, five in the second, and four in the hird. The optimal strategy has Lagrange multipliers $\mu = 0.059733$, $\lambda_1 = 0.020749$, $\lambda_2 = 0.022008$ and $\lambda_3 = 0.020791$. The cost of the journey is $J = 199.05$. Without speed limits, the cost of the journey is $J = 197.95$. The speed profile is shown in Figure 11-6.

□

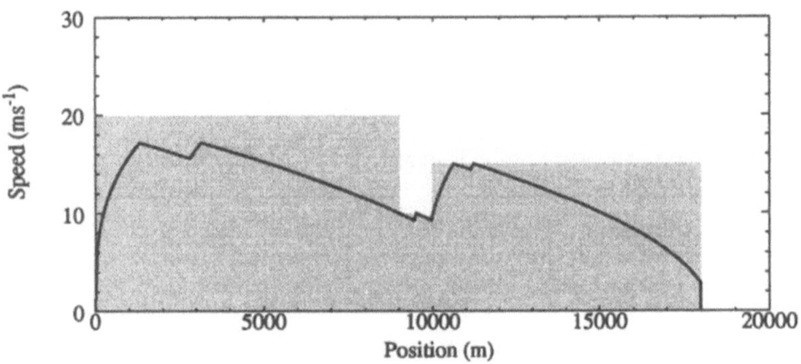

Figure 11-5: Speed $v(x)$ for Example 11.4

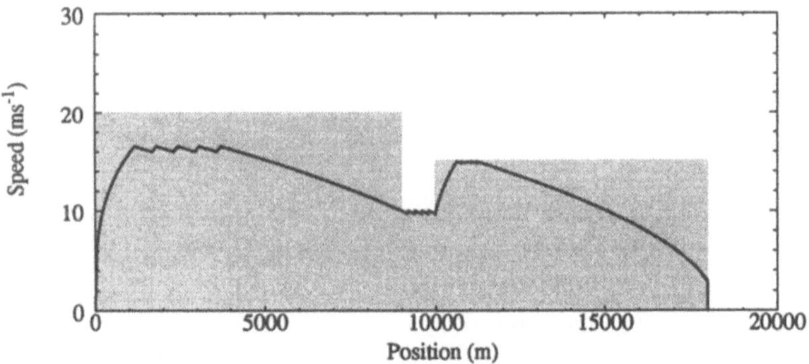

Figure 11-6: Speed $v(x)$ for Examples 11.4 and 11.5

## CHAPTER 12
# NON-CONSTANT GRADIENT

In this chapter we use the fuel-consumption model to formulate the train control problem on non-level track. Once again, an optimal strategy no longer exists. That is, there is no feasible strategy that minimises fuel consumption. We will show that for a given sequence of control settings, there exist optimal switching points that define a strategy of optimal type. This strategy minimises fuel consumption for the given control sequence. We find key equations that define necessary conditions for a strategy of optimal type, and present an algorithm for solution of these equations. On non-steep track an approximate speed-holding strategy is best. On steep track, speed-holding may be disrupted by segments of maximum power around steep inclines, or by segments of coasting around steep declines.

This material in this chapter was collectively published in the proceedings of CTAC-91 [13], in *Control Engineering Practice* [10] and in a University of South Australia School of Mathematics Report [36].

## 12.1 Notation

Because of the non-zero track gradient, it is convenient to use position as the independent variable. This requires some minor changes to our previous notation. In general terms, we use the notation of Chapter 8 and the equations of motion written in the form of (2.6) and (2.7).

Consider a fixed sequence $\{j(k+1)\}_{k=0, 1, ..., n}$ of control settings. This sequence defines a subset

$$S(\{j(k+1)\}_{k=0, 1, ..., n})$$

of control strategies each with $n + 1$ distinct control phases. Let us define

$$0 = x_0 \leq x_1 \leq ... \leq x_{n+1}$$

where $x_0$ is the starting point, $\{x_k\}_{k=1,2,\ldots,n}$ are the switching points, $x_{n+1}$ is the stopping point and $j(k+1)$ is the control setting on the interval $(x_k, x_{k+1})$. As before, we use $\xi_{k+1}$ to denote the length of the interval $(x_k, x_{k+1})$. The control strategy from the given subset and with the above switching points is denoted by

$$S(\{[j(k+1); (x_k, x_{k+1})]\}_{k=0,1,\ldots,n}).$$

We write $t_k = t(x_k)$ to denote the time when the train is at position $x_k$, and use $\tau_{k+1} = t_{k+1} - t_k$ to denote the time taken to traverse the interval $(x_k, x_{k+1})$. The points $\{t_k\}_{k+1,2,\ldots,n}$ are the switching times.

## 12.2  Solution of the Equations of Motion

Track gradient data is normally stored in digital form and hence the true gradient is represented as a piecewise constant gradient. For both theoretical and computational reasons it is convenient to use the gradient data in this form.

In this section we show that the true solutions to the equations of motion can be computed to any desired accuracy using a piecewise constant approximation to the true gradient.

For each strategy, (2.7) can be written in the general form

$$v\frac{dv}{dx} = u(x, v) \tag{12.1}$$

where we have

$$\frac{\partial u}{\partial x} = \frac{dg}{dx}$$

and

$$\frac{\partial u}{\partial v} < 0$$

for all $x \in [0, X]$ and all $v > 0$. Let $v(x)$ denote the solution to (12.1) on the interval $[0, X]$ subject to the initial condition $v(0) = v_0$. We establish bounds on this solution by solving a simpler equation. For each subdivision

$$0 = h_0 < h_1 < \ldots < h_{p+1} = X$$

we choose $h_q^m, h_q^M \in [h_q, h_{q+1}]$ such that

$$g(h_q^m) \leq g(x) \leq g(h_q^M)$$

for $x \in [h_q, h_{q+1}]$ and consider solutions $v_m(x)$ and $v_M(x)$ to the simpler equations

$$v_m \frac{dv_m}{dx} = u(h_q^m, v_m)$$

and

$$v_M \frac{dv_M}{dx} = u(h_q^M, v_M)$$

for $x \in [h_q, h_{q+1}]$ and each $q = 0, 1, ..., p$ subject to the initial conditions $v_m(0) = v_M(0) = v_0$. We can now state our first basic lemma.

## Lemma 12.1

Let $v_m(x)$, $v(x)$ and $v_M(x)$ be the functions defined above. Then $v_m(x) \leq v(x) \leq v_M(x)$ for all $x \in [0, X]$.

## Proof

Let $x \in [h_q, h_{q+1}]$. If we assume $v(x) < v_m(x)$ then it follows that

$$v \frac{dv}{dx} = u(x, v) > u(h_q^m, v_m) = v_m \frac{dv_m}{dx}.$$

Since $v(y)$ and $v_m(y)$ are continuous there is a neighbourhood $(a, c)$ of $x$ such that $v(y) < v_m(y)$ for all $y \in (a, c)$. If we let

$$b = \inf \{ a| \, v(y) < v_m(y) \text{ for all } y \in (a, c) \}$$

then the continuity of $v(y)$ and $v_m(y)$ implies $v(b) = v_m(b)$. Thus we have an interval $(b, c)$ with $v(b) = v_m(b)$ and $v(y) < v_m(y)$ for all $y \in (b, c)$. Now we can easily see that for each $y \in (b, c)$ we have

$$[v(y)]^2 - [v_m(y)]^2 = \int_b^y \left[ 2v \frac{dv}{dx} - 2v_m \frac{dv_m}{dx} \right] dx > 0$$

which is a contradiction. Thus $v_m(x) \le v(x)$ for all $x$. A similar argument shows that $v_M(x) \ge v(x)$ for all $x$.

$\square$

Let $g(x)$ be continuous on $[0, X]$. Since $[0, X]$ is compact it follows that $g(x)$ is uniformly continuous on this interval. If we choose $\varepsilon < 0$ then we can find $\delta > 0$ such that $|g(x) - g(y)| < \varepsilon$ whenever $|x - y| < \delta$. We suppose the subdivision $\{h_q\}$ is chosen so that $|h_{q+1} - h_q| < \delta$. We can now state our second basic lemma.

### Lemma 12.2

*For all $x \in [0, X]$ we have*

$$[v_M(x)]^2 - [v_m(x)]^2 < 2\varepsilon X.$$

### Proof

For each $y \in [h_q, h_{q+1}]$ we have

$$[v_M(y)]^2 - [v_m(y)]^2$$

$$= \sum_{r=0}^{q-1} \int_{h_r}^{h_{r+1}} \left[ 2v_M \frac{dv_M}{dx} - 2v_m \frac{dv_m}{dx} \right] dx + \int_{h_q}^{y} \left[ 2v_M \frac{dv_M}{dx} - 2v_m \frac{dv_m}{dx} \right] dx$$

$$\le \sum_{r=0}^{q-1} \int_{h_r}^{h_{r+1}} [2g(h_r^M) - 2g(h_r^m)] \, dx + \int_{h_q}^{y} [2g(h_q^M) - 2g(h_q^m)] \, dx$$

$$< 2\varepsilon X$$

$\square$

These two lemmas show that the true solution can be computed to any required accuracy by subdividing the given interval into a collection of subintervals and solving a simpler equation for which the gradient is constant on each subinterval.

## 12.3  Track Gradient Analysis and Terminology

Subdivide the interval $[0, X]$ by setting

$$0 = h_0 < h_1 < \ldots < h_{p+1} = X$$

and suppose that the track gradient is constant on each subinterval $(h_q, h_{q+1})$. We consider a fixed control sequence $\{j(k+1)\}_{k=0, 1, \ldots, n}$ and a general strategy

$$S(\{[j(k+1); (x_k, x_{k+1})]\}_{k=0, 1, \ldots, n}) \in S(\{j(k+1)\}_{k=0, 1, \ldots, n}).$$

Equation (2.7) is now written in the following way. When

$$x \in (x_k, x_{k+1}) \cap (h_q, h_{q+1})$$

we have

$$v\frac{dv}{dx} = \frac{Hf_{j(k+1)}}{v} + K_{j(k+1)} - r(v) + g_q$$

$$= u_{[k, q]}(v)$$

where $g_q$ is the acceleration due to the gradient on the interval $(h_q, h_{q+1})$. We could for example assume that $g_q = g(h_q)$ where $g(x)$ is the acceleration due to the known track gradient. Ultimately we will need to specify the locations of the variable points $\{x_k\}_{k=1, 2, \ldots, n}$ relative to the fixed points $\{h_q\}_{q=0, 1, \ldots, p+1}$. We use the notation $\{q(k)\}_{k=0, 1, \ldots, n}$ to denote the essentially unknown sequence with $q(0) = 0$ and $q(n+1) = p$, and with

$$h_{q(k)} \leq x_k < h_{q(k)+1}$$

for each $k = 1, 2, \ldots, n$.

## 12.4  The Speed Profiles

In this section we define functions that satisfy the equations of motion. For each $j = 0, 1, \ldots, m$ let $W_{[j, q]}$ be the unique solution to the equation

$$Hf_j - v[r(v) - g_q] = 0.$$

If the fuel supply rate is $f_j$ and if the acceleration due to the track gradient is $g_q$ then $W_{[j, q]}$ is the speed at which the net acceleration is zero. Choose $W$ such that

$W > W_{[j, q]}$ for all $[j, q]$. A direct integration of the equation of motion can now be used to define position as a function of speed.

For $k < n$ we set

$$x_{[k, q]}(v) = \begin{cases} \displaystyle\int_0^v \frac{w\,dw}{u_{[k, q]}(w)} & \text{for } v \in [0, W_{[j(k+1), q]}) \\[4mm] \displaystyle\int_v^W \frac{(-1)\,w\,dw}{u_{[k, q]}(w)} & \text{for } v \in (W_{[j(k+1), q]}, W] \end{cases}$$

We use $v_q = v(h_q)$ to denote the speed at $x = h_q$ and $V_k = v(x_k)$ to denote the speed at $x = x_k$. In general the value of $v_q$ determines which of the above definitions is used. Because the slope of the graph $y = vr(v)$ is positive and bounded when $v \in [0, W]$ it is easy to show that $x_{[k, q]}(v) \to \infty$ as $v \to W_{[j(k+1), q]}$. It now follows that if $v_q < W_{[j(k+1), q]}$ then $v_{q+1} < W_{[j(k+1), q]}$. Alternatively, if $v_q > W_{[j(k+1), q]}$ then $v_{q+1} > W_{[j(k+1), q]}$. Whichever of the above definitions is used, we denote the inverse function by $v_{[k, q]}$. When $k = n$ we define

$$x_{[n, q]}(v) = \int_v^W \frac{(-1)\,w\,dw}{u_{[n, q]}(w)}$$

for $v \in [0, W]$ and let $v_{[n, q]}$ denote the inverse function. We assume that $K > g_q$ for all $q$. For all $k$ we claim that the actual speed on the interval $(x_k, x_{k+1})$ is given by

$$v(x) = \begin{cases} v_{[k, q(k)]}(x - x_k + x_{[k, q(k)]}(V_k)) & x \in (x_k, h_{q(k)+1}) \\[2mm] v_{[k, q]}(x - h_q + x_{[k, q]}(v_r)) & x \in (h_r, h_{q+1}) \\[2mm] v_{[k, q(k+1)]}(x - h_{q(k+1)} + x_{[k, q(k+1)]}(v_{q(k+1)})) & \\[1mm] & x \in (h_{q(k+1)}, x_{k+1}) \end{cases}$$

Note that

$$x_{[k, q]}{}'(v) = \frac{v}{u_{[k, q]}(v)}$$

and

$$v_{[k, q]}{}'(x) = \frac{u_{[k, q]}(v_{[k, q]}(x))}{v_{[k, q]}(x)}$$

and also that the time taken to traverse the interval $(x_k, x_{k+1})$ is given by

$$
\tau_{k+1} = \int_{x_{[k,q(k)]}(V_k)}^{h_{q(k)+1} - x_k + x_{[k,q(k)]}(V_k)} \frac{d\rho}{v_{[k,q(k)]}(\rho)}
$$
$$
+ \sum_{q=q(k)+1}^{q(k+1)-1} \int_{x_{[k,q]}(v_q)}^{h_{q+1} - h_q + x_{[k,q]}(v_q)} \frac{d\rho}{v_{[k,q]}(\rho)}
$$
$$
+ \int_{x_{[k,q(k+1)]}(v_{q(k+1)})}^{x_{k+1} - h_{q(k+1)} + x_{[k,q(k+1)]}(v_{q(k+1)})} \frac{d\rho}{v_{[k,q(k+1)]}(\rho)} \tag{12.2}
$$

for $k < n$ and by

$$
\tau_{n+1} = \int_{x_{[n,q(n)]}(V_n)}^{h_{q(n)+1} - x_n + x_{[n,q(n)]}(V_n)} \frac{d\rho}{v_{[n,q(n)]}(\rho)}
$$
$$
+ \sum_{q=q(n)+1}^{q(n+1)-1} \int_{x_{[n,q]}(v_q)}^{h_{q+1} - h_q + x_{[n,q]}(v_q)} \frac{d\rho}{v_{[n,q]}(\rho)}
$$
$$
+ \int_{x_{[n,q(n+1)]}(v_{q(n+1)})}^{x_{[n,q(n+1)]}(0)} \frac{d\rho}{v_{[n,q(n+1)]}(\rho)} \tag{12.3}
$$

when $k = n$. Since $V_k = V_k(\xi_1, \xi_2, ..., \xi_k)$ it follows that $v_q = v_q(\xi_1, \xi_2, ..., \xi_k)$ for each $q$ with $q(k) < q \leq q(k+1)$ and hence $\tau_{k+1} = \tau_{k+1}(\xi_1, \xi_2, ..., \xi_k)$.


## 12.5  The Constraints

There are important constraints that must also be satisfied. Since the initial and final speeds are zero it is necessary that

$$
V_0 = V_{n+1} = 0.
$$

It is important to realise that the variable $\xi_{n+1}$ can be eliminated from the problem using the equation $V_{n+1} = 0$. This equation can be written in the form

$$
v_{[n,q(n+1)]}(x_{n+1} - h_{q(n+1)} + x_{[n,q(n+1)]}(v_{q[n+1]})) = 0
$$

and hence

$$
x_{n+1} - h_{q(n+1)} + x_{[n,q(n+1)]}(v_{q[n+1]}) = x_{[n,q(n+1)]}(0).
$$

Thus

$$\xi_{n+1} = h_{q(n+1)} + x_{[n,\,q(n+1)]}(0) - x_{[n,\,q(n+1)]}(v_{q(n+1)}) - \sum_{k=0}^{n-1} \xi_{k+1}$$

and hence the distance constraint

$$\sum_{k=0}^{n} \xi_{k+1} = X$$

can be rewritten in the form

$$h_{q(n+1)} + x_{[n,\,q(n+1)]}(0) - x_{[n,\,q(n+1)]}(v_{q(n+1)}) = X.$$

Since $x_{[n,\,q(n+1)]}(v_{q(n+1)})$ depends only on $\xi = (\xi_1, \xi_2, ..., \xi_n)$ we can define

$$x(\xi) = h_{q(n+1)} + x_{[n,\,q(n+1)]}(0) - x_{[n,\,q(n+1)]}(v_{q(n+1)})$$

in which case the distance constraint becomes

$$x(\xi) = X.$$

The time constraint is expressed simply as

$$\sum_{k=0}^{n} \tau_{k+1} = T.$$

Since $\tau_{k+1} = \tau_{k+1}(\xi_1, \xi_2, ..., \xi_{k+1})$ and since $\xi_{n+1}$ depends only on $\xi$ we can define

$$t(\xi) = \sum_{k=0}^{n} \tau_{k+1}$$

and the time constraint becomes

$$t(\xi) = T.$$

# 12.6 Necessary Conditions for a Strategy of Optimal Type

## 12.6.1 The Lagrangean Function

For the subset $S(\{j(k+1)\}_{k=0,1,...,n})$ of control strategies we define a Lagrangean function of the form

$$J(\xi, \lambda, \mu) = HJ(\xi) + \lambda\,[X - x(\xi)] + \mu\,[t(\xi) - T] \tag{12.4}$$

where $J(\xi)$ is the cost of the journey, $x(\xi)$ is the distance travelled, $t(\xi)$ is the time taken, and $\lambda$ and $\mu$ are Lagrange multipliers. To find necessary conditions for a minimum cost strategy we apply the Kuhn-Tucker conditions

$$\frac{\partial J}{\partial \xi_k} = 0$$

for all $k$ and the complementary slackness conditions

$$\lambda\,[X - x(\xi)] = 0$$

and

$$\mu\,[t(\xi) - T] = 0.$$

If we weaken the equality constraints to read $x(\xi) \geq X$ and $t(\xi) \leq T$ then we can also guarantee that $\lambda$ and $\mu$ are non-negative. It is intuitively obvious that the weakened problem has the same solution. We will assume that $\xi$ lies in an open set such that $V_k(\xi) \neq W_{[j(k+1),\,q(k)]}$ and such that $v_q(\xi) \neq W_{[j(k+1),\,q]}$ for each $q$ with $q(k) + 1 \leq q \leq q(k+1)$ and each $k = 1, 2, ..., n$.

Application of the Kuhn-Tucker conditions gives a set of complicated equations, which can be simplified to give key equations that must be satisfied by a strategy of optimal type.

## 12.6.2 The Key Equations

To explain the key equations, it is necessary to introduce some more terminology. For each $k = 1, 2, ..., n$ and for $q(k) < r \leq q(k+1)$ define

$$R_r = \frac{u_{[k,\,r-1]}(v_r)}{u_{[k,\,r]}(v_r)}.$$

For each $q$ with $q(k) \le q < q(k+1)$ define

$$\mathcal{R}_{[k,\,q]} = \prod_{r=q+1}^{q(k+1)} R_r$$

and let $\mathcal{R}_k = \mathcal{R}_{[k,\,q(k)]}$. When $x \le x_{k+1}$ the *effective speed* $\mathcal{V}_k(x)$ at the point $x$ is defined by $\mathcal{V}_k(x) = v(x)$ for $x \in [h_{q(k+1)}, x_{k+1}]$ and by

$$\frac{1}{\mathcal{V}_k(x)} = \frac{1}{\mathcal{R}_{[k,\,q]}}\left[\frac{1}{v(x)} - \frac{1}{v_{q+1}}\right] + \sum_{r=q+1}^{q(k+1)-1} \frac{1}{\mathcal{R}_{[k,\,r]}}\left[\frac{1}{v_r} - \frac{1}{v_{r+1}}\right] + \frac{1}{v_{q(k+1)}}$$

for $x \in [h_q, h_{q+1}]$ and $q < q(k+1)$. The effective speed at the point $x_k$ is denoted by $\mathcal{V}_k = \mathcal{V}_k(x_k)$. We will also use the notation

$$Q_k = \frac{u_{[k-1,\,q(k)]}(V_k)}{u_{[k,\,q(k)]}(V_k)}$$

for each $k = 1, 2, \ldots, n$.

The necessary conditions for a strategy of optimal type can now be stated in the following way. For the subset $S(\{j(k+1)\}_{k=0,1,\ldots,n})$ of control strategies there are non-negative constants $\lambda$ and $\mu$ such that

$$\frac{\mu}{\mathcal{V}_n} = \lambda, \tag{12.5}$$

$$\frac{\mu}{\mathcal{V}_{n-1}} - \frac{u_{[n-1,\,q(n-1)]}(V_{n-1})}{\mathcal{R}_{n-1}} = \frac{\mu}{V_n} \tag{12.6}$$

and such that

$$\frac{[\mu + Hf_{j(k+1)}]}{\mathcal{V}_k} - \frac{u_{[k,\,q(k)]}(V_k)}{\mathcal{R}_k}$$

$$= \frac{[\mu + Hf_{j(k+1)}]}{V_{k+1}} - u_{[k,\,q(k+1)]}(V_{k+1}) \tag{12.7}$$

for each $k = 1, 2, \ldots, n-2$. These equations will be called the *key equations*. We emphasise that the key equations provide necessary conditions for the strategies of optimal type but in general may not provide sufficient conditions. On the other hand, there are many specific situations where the key equations will also be sufficient. For this reason, in a slight abuse of terminology, we will refer to the strategies obtained by solving the key equations as strategies of optimal type. A detailed derivation of the key equations is given below.

### 12.6.3 The Kuhn-Tucker Conditions

Direct application of the Kuhn-Tucker conditions gives a set of complicated equations which can be simplified by an essentially inductive argument. It is convenient to begin with an outline the argument.

The condition

$$\frac{\partial \mathcal{J}}{\partial \xi_n} = 0$$

can be simplified to give

$$(-1)\,\lambda + \mu\,\frac{1}{\mathcal{V}_n} = 0. \tag{12.8}$$

When the condition

$$\frac{\partial \mathcal{J}}{\partial \xi_{n-1}} = 0$$

is combined with (12.8) we can deduce that

$$\mu\,(1 - Q_{n-1})\,\mathcal{R}_{n-1}\left[\frac{1}{\mathcal{V}_{n-1}} - \frac{1}{V_n}\right] = \frac{(-1)\,Hf_{j(n-1)}}{V_{n-1}}. \tag{12.9}$$

In similar fashion we can combine the condition

$$\frac{\partial \mathcal{J}}{\partial \xi_{n-2}} = 0$$

with (12.8) and (12.9) to deduce that

$$(1 - Q_{n-2})\,\mathcal{R}_{n-2}\left\{[\mu + Hf_{j(n-1)}]\left[\frac{1}{\mathcal{V}_{n-2}} - \frac{1}{V_{n-1}}\right]\right\}$$
$$+ u_{[n-2,\,q(n-1)]}(V_{n-1}) = \frac{[Hf_{j(n-1)} - f_{j(n-2)}]}{V_{n-2}}.$$

If we assume that

$$(1 - Q_h) \, \mathcal{R}_h \left\{ [\mu + H f_{j(h+1)}] \left[ \frac{1}{\mathcal{V}_h} - \frac{1}{V_{h+1}} \right] + u_{[h,\, q(h+1)]}(V_{h+1}) \right\}$$

$$= \frac{[H f_{j(h+1)} - f_{j(h)}]}{V_h} \qquad\qquad (12.10)$$

for each $h$ with $k < h < n - 2$ then the additional equation

$$\frac{\partial \mathcal{J}}{\partial \xi_k} = 0$$

allows us to deduce that

$$(1 - Q_k) \, \mathcal{R}_k \left\{ [\mu + H f_{j(k+1)}] \left[ \frac{1}{\mathcal{V}_k} - \frac{1}{V_{k+1}} \right] + u_{[k,\, q(k+1)]}(V_{k+1}) \right\}$$

$$= \frac{[H f_{j(k+1)} - f_{j(k)}]}{V_k} . \qquad\qquad (12.11)$$

Therefore the hypothesis is established for $h = k$. We can now deduce that (12.10) is valid for all $h = 1, 2, \ldots, n - 2$. Equation (12.8) is equivalent to (12.5). Using

$$Q_{n-1} = \frac{u_{[n-2,\, q(n-1)]}(V_{n-1})}{u_{[n-1,\, q(n-1)]}(V_{n-1})}$$

and

$$u_{[n-1,\, q(n-1)]}(V_{n-1}) - u_{[n-2,\, q(n-1)]}(V_{n-1}) = (-1) \frac{H f_{j(n-1)}}{V_{n-1}}$$

it can be seen that (12.9) and (12.6) are equivalent. Similarly, using

$$Q_k = \frac{u_{[k-1,\, q(k)]}(V_k)}{u_{[k,\, q(k)]}(V_k)}$$

and

$$u_{[k,\, q(k)]}(V_k) - u_{[k-1,\, q(k)]}(V_k) = \frac{H [f_{j(k+1)} - f_{j(k)}]}{V_k}$$

we can see that (12.11) and (12.7) are equivalent.

## 12.7 Derivation of the Key Equations

### 12.7.1 Calculation of some Useful Derivatives

To apply the Kuhn-Tucker conditions it is necessary to calculate some partial derivatives. For $h \leq k \leq n$ and $(h_q, h_{q+1}) \subset (x_k, x_{k+1})$ we have

$$V_k = v_{[k-1, q(k)]}(x_k - h_{q(k)} + x_{[k-1, q(k)]}(v_{q(k)}))$$

and hence

$$\frac{\partial V_k}{\partial \xi_k} = v_{[k-1, q(k)]}{}'(x_k - h_{q(k)} + x_{[k-1, q(k)]}(v_{q(k)}))$$

$$= \frac{u_{[k-1, q(k)]}(V_k)}{V_k}.$$

In similar fashion we have

$$v_{q+1} = v_{[k, q]}(h_{q+1} - h_q + x_{[k, q]}(v_q))$$

and hence

$$\frac{\partial v_{q+1}}{\partial \xi_k} = v_{[k, q]}{}'(h_{q+1} - h_q + x_{[k, q]}(v_q))x_{[k, q]}{}'(v_q)\frac{\partial v_q}{\partial \xi_k}$$

$$= \frac{u_{[k, q]}(v_{q+1})}{v_{q+1}}\frac{v_q}{u_{[k, q]}(v_q)}\frac{\partial v_q}{\partial \xi_k}.$$

By repeated application of this process we obtain

$$\frac{\partial v_{q+1}}{\partial \xi_k} = \frac{u_{[k, q]}(v_{q+1})}{v_{q+1}}\left[\prod_{r = q(k)+2}^{q} R_r\right]\frac{v_{q(k)+1}}{u_{[k, q(k)+1]}(v_{q(k)+1})}\frac{\partial v_{q(k)+1}}{\partial \xi_k}.$$

Since

$$v_{q(k)+1} = v_{[k, q(k)]}(h_{q(k)+1} - x_k + x_{[k, q(k)]}(V_k))$$

it follows that

$$
\begin{aligned}
\frac{\partial v_{q(k)+1}}{\partial \xi_k}
&= v_{[k,\,q(k)]}'(h_{q(k)+1} - x_k + x_{[k,\,q(k)]}(V_k))\left[-1 + x_{[k,\,q(k)]}'(V_k)\frac{\partial V_k}{\partial \xi_k}\right] \\[2mm]
&= \frac{u_{[k,\,q(k)]}(v_{q(k)+1})}{v_{q(k)+1}}\left[-1 + \frac{V_k}{u_{[k,\,q(k)]}(V_k)}\,\frac{u_{[k-1,\,q(k)]}(V_k)}{V_k}\right] \\[2mm]
&= \frac{u_{[k,\,q(k)]}(v_{q(k)+1})}{v_{q(k)+1}}\,(-1 + Q_k)
\end{aligned}
$$

and hence

$$
\frac{\partial v_{q+1}}{\partial \xi_k} = \frac{u_{[k,\,q]}(v_{q+1})}{v_{q+1}}\,(-1 + Q_k)\prod_{r=q(k)+1}^{q} R_r.
$$

When $h < k$ a similar process can be used to show that

$$
\frac{\partial v_{q+1}}{\partial \xi_h} = \frac{u_{[k,\,q]}(v_{q+1})}{v_{q+1}}\left[(-1 + Q_k) + \sum_{i=h}^{k-1}(-1 + Q_i)\prod_{j=i}^{k-1} Q_{j+1}\mathcal{R}_j\right]
$$

$$
\times \prod_{r=q(k)+1}^{q} R_r.
$$

It can also be shown that

$$
\begin{aligned}
\frac{\partial}{\partial \xi_n}\left[x_{[n,\,q(n+1)]}(v_{q(n+1)})\right]
&= x_{[n,\,q(n+1)]}'(v_{q(n+1)})\frac{\partial v_{q(n+1)}}{\partial \xi_n} \\[2mm]
&= (-1 + Q_n)\,\mathcal{R}_n
\end{aligned}
$$

and, for $h < n$,

$$
\frac{\partial}{\partial \xi_h}\left[x_{[n,\,q(n+1)]}(v_{q(n+1)})\right]
$$

$$
= x_{[n,\,q(n+1)]}'(v_{q(n+1)})\frac{\partial v_{r(n+1)}}{\partial \xi_h}
$$

$$
= \left[(-1 + Q_n) + \sum_{i=h}^{n-1}(-1 + Q_i)\prod_{j=i}^{n-1} Q_{j+1}\mathcal{R}_j\right]\mathcal{R}_n.
$$

A similar argument gives

$$
\frac{\partial V_{k+1}}{\partial \xi_k} = \frac{u_{[k,\,q(k+1)]}(V_{k+1})}{V_{k+1}}\,[1 + (-1 + Q_k)\,\mathcal{R}_k]
$$

and, for $h < k$,

$$\frac{\partial V_{k+1}}{\partial \xi_h} = \frac{u_{[k,\,q(k+1)]}(V_{k+1})}{V_{k+1}}$$

$$\times \left\{ 1 + \left[ (-1 + Q_k) + \sum_{i=h}^{k-1} (-1 + Q_i) \prod_{j=i}^{k-1} Q_{j+1} \mathcal{R}_j \right] \mathcal{R}_k \right\}. \qquad (12.12)$$

From (12.3) it follows that

$$\frac{\partial \tau_{n+1}}{\partial \xi_n} = (1 - Q_n) \frac{\mathcal{R}_n}{\mathcal{V}_n} - \frac{1}{V_n}$$

and in general, for $h < n$,

$$\frac{\partial \tau_{n+1}}{\partial \xi_h} = \left[ (1 - Q_n) + \sum_{i=h}^{n-1} (1 - Q_i) \prod_{j=i}^{n-1} Q_{j+1} \mathcal{R}_j \right] \frac{\mathcal{R}_n}{\mathcal{V}_n} - \frac{1}{V_n}. \qquad (12.13)$$

When $k < n$ we use (12.2) to show that

$$\frac{\partial \tau_{k+1}}{\partial \xi_{k+1}} = \frac{1}{V_{k+1}} \qquad (12.14)$$

and

$$\frac{\partial \tau_{k+1}}{\partial \xi_k} = (1 - Q_k) \, \mathcal{R}_k \left[ \frac{1}{\mathcal{V}_k} - \frac{1}{V_{k+1}} \right] + \frac{1}{V_{k+1}} - \frac{1}{V_k}.$$

When $h < k$ a similar argument gives

$$\frac{\partial \tau_{k+1}}{\partial \xi_h} = \left[ (1 - Q_k) + \sum_{i=h}^{k-1} (1 - Q_i) \prod_{j=i}^{k-1} Q_{j+1} \mathcal{R}_j \right] \mathcal{R}_k \left[ \frac{1}{\mathcal{V}_k} - \frac{1}{V_{k+1}} \right]$$

$$+ \frac{1}{V_{k+1}} - \frac{1}{V_k}. \qquad (12.15)$$

### 12.7.2  The Kuhn-Tucker Equations

We make the inductive assumption that the equation

$$\frac{\partial \mathcal{J}}{\partial \xi_h} = 0 \tag{12.16}$$

can be reduced to (12.10) for each $h$ with $k < h < n - 2$. Since (12.16) with $h = k$ can be written in the expanded form

$$
\begin{aligned}
&\frac{Hf_{j(k)}}{V_k} + Hf_{j(k+1)}\left\{ (1 - Q_k)\,\mathcal{R}_k\left[\frac{1}{\mathcal{V}_k} - \frac{1}{V_{k+1}}\right] + \frac{1}{V_{k+1}} - \frac{1}{V_k} \right\} \\[2mm]
&+ \sum_{h=k+1}^{n-2} Hf_{j(k+1)}\left\{ \left[(1 - Q_h) + \sum_{i=k}^{h-1}(1 - Q_i)\prod_{j=i}^{h-1} Q_{j+1}\mathcal{R}_j\right] \right. \\[2mm]
&\qquad\qquad\qquad\qquad \left. \times \mathcal{R}_h\left[\frac{1}{\mathcal{V}_h} - \frac{1}{V_{h+1}}\right] + \frac{1}{V_{h+1}} - \frac{1}{V_h} \right\} \\[2mm]
&- \lambda\left[(1 - Q_n) + \sum_{i=k}^{n-1}(1 - Q_i)\prod_{j=i}^{n-1} Q_{j+1}\mathcal{R}_j\right]\mathcal{R}_n \\[2mm]
&+ \mu\left( \frac{1}{V_k} + \left\{ (1 - Q_k)\mathcal{R}_k\left[\frac{1}{\mathcal{V}_k} - \frac{1}{V_{k+1}}\right] + \frac{1}{V_{k+1}} - \frac{1}{V_k} \right\} \right. \\[2mm]
&\qquad + \sum_{h=k+1}^{n-2}\left\{ \left[(1 - Q_h) + \sum_{i=k}^{h-1}(1 - Q_i)\prod_{j=i}^{h-1} Q_{j+1}\mathcal{R}_j\right] \right. \\[2mm]
&\qquad\qquad\qquad\qquad \left. \times \mathcal{R}_h\left[\frac{1}{\mathcal{V}_h} - \frac{1}{V_{h+1}}\right] + \frac{1}{V_{h+1}} - \frac{1}{V_h} \right\} \\[2mm]
&\qquad \left. + \left\{ \left[(1 - Q_n) + \sum_{i=k}^{n-1}(1 - Q_i)\prod_{j=i}^{n-1} Q_{j+1}\mathcal{R}_j\right]\frac{\mathcal{R}_n}{\mathcal{V}_n} - \frac{1}{V_n} \right\} \right) = 0
\end{aligned}
$$

it follows, by applying (12.8) and by making a minor rearrangement, that

$$[\mu + Hf_{j(k+1)}]\,(1 - Q_k)\,\mathcal{R}_k\left[\frac{1}{\mathcal{V}_k} - \frac{1}{V_{k+1}}\right]$$

$$+ \sum_{h=k+1}^{n-2} [\mu + Hf_{j(h+1)}]\left[(1 - Q_h) + \sum_{i=k}^{h-1}(1 - Q_i)\prod_{j=i}^{h-1}Q_{j+1}\mathcal{R}_j\right]$$

$$\times \mathcal{R}_h\left[\frac{1}{\mathcal{V}_h} - \frac{1}{V_{h+1}}\right]$$

$$+ \mu\left[(1 - Q_{n-1}) + \sum_{i=k}^{n-2}(1 - Q_i)\prod_{j=i}^{n-2}Q_{j+1}\mathcal{R}_j\right]\mathcal{R}_{n-1}\left[\frac{1}{\mathcal{V}_{n-1}} - \frac{1}{V_n}\right]$$

$$+ \frac{Hf_{j(k)}}{V_k} + \sum_{h=k}^{n-2} Hf_{j(h+1)}\left[\frac{1}{V_{h+1}} - \frac{1}{V_h}\right] = 0. \tag{12.17}$$

From (12.9) we can see that

$$\mu\left[(1 - Q_{n-1}) + \sum_{i=k}^{n-2}(1 - Q_i)\prod_{j=i}^{n-2}Q_{j+1}\mathcal{R}_j\right]\mathcal{R}_{n-1}\left[\frac{1}{\mathcal{V}_{n-1}} - \frac{1}{V_n}\right]$$

$$= (-1)\left\{1 + \frac{Q_{n-1}}{1 - Q_{n-1}}\left[(1 - Q_{n-2}) + \sum_{i=k}^{n-3}(1 - Q_i)\prod_{j=i}^{n-3}Q_{j+1}\mathcal{R}_j\right]\right.$$

$$\left.\times \mathcal{R}_{n-1}\right\}\frac{Hf_{j(n-1)}}{V_{n-1}}$$

$$= (-1)\frac{Hf_{j(n-1)}}{V_{n-1}} + \left[(1 - Q_{n-2}) + \sum_{i=k}^{n-3}(1 - Q_i)\prod_{j=i}^{n-3}Q_{j+1}\mathcal{R}_j\right]$$

$$\times \mathcal{R}_{n-2}u_{[n-2,\,q(n-1)]}(V_{n-1}). \tag{12.18}$$

By the inductive hypothesis we deduce that

$$\left\{[\mu + Hf_{j(h+1)}]\left[\frac{1}{\mathcal{V}_h} - \frac{1}{V_{h+1}}\right] + u_{[h,\,q(h+1)]}(V_{h+1})\right\}$$

$$\times \left[(1 - Q_h) + \sum_{i=k}^{h-1}(1 - Q_i)\prod_{j=i}^{h-1}Q_{j+1}\mathcal{R}_j\right]\mathcal{R}_h$$

$$= \left\{ 1 + \frac{Q_h}{1-Q_h}\left[ (1-Q_{h-1}) + \sum_{i=k}^{h-2}(1-Q_i)\prod_{j=i}^{h-2}Q_{j+1}\mathcal{R}_j\right]\mathcal{R}_{h-1}\right\}$$

$$\times \frac{H\,[f_{j(h+1)} - f_{j(h)}]}{V_h}$$

$$= \frac{H\,[f_{j(h+1)} - f_{j(h)}]}{V_h} + \left[ (1-Q_{h-1}) + \sum_{i=k}^{h-2}(1-Q_i)\prod_{j=i}^{h-2}Q_{j+1}\mathcal{R}_j\right]$$

$$\times \mathcal{R}_{h-1}u_{[h-1,\,q(h)]}(V_h)$$

for all $h$ with $k+1 < h \le n-2$. The inductive hypothesis also tells us that

$$\left\{ [\mu + Hf_{j(k+2)}]\left[\frac{1}{V_{k+1}} - \frac{1}{V_{k+2}}\right] + u_{[k+1,\,q(k+2)]}(V_{k+2})\right\}$$

$$\times \left\{ [\,(1-Q_{k+1}) + (1-Q_k)\,Q_{k+1}\mathcal{R}_k]\,\mathcal{R}_{k+1}\right\}$$

$$= \left\{ 1 + \frac{Q_{k+1}}{1-Q_{k+1}}(1-Q_k)\,\mathcal{R}_k\right\}\frac{H\,[f_{j(k+2)} - f_{j(k+1)}]}{V_{k+1}}$$

$$= \frac{H\,[f_{j(k+2)} - f_{j(k+1)}]}{V_{k+1}} + (1-Q_k)\,\mathcal{R}_k u_{[k,\,q(k+1)]}(V_{k+1}). \qquad (12.19)$$

By adding (12.18), for all $h$ with $k+1 < h \le n-2$, and (12.19) we obtain

$$\sum_{h=k+1}^{n-2}\left\{ [\mu + Hf_{j(h+1)}]\left[\frac{1}{V_h} - \frac{1}{V_{h+1}}\right] + u_{[h,\,q(h+1)]}(V_{h+1})\right\}$$

$$\times \left[ (1-Q_h) + \sum_{i=k}^{h-1}(1-Q_i)\prod_{j=i}^{h-1}Q_{j+1}\mathcal{R}_j\right]\mathcal{R}_h$$

$$+\mu\left[ (1-Q_{n-1}) + \sum_{i=k}^{n-2}(1-Q_i)\prod_{j=i}^{n-2}Q_{j+1}\mathcal{R}_j\right]\mathcal{R}_{n-1}\left[\frac{1}{V_{n-1}} - \frac{1}{V_n}\right]$$

$$= (-1)\frac{Hf_{j(n-1)}}{V_{n-1}} + \sum_{h=k+1}^{n-2}\frac{H\,[f_{j(h+1)} - f_{j(h)}]}{V_h}$$

$$+ (1-Q_k)\,\mathcal{R}_k u_{[k,\,q(k+1)]}(V_{k+1})$$

$$+ \sum_{h=k+1}^{n-2} \left[ (1-Q_h) + \sum_{i=k}^{h-1} (1-Q_i) \prod_{j=i}^{h-1} Q_{j+1} \mathcal{R}_j \right] \mathcal{R}_h u_{[k, q(k+1)]}(V_{k+1})$$

from which it follows that

$$\sum_{h=k+1}^{n-2} [\mu + Hf_{j(h+1)}] \left[ (1-Q_h) + \sum_{i=k}^{h-1} (1-Q_i) \prod_{j=i}^{h-1} Q_{j+1} \mathcal{R}_j \right]$$

$$\times \mathcal{R}_h \left[ \frac{1}{\mathcal{V}_h} - \frac{1}{V_{h+1}} \right]$$

$$+ \mu \left[ (1-Q_{n-1}) + \sum_{i=k}^{n-2} (1-Q_i) \prod_{j=i}^{n-2} Q_{j+1} \mathcal{R}_j \right] \mathcal{R}_{n-1} \left[ \frac{1}{\mathcal{V}_{n-1}} - \frac{1}{V_n} \right]$$

$$+ \frac{Hf_{j(k+1)}}{V_{k+1}} + \sum_{h=k+1}^{n-2} Hf_{j(h+1)} \left[ \frac{1}{V_{h+1}} - \frac{1}{V_h} \right]$$

$$- (1-Q_k) \mathcal{R}_k u_{[k, q(k+1)]}(V_{k+1})$$

$$= 0$$

By comparing this equation with (12.17) we deduce (12.11). Thus the hypothesis is established for $h = k$ and (12.10) is valid for all $h = 1, 2, ..., n-2$.


## 12.8  An Alternative Form for the Key Equations

To solve the key equations it is convenient to rewrite them in an alternative form. We make some additional definitions. For $v > 0$ let

$$E_\mu(v) = \frac{\mu}{v} + r(v)$$

be the *energy density* function and define a function $\{\varepsilon_\mu\}_k (x)$ in the region $x \le x_{k+1}$ by the formula

$$\{\varepsilon_\mu\}_k (x) = E_\mu [v(x)]$$

for $x \in [h_{q(k+1)}, x_{k+1}]$ and by

$$\{\mathcal{E}_{\mu}\}_k(x) = \frac{1}{\mathcal{R}_{[k,q]}} [E_{\mu}[v(x)] - E_{\mu}(v_{q+1})]$$

$$+ \sum_{r=q+1}^{q(k+1)-1} \frac{1}{\mathcal{R}_{[k,r]}} [E_{\mu}(v_r) - E_{\mu}(V_{r+1})] + E_{\mu}(v_{q(k+1)})$$

for $x \in [h_q, h_{q+1}]$ and $q < q(k+1)$. The function $\{\mathcal{E}_{\mu}\}_k(x)$ is a gradient weighted average of $E_{\mu}(v)$ over the interval $(x, x_{k+1})$, and can be regarded as the *effective energy density* for the control strategy.

The key equations can now be rewritten as follows. There are non-negative constants $\lambda$ and $\mu$ such that $x_n < X$ is the solution to

$$\frac{\mu}{\mathcal{V}_n(x)} = \lambda, \tag{12.20}$$

$x_{n-1} < x_n$ is the solution to

$$\{\mathcal{E}_{\mu}\}_{n-1}(x) = \frac{\mu}{V_n} + g_{q(n)} \tag{12.21}$$

and in general $x_k < x_{k+1}$ is the solution to

$$\{\mathcal{E}_{\mu}\}_k(x) = E_{\mu}(V_{k+1}) \tag{12.22}$$

for each $k = 1, 2, ..., n - 2$. On level track we note that $\mathcal{V}_k(x) = v$ and $\{\mathcal{E}_{\mu}\}_k(x) = E_{\mu}(v)$ where $v = v(x)$. Since $g_{q(n)} = 0$ on level track it can be seen that our key equations generalise the key equations (8.1) and (8.2).

## 12.9  The Strategies of Optimal Type

### 12.9.1  The Properties of the Effective Energy Density Function

In this section we show that the effective energy density function $\{\mathcal{E}_{\mu}\}_k(x)$ is continuous and has a continuous derivative in the region $x \le x_{k+1}$. For convenience we write $v = v(x)$ throughout this section. When $x \in (h_q, h_{q+1})$ it is obvious that these properties are true and that

$$\{\mathcal{E}_{\mu}\}_k'(x) = \frac{1}{\mathcal{R}_{[k,q]}} E_{\mu}'(v) \frac{u_{[k,q]}(v)}{v}. \tag{12.23}$$

When $x = h_q$ we have the following results.

### Lemma 12.3

$\{\mathcal{E}_\mu\}_k (x)$ *is continuous at* $x = h_q$.

### Proof

For $x \in (h_q, h_{q+1})$ we have

$$\{\mathcal{E}_\mu\}_k (x) = \frac{1}{\mathcal{R}_{[k,q]}} [E_\mu(v) - E_\mu(v_{q+1})] + \{\mathcal{E}_\mu\}_k (h_{q+1})$$

$$\rightarrow \frac{1}{\mathcal{R}_{[k,q]}} [E_\mu(v_q) - E_\mu(v_{q+1})] + \{\mathcal{E}_\mu\}_k (h_{q+1})$$

$$= \{\mathcal{E}_\mu\}_k (h_q)$$

as $x \downarrow h_q$. For $x \in (h_{q-1}, h_q)$ we have

$$\{\mathcal{E}_\mu\}_k (x) = \frac{1}{\mathcal{R}_{[k,q-1]}} [E_\mu(v) - E_\mu(v_q)] + \{\mathcal{E}_\mu\}_k (q)$$

$$\rightarrow \{\mathcal{E}_\mu\}_k (h_q)$$

as $x \uparrow h_q$.

$\square$

### Lemma 12.4

$\{\mathcal{E}_\mu\}_k {}'(x)$ *is continuous at* $x = h$.

### Proof

For $x \in (h_q, h_{q+1})$ we have

$$\{\mathcal{E}_\mu\}_k {}'(x) = \frac{1}{\mathcal{R}_{[k,q]}} E_\mu{}'(v) \frac{F_{[k,q]}(v)}{v}$$

$$\to \frac{1}{\mathcal{R}_{[k,\,q]}} E_\mu{}'(v_q) \frac{u_{[k,\,q]}(v_q)}{v_q}$$

as $x \downarrow h_q$. For $x \in (h_{q-1}, h_q)$ we have

$$\{\mathcal{E}_\mu\}_k{}'(x) = \frac{1}{\mathcal{R}_{[k,\,q-1]}} E_\mu{}'(v) \frac{u_{[k,\,r-1]}(v)}{v}$$

$$\to \frac{1}{\mathcal{R}_{[k,\,q-1]}} E_\mu{}'(v_q) \frac{u_{[k,\,q-1]}(v_q)}{v_q}$$

$$\to \frac{1}{\mathcal{R}_{[k,\,q]}} E_\mu{}'(v_q) \frac{u_{[k,\,q]}(v_q)}{v_q}$$

as $x \uparrow h_q$.

$\square$

### 12.9.2 The Structure of a Strategy of Optimal Type

From (12.23) and the convexity of the graph $y = E_\mu(v)$ it can be seen that the function $\{\mathcal{E}_\mu\}_k(x)$ has a local turning point only when $v = w_\mu$, where $w_\mu$ is the unique minimum turning point for the graph $y = E_\mu(v)$. Note that the critical speed $w_\mu$ satisfies the equation

$$w_\mu^2 r'(w_\mu) = \mu$$

We will show that $w_\mu$ defines an approximate holding speed. These observations are crucial to our understanding of the solutions to the key equations. The detailed solution algorithm is discussed in the next section.

If power is applied to the train on the interval $(x_k, x_{k+1})$ and if we assume that the speed increases throughout the interval then there is precisely one solution $x_k < x_{k+1}$ to (12.22) with

$$v(x_k) = V_k < w_\mu < V_{k+1} = v(x_{k+1}).$$

If the coast control is used and we assume that the speed decreases throughout the interval then once again there is precisely one solution $x_k < x_{k+1}$, but with

$$v(x_k) = V_k > w_\mu > V_{k+1} = v(x_{k+1}).$$

In each of the above cases the interval $(x_k, x_{k+1})$ is uniquely determined.

Therefore when the gradient is not too large we can expect the speed to oscillate about the critical value $w_\mu$ as the control is switched between power and coast. As the number of phases is increased it is intuitively reasonable to suggest that the length of each phase is reduced and hence the critical speed $w_\mu$ can be interpreted as an approximate holding speed.

Thus for non-steep track the strategy of optimal type will contain an initial phase of maximum power, followed by an approximate speed-hold phase near the critical speed $w_\mu$, a semi-final *coast* phase and a final brake phase.

In practice the track gradient may be sufficiently large to disrupt this general pattern. Thus we may have situations where maximum power is applied to the train but the speed decreases. On the other hand there may be situations where the coast control is used and the speed increases. When these situations occur we will say that the track is *steep*. Note that this definition depends on the speed of the train, and whether or not a section of track is classified as steep will depend on the nature of the strategy. In such cases it is possible that there will be more than one solution to the key equations and consequently it is possible that the interval $(x_k, x_{k+1})$ will not be uniquely determined. We reiterate that the key equations give necessary conditions for a strategy of optimal type, but these conditions may not be sufficient.

Nevertheless, on steep track, the strategy of optimal type will have the same form as on non-steep track except that the approximate speed-hold phase may be disrupted by phases of maximum power around the steep inclines and by phases of coast around the steep declines. These disruptions can be interpreted as necessary adjustments to keep the speed as near as possible to the desired holding speed. The theoretical difficulties associated with steep track are not yet fully resolved. Nevertheless, the understanding developed in this chapter has allowed our solution procedure to be used in practice on tracks with steep gradients.

In Chapter 3 we showed that any segment of non-negative measurable control can be approximated by a sequence of *coast–power* pairs. Thus we can construct a strategy of optimal type with almost minimum fuel consumption.


## 12.10  An Algorithm for Solving the Key Equations

In this section we will describe an algorithm for solution of the key equations. This algorithm was first described in a paper presented at the 1991 Computational Techniques and Applications Conference [13].

We suppose the control sequence $\{j(k+1)\}_{k=0,1,\ldots,n}$ is given and that we seek a strategy of optimal type from the subset $S(\{j(k+1)\}_{k=0,1,\ldots,n})$. We assume that $n$ is even and that

$$j(k+1) = \begin{cases} m & \text{if } k \text{ is even} \\ 0 & \text{if } k \text{ is odd} \end{cases}$$

for $k < n$. We stress that this assumption is not a necessary part of the proposed algorithm, but note that in seeking a strategy that minimises fuel consumption it is sufficient to restrict our attention to control subsets of this form. This point was explained in Chapter 3.

We assume that the track is not steep. That is, we assume that the speed increases during each power phase and decreases during each coast or brake phase. Although the proposed algorithms can still be used on steep track the existence and uniqueness of solutions is a more difficult issue.

The speed profile can be calculated in the following way. In general we consider an interval $[x_k, x_{k+1}]$ and suppose that $V_{k+1}$ and $x_{k+1}$ are known. In the first case we suppose that $j(k+1) = m$. For each $x \in [h_{q(k+1)}, x_{k+1}]$ we define $v = V_{[k, q(k+1)]}(x)$ as the solution to the equation

$$\int_v^{V_{k+1}} \frac{w\,dw}{u_{[k, q(k+1)]}(w)} = x_{k+1} - x$$

and in particular we note that $v_{q(k+1)} = V_{[k, q(k+1)]}(h_{q(k+1)})$. We now use a recursive definition for $v_q$ when $q < q(k+1)$. We define $v = v_r$ as the solution to the equation

$$\int_v^{V_{r+1}} \frac{w\,dw}{u_{[k, r]}(w)} = h_{r+1} - h_r$$

for each $r = q(k+1) - 1, q(k+1) - 2, ..., q$. For $x \in [h_q, h_{q+1}]$ we define $v = V_{[k, q]}(x)$ as the solution to the equation

$$\int_v^{V_{q+1}} \frac{w\,dw}{u_{[k, q]}(w)} = h_{q+1} - x.$$

Note that $v_q = V_{[k, q]}(h_q)$, $v_{q+1} = V_{[k, q]}(h_{q+1})$ and that $v_q < v < v_{q+1}$ for $x \in (h_q, h_{q+1})$. A similar calculation procedure applies when $j(k+1) = 0$ or $j(k+1) = -1$ but in each of these cases $v_q > v > v_{q+1}$ for $x \in (h_q, h_{q+1})$.

To solve the key equations (12.20), (12.21) and (12.22) we begin by noting the useful recursive relationships

$$\frac{1}{\mathcal{V}_k(h_q)} = \frac{1}{\mathcal{R}_{[k,q]}} \left[ \frac{1}{v_q} - \frac{1}{v_{q+1}} \right] + \frac{1}{\mathcal{V}_k(h_{q+1})} \tag{12.24}$$

and

$$\{\mathcal{E}_\mu\}_k(h_q) = \frac{1}{\mathcal{R}_{[k,q]}} [E_\mu(v_q) - E_\mu(v_{q+1})] + \{\mathcal{E}_\mu\}_k(h_{q+1}). \tag{12.25}$$

The key equations can now be solved as follows. We assume that the parameters $\lambda$ and $\mu$ are known.

Consider the interval $[x_k, x_{n+1}]$. We know that $x_{n+1} = X$ and that $V_{n+1} = 0$, but $x_n$ and $V_n$ are unknown. For $k = n$ we use the recursive calculation (12.24) to find $q = q(n)$ such that

$$\frac{\mu}{\mathcal{V}_k(h_q)} \le \lambda < \frac{\mu}{\mathcal{V}_k(h_{q+1})}$$

Now we find $v = V_n \in (v_{q(n)+1}, v_{q(n)}]$ such that

$$\mu \left[ \frac{1}{\mathcal{R}_n} \left\{ \frac{1}{v} - \frac{1}{v_{q(n)+1}} \right\} + \frac{1}{\mathcal{V}_n(h_{q(n)+1})} \right] = \lambda$$

and finally calculate $x = x_n \in [h_{q(n)}, h_{q(n)+1})$ from

$$x = h_{q(n)+1} - \int_{V_n}^{v_{q(n)+1}} \frac{w\,dw}{u_{[k,q(n)]}(w)}.$$

This completes the first stage of the calculation.

Now consider the interval $[x_{n-1}, x_n]$. Both $x_n$ and $V_n$ are now known but $x_{n-1}$ and $V_{n-1}$ are unknown. For $k = n-1$ we use the recursive calculation (12.25) to find $q = q(n-1)$ such that

$$\{\mathcal{E}_\mu\}_{n-1}(h_q) \ge \frac{\mu}{V_n} + g_{q(n)} > \{\mathcal{E}_\mu\}_{n-1}(h_{q+1}).$$

Now we find $v = V_{n-1} \in (v_{q(n-1)+1}, v_{q(n-1)}]$ such that

$$\frac{1}{\mathcal{R}_{n-1}} [E_\mu(v) - E_\mu(v_{q(n-1)+1})] + \{\mathcal{E}_\mu\}_{n-1}(h_{q(n-1)+1}) = \frac{\mu}{V_n} + g_{q(n)}$$

and finally calculate $x = x_{n-1} \in [h_{q(n-1)}, h_{q(n-1)+1})$ from

$$x = h_{q(n-1)+1} - \int_{V_{n-1}}^{v_{q(n-1)+1}} \frac{w\,dw}{u_{[n-1,\,q(n-1)]}(w)}.$$

Note that $V_{n-1} > w_\mu$. This completes the second stage of the calculation.

In general we consider an interval $[x_k, x_{k+1}]$ where $x_{k+1}$ and $V_{k+1}$ are known from the previous stage but where $x_k$ and $V_k$ are unknown. We use the recursive calculation (12.25) to find $q = q(k)$ such that

$$\{\mathcal{E}_\mu\}_k(h_q) \geq E_\mu(V_{k+1}) > \{\mathcal{E}_\mu\}_k(h_{q+1}).$$

If we assume in the first instance that $j(k+1) = m$ then $v_q < v_{q+1}$. Now we find $v = V_k \in [v_{q(k)+1}, v_{q(k)})$ such that

$$\frac{1}{\mathcal{R}_k}[E_\mu(v) - E_\mu(v_{q(k)+1})] + \{\mathcal{E}_\mu\}_k(h_{q(k)+1}) = E_\mu(V_{k+1}).$$

Note that $V_k < w_\mu < V_{k+1}$. Alternatively, if we assume in the second instance that $j(k+1) = 0$ then $v_q > v_{q+1}$ and $v = V_k \in [v_{q(k)+1}, v_{q(k)})$. Finally we calculate $x = x_k \in [h_{q(k)}, h_{q(k)+1})$ from

$$x = h_{q(k)+1} - \int_{V_k}^{v_{q(k)+1}} \frac{w\,dw}{u_{[k,\,q(k)]}(w)}.$$

This calculation determines a strategy of optimal type. It is necessary to adjust the parameters $\lambda$ and $\mu$ to obtain a feasible strategy. These adjustment procedures are similar to those described in Sections 8.5 and 8.6.

The task of determining $\lambda$ and $\mu$ is an iterative process. For level track, the key equation (12.22) reduces to the form

$$E_\mu(v) = \lambda.$$

For $\lambda > E_\mu(w_\mu)$ this equation has precisely two solutions $V$ and $W$ with $V < w_\mu < W$. Thus we obtain an approximate speed-holding strategy with the speed oscillating between $V$ and $W$. Since $w_\mu^2\, r'(w_\mu) = \mu$ and since $w_\mu > X/T$ we can begin the iterative process with $\mu = \mu_{est}$ where

$$\mu_{est} = \left(\frac{X}{T}\right)^2 r'\left(\frac{X}{T}\right) + \varepsilon$$

and $\varepsilon$ is a small positive number. Since the value $\lambda$ must be greater than $E_\mu(w_\mu)$ we can begin with $\lambda = \lambda_{est}$ where

$$\lambda_{est} = E_{\mu_{est}}(w_{\mu_{est}}) + \delta$$

and $\delta$ is a small positive number. As $\delta$ decreases, the values of $V$ and $W$ move closer together and the length of each phase decreases. Thus we adjust $\delta$ until we obtain $x_0 = 0$. This means that the distance constraint is satisfied. It is also necessary to satisfy the time constraint. We calculate

$$T_\mu = \int_0^X \frac{1}{v(x)} dx.$$

If $T_\mu > T$ then $\mu$ must be increased. If $T_\mu < T$ then $\mu$ must be decreased. The whole process is now repeated. On level track it is easy to see that $\delta$ and $\varepsilon$ are both positive. On non-level track, this may not be the case. The adjustment of $\lambda$ and $\mu$ is illustrated for level track in the following example.

### Example 12.1  Adjustment of $\lambda$ and $\mu$

We illustrate the adjustment of $\lambda$ and $\mu$ used to calculate Example 8.5. We calculate $X/T = 12$ and choose $\varepsilon = 0.05$ and $\delta = 10^{-6}$.

**Table 12-1: Adjustment of $\lambda$ and $\mu$ for Example 8.5**

| $\varepsilon$ $(\times 10^{-2})$ | $\delta$ $(\times 10^{-6})$ | $w_\mu$ | $\lambda$ $(\times 10^{-2})$ | $\mu$ $(\times 10^{-2})$ | $x_0$ | $t_0$ | Comments |
|---|---|---|---|---|---|---|---|
| 0.5 | 1.0 | 12.796 | 1.8716 | 3.0056 | 8531.90 | 420.87 | Distance travelled too small. Increase $\delta$. |
| 0.5 | 6.0 | 12.796 | 1.8721 | 3.0056 | 2962.62 | −13.25 | Speed too low. Increase $\varepsilon$. |
| 1.5 | 0.6 | 14.156 | 1.9457 | 4.0056 | 7797.27 | 418.50 | Distance travelled too small. Increase $\delta$. |
| 1.5 | 5.0 | 14.156 | 1.9462 | 4.0056 | 1527.83 | −23.28 | Speed too low. Increase $\varepsilon$. |
| 3.3525 | 3.3285 | 16.170 | 2.0680 | 5.8581 | −0.083 | 0.018 | Close enough. |

$\square$

The Scheduling and Control Group has developed a prototype computer program that is designed to do the above calculations on board a long-haul freight train in real time.


## 12.11  Examples for Non-Steep Track

The following examples are based on data obtained from train models used by the Scheduling and Control Group at the University of South Australia. Length is measured in metres and time is measured in seconds. We consider a journey with $X = 18000$ and $T = 1500$. We assume that

$$r(v) = a + bv + cv^2$$

where $a = 1.5 \times 10^{-2}$, $b = 3 \times 10^{-5}$ and $c = 6 \times 10^{-6}$. We take $H = 1.5$ and $K = 1$, and assume only two allowable rates of fuel supply with $f_0 = 0$ and $f_1 = 1$. The strategies of optimal type in these examples all contain an approximate speed-hold segment. This segment is constructed using alternate phases of coast and maximum power. For each $x \in [0, X]$ we will take

$$g(x) = \alpha x (X - 2x) (X - x)$$

where $\alpha$ is a constant. Outside the interval $[0, X]$ we assume that $g(x) = 0$. In all examples the value of $\alpha$ is so small that the gradient would not be readily apparent to the naked eye. The height of the track is determined by the formula

$$h(x) \approx -\frac{\alpha}{9.8} \int_0^x \xi(X - 2\xi)(X - \xi)\, d\xi$$

$$= -\frac{\alpha}{19.6} x^2 (X - x)^2$$

and so in the case where $|\alpha| = 2 \times 10^{-14}$ we have $|h(x)| < 6.7$ for all $x$. Thus in a total journey of 18 kilometres there is a rise or fall of only 6.7 metres. The strategies of optimal type in these examples all contain an approximate speed-hold segment. This segment is constructed using alternate phases of coast and maximum power. Nevertheless we will show that even these small gradients have a dramatic effect on the position of the switching points and in particular we will show that the extent of the semi-final coast phase is drastically changed.

There are some important general observations that should be made at this stage. There is only a small decrease in the fuel consumption when a strategy of optimal type using only one *coast–power* pair is replaced by a strategy of optimal type using nine *coast–power* pairs. Thus a seemingly rudimentary approximation to the idealised minimum cost strategy can be very energy-efficient. Hence in practice we do not need a large number of *coast–power* control pairs to approximate a true speed-holding strategy. Indeed, the examples confirm in practical terms that *coast–power* control is almost as close as we please to continuous control.

In our examples we have used simple gradient profiles. There are two critical factors that determine an energy-efficient strategy. First, it is necessary to keep the train speed close to the selected holding speed during the approximate speed-hold phase. Second, it is important to choose the correct switching point to begin the semi-final coast phase. We have found that selection of a non-optimal holding speed does not necessarily cause a large increase in fuel consumption as long as appropriate switching points are used, but we show that manual selection of the switching points is difficult because of extreme sensitivity to small changes in gradient. In our simple examples, selection of the correct switching point to begin the semi-final *coast* phase is extremely important. With metropolitan railcars on apparently flat track with small distances between stations we have found that even the most experienced drivers could not choose this point effectively. This was demonstrated in our Metromiser trials, where audited fuel savings of 14% and improved timekeeping were achieved when Metromiser was used to select the point where the semi-final coast phase begins.

### 12.11.1 Level Track Strategies Applied to Track with Small Gradients

We consider the level track strategy on a track with small gradients. The gradients are so small that they would not be apparent to the naked eye. The examples will show that the level track strategy should not be used. There are three alternative ways in which the level track strategy could be implemented. The most natural way is to select the same switching points, but it is also possible to select the same switching times or the same switching speeds. In all cases, the strategies can be shown to be inappropriate.

***Example 12.2  Level Track Switching Points over a Small Valley***

We consider a small valley. To define a track with an initial downhill section and a final uphill section we let $\alpha = 2 \times 10^{-14}$. We use nine *coast–power* pairs and the same switching points as we used in Example 8.5. The strategy is not feasible, and the final phase is degenerate. The train stops at time $t = 1363.73$ and at position $x = 16988.21$. The cost of the journey is $J = 190.58$.

□

### *Example 12.3  Level Track Switching Points over a Small Hill*

We now consider a small hill. To define a track with an initial uphill section and a final downhill section we let $\alpha = (-2) \times 10^{-14}$. We use the same strategy and switching points as the previous example. The strategy is not feasible and is not energy-efficient. The train stops at time $t = 1461.125$ and at position $x = 18028.070$. The cost of the journey is $J = 218.18$.

□

The speed profiles for Examples 12.2 and 12.3 are compared to the speed profile for Example 8.5 in Figures 12-1.

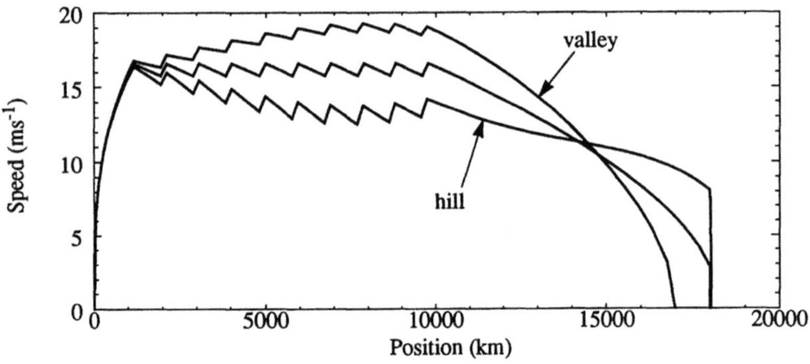


Figure 12-1: Speed $v(x)$ for Examples 8.5, 12.2 and 12.3

### *Example 12.4  Level Track Switching Times over a Small Valley*

Once again, we consider a small valley. We let $\alpha = 2 \times 10^{-14}$, and use nine *coast–power* pairs and the same switching times as we used in Example 8.5. The strategy is not feasible, and the final phase is degenerate. The train stops at time $t = 1463.23$ and at position $x = 17815.32$. The cost of the journey is $J = 201.89$.

□

### *Example 12.5  Level Track Switching Times over a Small Hill*

We now consider a small hill, with $\alpha = -2 \times 10^{-14}$. We use the same strategy and switching times as the previous example. The strategy is not feasible. The train stops at time $t = 1503.97$ and at position $x = 17041.58$. The cost of the journey is $J = 201.89$.

$\square$

The speed profiles for Examples 12-4 and 12-5 are compared to the speed profile for Example 8.5 in Figure 12-2.

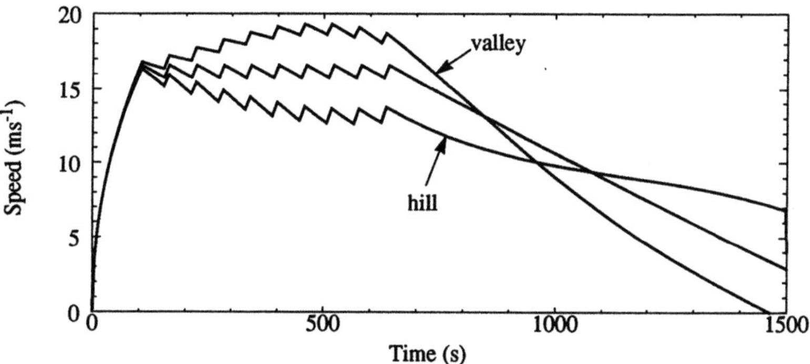


Figure 12-2: Speed $v(t)$ for Examples 8.5, 12.4 and 12.5

### 12.11.2  Strategies of Optimal Type on Track with Small Gradients

### *A Small Valley*

As before we let $\alpha = 2 \times 10^{-14}$. We consider a strategy with one *coast–power* pair and a strategy with nine *coast–power* pairs. In each case we use the algorithms described in Section 12.10 to calculate a strategy of optimal type. The holding speed is given by $v \approx 15$ and the speed-hold segment extends from $(x, t) \approx (865, 85)$ to $(x, t) \approx (13700, 920)$. For a non-level track the critical speeds that determine the speed-holding interval $[V, W]$ are not uniquely defined. We set

$$V = \min \left\{ V_{2i} \ \text{for} \ i \in \left[ 1, \frac{n-1}{2} \right] \right\}$$

and

$$W = \max \left\{ V_{2i-1} \text{ for } i \in \left[ 1, \frac{n-1}{2} + 1 \right] \right\}$$

We find that $[V, W] = [11.296, 18.509]$ in the example with one coast-power pair and $[V, W] = [14.466, 15.642]$ in the example with nine coast-power pairs. For practical purposes we note that the fuel consumption is reduced only marginally by the more elaborate strategy.

### Example 12.6  One Coast-Power Pair over a Small Valley

If we set $\lambda = 2.23205 \times 10^{-2}$ and $\mu = 4.39305 \times 10^{-2}$ then the critical speed is given by $w_\mu = 14.622$. The fuel consumption is given by $J = 200.66$.

$\square$

### Example 12.7  Nine Coast-Power Pairs over a Small Valley

If we set $\lambda = 2.25335 \times 10^{-2}$ and $\mu = 4.76702 \times 10^{-2}$ then the critical speed is given by $w_\mu = 15.047$. The fuel consumption is given by $J = 200.18$.

$\square$

The speed profiles for Examples 12.6 and 12.7 are shown in Figure 12-3.

### A Small Hill

As before we let $\alpha = -2 \times 10^{-14}$. Once again we consider a strategy with one *coast–power* pair and a strategy with nine *coast–power* pairs. The holding speed is given by $v \approx 18.8$ and the speed-hold segment extends from $(x, t) \approx (1890, 145)$ to $(x, t) \approx (5300, 325)$. We find that $[V, W] = [17.737, 19.926]$ in the example with one *coast–power* pair and $[V, W] = [18.665, 19.013]$ in the example with nine *coast–power* pairs. Once again, the fuel consumption is reduced only marginally by the more elaborate strategy.

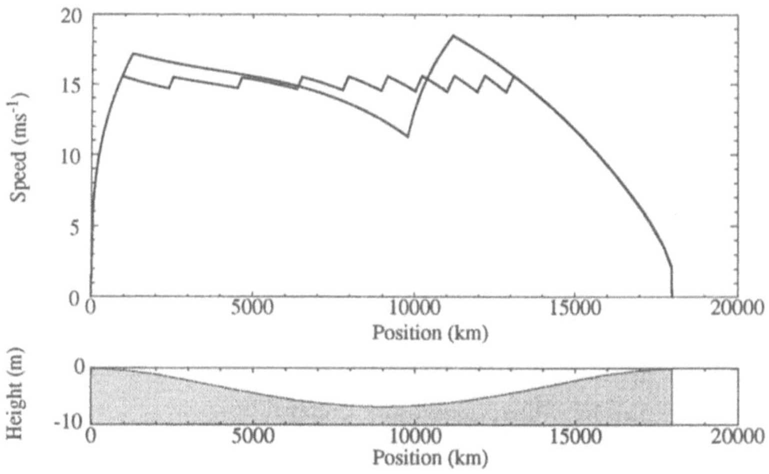

Figure 12-3: Speed and track height for Examples 12.6 and 12.7

### Example 12.8  One Coast-Power Pair over a Small Hill

If we set $\lambda = 1.60102 \times 10^{-2}$ and $\mu = 9.04583 \times 10^{-2}$ then the critical speed is given by $w_\mu = 18.809$. The fuel consumption is given by $J = 209.54$.

$\square$

### Example 12.9  Nine Coast-Power Pairs over a Small Hill

If we set $\lambda = 1.60209 \times 10^{-2}$ and $\mu = 9.08786 \times 10^{-2}$ then the critical speed is given by $w_\mu = 18.839$. The fuel consumption is given by $J = 209.52$.

$\square$

The speed profiles for Examples 12.8 and 12.9 are shown in Figure 12-4.

### A Brief Comparison of the Strategies of Optimal Type

The most significant difference in the strategies of optimal type described in the above examples is the length of the final coast phase. In a total journey of 18 kilometres we have a final coast phase of 4.3 kilometres when travelling over the small valley and a final coast phase of 12.7 kilometres when travelling over the small hill. Since the rise and fall of the track in each case is less than 6.7 metres it is clear that this

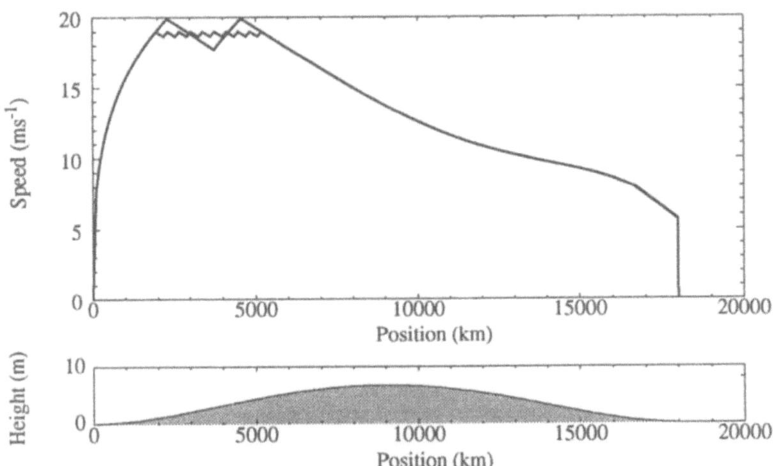


Figure 12-4: Speed and track height for Examples 12.8 and 12.9

difference could not be determined precisely by the driver alone. Because there is no fuel consumption while coasting we can see that an error in estimating the length of the semi-final coast phase can have a significant effect on the cost.

We have also shown that ad hoc strategies, even those with the correct structure, can result in non-feasible journeys or journeys where fuel consumption is increased significantly. For example the strategy of Example 12.3 gives a 5% increase in fuel consumption over the strategy of Example 12.9.

With more complicated track gradient profiles there may be a number of extended coast or power phases. Selection of appropriate switching points will be important for each of these phases. In practice there are many sections of track on a typical long-haul journey where the gradients are classified as steep. Even in an apparently flat continent such as Australia this is the case. Under these conditions the predominant speed-hold mode is interrupted by segments of coast and power. Selection of switching points is particularly important in defining the extent of each of these segments. Inappropriate selection will mean that excessive fuel is consumed in returning the train to the selected holding speed.

## 12.12  Examples for Steep Track

We use the same train parameters as before.


***Example 12.10  A Strategy of Optimal Type on Steep Track (Cheng's Climb)***

We consider a journey with $X = 50000$ and $T = 3000$. For $x \in [0, X]$ we take

$$
g(x) = \begin{cases} 0 & 0 < x < 20000 \\ -0.1 & 20000 < x < 25000 \\ 0 & 25000 < x < X \end{cases}
$$

and outside the interval $[0, X]$ we assume that $g(x) = 0$. The total rise is about 51 metres. We consider a strategy of optimal type with nine *coast–power* pairs. If we set $\lambda = 2.3030148 \times 10^{-2}$ and $\mu = 0.1003085$ then the critical speed is given by $w_\mu = 19.495$. The fuel consumption is given by $J = 924.39$.

$\square$

The speed and track height profiles for Example 12.10 are shown in Figure 12-5.

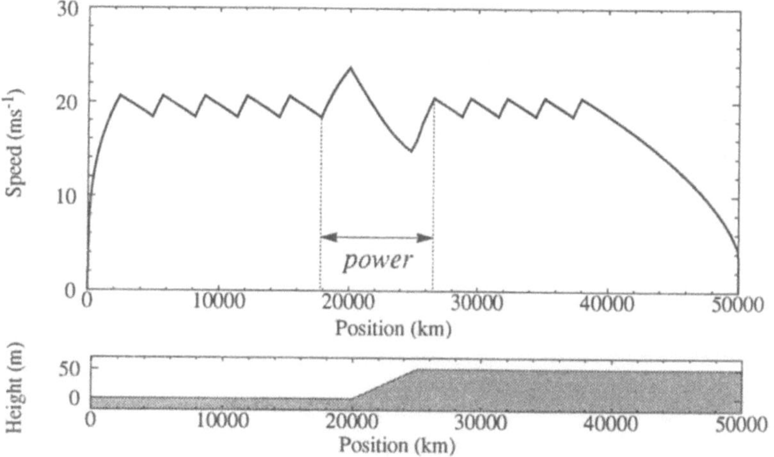


Figure 12-5: Speed and track height for Example 12.10

***Example 12.11  A Strategy of Optimal Type on Steep Track (Benjamin's Bank)***

We now consider a journey with $X = 50000$ and $T = 3000$. For $x \in [0, X]$ we take

$$g(x) = \begin{cases} 0 & 0 < x < 20000 \\ 0.05 & 20000 < x < 25000 \\ 0 & 25000 < x < X \end{cases}$$

and outside the interval $[0, X]$ we assume $g(x) = 0$. The total fall is about 25 metres. We consider a strategy of optimal type with nine *coast–power* pairs. If we set $\lambda = 2.2842050 \times 10^{-2}$ and $\mu = 9.679438 \times 10^{-2}$ then the critical speed is given by $w_\mu = 19.255$. The fuel consumption is given by $J = 424.48$.

☐

The speed and track height profiles for Example 12.11 are shown in Figure 12-6.

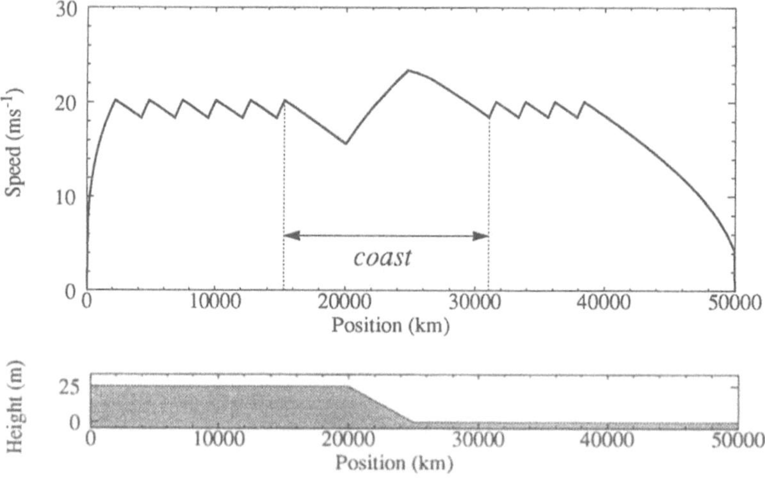


Figure 12-6: Speed and track height for Example 12.11

***Example 12.12  Strategy of Optimal Type on Steep Track (Milroy's Mountain)***

We now consider a journey with $X = 50000$ and $T = 3000$. For $x \in [0, X]$ we take

$$
g(x) = \begin{cases}
0 & 0 < x < 15000 \\
-0.1 & 15000 < x < 19000 \\
0 & 19000 < x < 20000 \\
0.05 & 20000 < x < 25000 \\
0 & 25000 < x < X
\end{cases}
$$

and outside the interval $[0, X]$ we assume $g(x) = 0$. We consider a strategy of optimal type with nineteen *coast–power* pairs. If we set $\lambda = 2.2895473 \times 10^{-2}$ and $\mu = 9.802139986 \times 10^{-2}$ then the critical speed is given by $w_\mu = 19.34$. The fuel consumption is given by $J = 691.36$.

$\square$

The speed and track height profiles for Example 12.12 are shown in Figure 12-5.

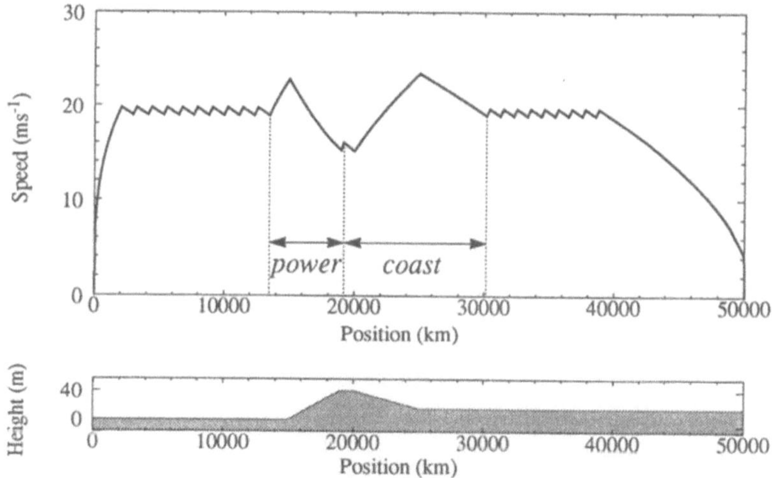


Figure 12-7: Speed and track height for Example 12.12

## CHAPTER 13

# CONTINUOUSLY VARYING GRADIENT

In the previous chapter we assumed a piecewise constant gradient and derived key equations for a strategy of optimal type. We also developed an algorithm for calculating the optimal switching points. In this chapter we show that the assumption of piecewise constant track gradient is unnecessary. Using an intuitive limit procedure applied to the key equations of the previous chapter we derive a more general form for the key equations. The general form of these equations is then derived rigorously by applying a standard perturbation argument to the equations of motion. We use the same problem formulation and essentially the same notation as in the previous chapter.

This material in this chapter was originally published in the *Journal of the Australian Mathematical Society, Series B* [12].

## 13.1  Some Additional Notation

For a control strategy $S(\{ [j(k+1); (x_k, x_{k+1})] \})_{k=0, 1, \ldots, n}$ the equations of motion will be written in the form

$$\frac{dt}{dx} = \frac{1}{v}$$

and

$$v\frac{dv}{dx} = \frac{Hf_{j(k+1)}}{v} + K_{j(k+1)} - r(v) + g(x)$$

$$= u_k(x, v)$$

for $x \in (x_k, x_{k+1})$.

It is necessary to define some additional gradient weighted average procedures. For this reason we will modify and extend the notation of the previous chapter in the following way.

For each measurable function $f : \Re_+ \to \Re$ we define a special gradient weighted average value at the point $x \in (h_q, h_{q+1}) \subset (x_k, x_{k+1})$ by the formula

$$\mathcal{A}_k^p [f(v)] (x)$$

$$= f[v(x)] \frac{1}{\mathcal{R}_{[k,q]}} + \sum_{r=q}^{q(k+1)-1} f(v_{r+1}) \left[ \frac{1}{\mathcal{R}_{[k,r+1]}} - \frac{1}{\mathcal{R}_{[k,r]}} \right]$$

where $v = v(x)$ denotes the speed at the point $x$. We use the notation

$$\mathcal{A}_k^p [f]$$

to indicate that the average is taken over the control interval $(x_k, x_{k+1})$ and that the interval $[0, X]$ is divided into $p + 1$ subintervals $(h_q, h_{q+1})$ of constant gradient. We say that

$$\mathcal{A}_k^p [f(v)] (x)$$

is the *effective value* of $f(v)$ at the point $x$. Somewhat perversely, the *effective speed* $\mathcal{V}_k^p(x)$ is defined by

$$\frac{1}{\mathcal{V}_k^p(x)} = \mathcal{A}_k^p \left[ \frac{1}{v} \right](x).$$

The *effective energy density* $\{\mathcal{E}_\mu\}_k^p (x)$ is defined by

$$\{\mathcal{E}_\mu\}_k^p (x) = \mathcal{A}_k^p [E_\mu(v)] (x)$$

where $E_\mu(v)$ is the energy density defined in the previous chapter. In the previous chapter we showed that the following conditions on the switching points $\{x_k\}$ are necessary for a strategy of optimal type. There are non-negative constants $\lambda$ and $\mu$ such that $x_n < X$ is a solution to

$$\frac{\mu}{\mathcal{V}_k^p(x)} = \lambda, \tag{13.1}$$

$x_{n-1} < x_n$ is a solution to

$$\{\mathcal{E}_\mu\}_{n-1}^p (x) = \frac{\mu}{V_n} + g_{r(n)} \tag{13.2}$$

and in general, $x_k < x_{k+1}$ is a solution to

$$\{\mathcal{E}_{\mu}\}_k^p(x) = E_{\mu}(V_{k+1}) \tag{13.3}$$

for each $k = 1, 2, \ldots, n-2$.

## 13.2  A General Form for the Key Equations

To explain the generalised form for the key equations it is necessary to introduce some relevant terminology. For each $k = 1, 2, \ldots, n$ and for each interval $[a, b] \subset [x_k, x_{k+1}]$ we define

$$\mathcal{D}_k(a, b) = \int_a^b \frac{1}{u_k(x, v(x))} \frac{\partial u_k}{\partial x}(x, v(x))dx$$

$$= \int_a^b \frac{1}{u_k(x, v(x))} dg(x)$$

and write $\mathcal{D}_k(x) = \mathcal{D}_k(x, x_{k+1})$ and $\mathcal{D}_k = \mathcal{D}_k(x_k)$. To describe the necessary conditions for a strategy of optimal type we use a similar notation to the notation used in the previous section for piecewise constant gradient.

For each measurable function $f : \mathfrak{R}_+ \to \mathfrak{R}$ and piecewise continuous gradient $g(x)$ we define the *effective value* at the point $x \in [x_k, x_{k+1}]$ by the formula

$$\mathcal{A}_k[f(v)](x) = f[v(x)] e^{\mathcal{D}_k(x)} + \int_x^{x_{k+1}} f[v(\xi)] de^{\mathcal{D}_k(\xi)}.$$

In particular the *effective speed* $V_k(x)$ is defined by

$$\frac{1}{V_k(x)} = \mathcal{A}_k\left[\frac{1}{v}\right](x) \tag{13.4}$$

and we write $V_k = V_k(x_k)$. The *effective energy density* is defined by

$$\{\mathcal{E}_{\mu}\}_k(x) = \mathcal{A}_k[E_{\mu}(v)](x).$$

In the case of a piecewise constant gradient, these definitions reduce to the definitions used in the previous section.

We will show that the following conditions on the switching points $\{x_k\}$ are necessary for a strategy of optimal type. There are non-negative constants $\lambda$ and $\mu$ such that $x_n < X$ is a solution to

$$\frac{\mu}{\mathcal{V}_n(x)} = \lambda, \tag{13.5}$$

$x_{n-1} < x_n$ is a solution to

$$\{\mathcal{E}_\mu\}_{n-1}(x) = \frac{\mu}{V_n} + g_{r(n)} \tag{13.6}$$

and in general, $x_k < x_{k+1}$ is a solution to

$$\{\mathcal{E}_\mu\}_k(x) = E_\mu(V_{k+1}) \tag{13.7}$$

for each $k = 1, 2, \ldots, n-2$.

## 13.3 Necessary Conditions for a Strategy of Optimal Type

### 13.3.1 An Intuitive Derivation of the Key Equations

For each $k = 1, 2, \ldots, n$ and for $q(k) < r \le q(k+1)$ it is easy to see that

$$\log R_r = \log u_{[k, r-1]}(v_s) - \log u_{[k, r]}(v_r)$$

$$\approx \frac{(-1)}{u_{[k, r]}(v_r)} \Delta g_r$$

where $\Delta g_r = g_r - g_{r-1}$. Therefore

$$\log \mathcal{R}_{[k, q]} \approx (-1) \sum_{r = q+1}^{q(k+1)} \frac{1}{u_{[k, r]}(v_r)} \Delta g_r.$$

If $q$ is chosen so that $x \in [h_q, h_{q+1}]$ and if we take the limit as $\Delta h_r \to 0$ then we obtain

$$\log \mathcal{R}_k(x) = \lim_{\Delta h_r \to 0} \log \mathcal{R}_{[k,\,q]}$$

$$= - \int_x^{x_{k+1}} \frac{1}{u_k(\xi,\, v(\xi))} dg(\xi)$$

$$= -\mathcal{D}_k(x)$$

and hence

$$\mathcal{R}_k(x) = e^{-\mathcal{D}_k(x)}.$$

Now since

$$\frac{1}{\mathcal{V}_k^p(x)} = \mathcal{A}_k^p\!\left[\frac{1}{v}\right]\!(x)$$

$$= \frac{1}{v(x)} \frac{1}{\mathcal{R}_{[k,\,q]}} + \sum_{r=q}^{q(k+1)-1} \frac{1}{v_{r+1}}\left[\frac{1}{\mathcal{R}_{[k,\,r+1]}} - \frac{1}{\mathcal{R}_{[k,\,r]}}\right]$$

it follows that

$$\lim_{\Delta h_r \to 0} \frac{1}{\mathcal{V}_k^p(x)} = \frac{1}{v(x)} e^{\mathcal{D}_k(x)} + \int_x^{x_{k+1}} \frac{1}{v(\xi)} de^{\mathcal{D}_k(\xi)}$$

$$= \mathcal{A}_k\!\left[\frac{1}{v}\right]\!(x)$$

$$= \frac{1}{\mathcal{V}_k(x)}.$$

We also have

$$\{\mathcal{E}_\mu\}_k^p(x) = \mathcal{A}_k^p\,[E_\mu(v)]\,(x)$$

$$= E_\mu\,[v(x)]\,\frac{1}{\mathcal{R}_{[k,\,q]}} + \sum_{r=q}^{q(k+1)-1} E_\mu(v_{r+1})\left[\frac{1}{\mathcal{R}_{[k,\,r+1]}} - \frac{1}{\mathcal{R}_{[k,\,r]}}\right]$$

and hence

$$\lim_{\Delta h_r \to 0} \{\mathcal{E}_\mu\}_k^p(x) = E_\mu[v(x)]\, e^{\mathcal{D}_k(x)} + \int_x^{x_{k+1}} E_\mu[v(\xi)]\, de^{\mathcal{D}_k(\xi)}$$

$$= \mathcal{A}_k[E_\mu(v)](x)$$

$$= \{\mathcal{E}_\mu\}_k(x).$$

The key equations (13.5), (13.6) and (13.7) are now obtained from (13.1), (13.2) and (13.3) using the above limit procedure.

### 13.3.2  Lagrangean Function and Kuhn-Tucker Equations

As in the previous chapter, we define a Lagrangean function of the form

$$\mathcal{J}(\xi, \lambda, \mu) = HJ(\xi) + \lambda[X - x(\xi)] + \mu[t(\xi) - T] \tag{13.8}$$

where $J(\xi)$ is the cost of the journey, $x(\xi)$ is the position, $t(\xi)$ is the time taken, and $\lambda$ and $\mu$ are the Lagrange multipliers. To find the minimum cost strategy for the given sequence of throttle settings we apply the Kuhn-Tucker conditions as before.

To apply the Kuhn-Tucker conditions it is necessary to calculate a number of partial derivatives. Although these partial derivatives have exactly the same form as the partial derivatives calculated in Section 12.7.1, the derivation uses perturbation theory and is more difficult. Details of the derivation are given in Sections 13.3.3 and 13.3.4. Direct application of the Kuhn-Tucker conditions gives a set of complicated equations that can be simplified using the essentially inductive argument applied in Section 12.6.3. Because the Lagrangeans (13.8) and (12.4) have the same form, and because the partial derivatives have the same form, the inductive argument is the same.

For convenience, we will use the notation

$$Q_k = \frac{u_{k-1}(x_k, V_k)}{u_k(x_k, V_k)}$$

for $k = 1, 2, \ldots, n$. The condition

$$\frac{\partial \mathcal{J}}{\partial \xi_n} = 0$$

can be simplified to give

$$(-1)\,\lambda + \mu\,\frac{1}{\mathcal{V}_n} = 0. \tag{13.9}$$

When the condition

$$\frac{\partial \mathcal{J}}{\partial \xi_{n-1}} = 0$$

is combined with (13.9) we deduce that

$$\mu\,(1 - Q_{n-1})\,e^{-\mathcal{D}_{n-1}}\left[\frac{1}{\mathcal{V}_{n-1}} - \frac{1}{V_n}\right] = \frac{(-1)\,Hf_{j(n-1)}}{V_{n-1}}. \tag{13.10}$$

In similar fashion we can combine the condition

$$\frac{\partial \mathcal{J}}{\partial \xi_{n-2}} = 0$$

with (13.9) and (13.10) to deduce that

$$(1 - Q_{n-2})\,e^{-\mathcal{D}_{n-2}}\left\{ [\mu + Hf_{j(n-1)}]\left[\frac{1}{\mathcal{V}_{n-2}} - \frac{1}{V_{n-1}}\right] \right\}$$

$$+ u_{n-2}(x_{n-1}, V_{n-1}) = \frac{[Hf_{j(n-1)} - f_{j(n-2)}]}{V_{n-2}}.$$

If we assume that

$$(1 - Q_h)\,e^{-\mathcal{D}_h}\left\{ [\mu + Hf_{j(h+1)}]\left[\frac{1}{\mathcal{V}_h} - \frac{1}{V_{h+1}}\right] + u_h(x_{h+1}, V_{h+1}) \right\}$$

$$= \frac{[Hf_{j(h+1)} - f_{j(h)}]}{V_h} \tag{13.11}$$

for each $h$ with $k < h < n - 2$ then the additional equation

$$\frac{\partial \mathcal{J}}{\partial \xi_k} = 0$$

allows us to deduce that

$$(1 - Q_k) \, e^{-\mathcal{D}_k} \left\{ [\mu + H f_{j(k+1)}] \left[ \frac{1}{\mathcal{V}_k} - \frac{1}{V_{k+1}} \right] + u_k(x_{k+1}, V_{k+1}) \right\}$$

$$= \frac{[H f_{j(k+1)} - f_{j(k)}]}{V_k}. \tag{13.12}$$

Therefore the hypothesis is established for $h = k$. We can now deduce that (13.11) is valid for all $h = 1, 2, \ldots, n - 2$. It is easy to show that (13.9), (13.10) and (13.12) are equivalent to (13.5)–(13.7).

### 13.3.3  Some Results from Perturbation Theory

Let $I = [x_1, x_2] \times [v_1, v_2] \subseteq \mathfrak{R}^2$ and let $\Phi : I \to \mathfrak{R}$ be a real valued function on $I$ with $\Phi(x, v)$ continuous and

$$\frac{\partial \Phi}{\partial v}(x, v)$$

well defined and continuous for $(x, v) \in I$. For the differential equation

$$\frac{dv}{dx} = \Phi(x, v) \tag{13.13}$$

the following standard results can be found in the literature [37]. The notation in this section, 13.3.3, is not intended to relate specifically to the train control problem.

### *Theorem 13.1  (Uniqueness Theorem)*

*There is exactly one solution $v(x) = f[a, V](x)$ satisfying the differential equation (13.13) and passing through the point $(a, V) \in I$.*

$\square$

### *Corollary 13.2*

*Let the solutions $f[a, V](x)$ be defined on some interval $[y_1, y_2]$ for each $V \in [w_1, w_2]$ and suppose that the point $(a, V_a) \in \text{int } J$, where*

$$J = [y_1, y_2] \times [w_1, w_2] \subseteq I.$$

*If we define*

$$f_\Delta(x) = f[a, V_a + \Delta](x)$$

*for each sufficiently small $\Delta$ then $f_\Delta \to f_0$ uniformly on $[y_1, y_2]$ as $\Delta \to 0$.*

$\square$

We use these two results to prove two useful corollaries.

### Corollary 13.3

*Let the solutions $f[a, V](x)$ be defined on the interval $[y_1, y_2]$ for each $V \in [w_1, w_2]$ and suppose that the point $(a, V_a) \in \operatorname{int} J$. If we define $V_x = f_0(x)$ and if $(x, V_x) \in \operatorname{int} J$ for all $x \in [a, b]$ then*

$$\log\left[\frac{\partial V_b}{\partial V_a}\right] = \int_a^b \frac{\partial \Phi}{\partial v}(x, f[a, V_a](x))\,dx.$$

### Proof

We write $V_x(\Delta) = f_\Delta(x)$ and use the notation $\Delta f(x) = f_\Delta(x) - f_0(x)$. Now

$$\begin{aligned}
\frac{d}{dx}[\Delta f(x)] &= \frac{d}{dx}[f_\Delta(x) - f_0(x)] \\
&= \Phi(x, f_\Delta(x)) - \Phi(x, f_0(x)) \\
&= \int_0^{\Delta f(x)} \frac{\partial \Phi}{\partial v}(x, f_0(x) + w)\,dw.
\end{aligned}$$

From Theorem 13.1 we know that $\Delta f(x) \neq 0$ for all $x \in [y_1, y_2]$ when $\Delta \neq 0$, and hence

$$\frac{\frac{d}{dx}[\Delta f(x)]}{\Delta f(x)} = \frac{1}{\Delta f(x)} \int_0^{\Delta f(x)} \frac{\partial \Phi}{\partial v}(x, f_0(x) + w)\,dw.$$

By integrating from $x = a$ to $x = b$ we obtain

$$\log\left[\frac{\Delta f(b)}{\Delta f(a)}\right] = \int_a^b \left\{\frac{1}{\Delta f(x)} \int_0^{\Delta f(x)} \frac{\partial \Phi}{\partial v}(x, f_0(x) + w)\,dw\right\} dx.$$

Since $\Delta f(x) = V_x(\Delta) - V_x(0) = \Delta V_x$ this becomes

$$\log\left[\frac{\Delta V_b}{\Delta V_a}\right] = \int_a^b \left\{ \frac{1}{\Delta f(x)} \int_0^{\Delta f(x)} \frac{\partial \Phi}{\partial v}(x, f_0(x) + w)\,dw \right\} dx$$

and since $\Delta f(x) \to 0$ uniformly on $[y_1, y_2]$ as $\Delta V_a = \Delta \to 0$ the result follows by taking the limit as $\Delta \to 0$.

$\square$

We will normally rewrite the result of Corollary 13.3 in a more convenient form. From the formula

$$\frac{d}{dx}[\Phi(x, f_0(x))] = \frac{\partial \Phi}{\partial x}(x, f_0(x)) + \frac{\partial \Phi}{\partial v}(x, f_0(x))\Phi(x, f_0(x))$$

we have by formal manipulation

$$\frac{\partial \Phi}{\partial v}(x, f_0(x)) = \frac{\dfrac{d}{dx}[\Phi(x, f_0(x))] - \dfrac{\partial \Phi}{\partial x}(x, f_0(x))}{\Phi(x, f_0(x))}$$

and hence, in a formal sense,

$$\int_a^b \frac{\partial \Phi}{\partial v}(x, f_0(x))\,dx = \log\left[\frac{\Phi(b, f_0(b))}{\Phi(a, f_0(a))}\right] - \mathcal{D}(a, b) \tag{13.14}$$

where

$$\mathcal{D}(a, b) = \int_a^b \frac{1}{\Phi(x, f_0(x))}\frac{\partial \Phi}{\partial x}(x, f_0(x))\,dx.$$

Therefore we will normally write the result of Corollary 13.3 in the form

$$\frac{\partial V_b}{\partial V_a} = \frac{\Phi(b, V_b)}{\Phi(a, V_a)}\, e^{-\mathcal{D}(a, b)} \tag{13.15}$$

where the right hand side is strictly defined by (13.14).

### *Corollary 13.4*

*Let $\delta(\xi)$ and $\Delta(\xi)$ be real valued continuously differentiable functions with $\delta(0) = 0$ and $\Delta(0) = 0$. Define $x(\xi) = x + \delta$, $V(\xi) = v + \Delta$ and*

$$V_x(\delta, \Delta) = f[a, V + \Delta](x + \delta)$$

*If $(x, V_x) \in \mathrm{int}\ J$ for all $x \in [a, b]$ then*

$$\left.\frac{\partial V_b}{\partial \xi}\right|_{\xi = 0} = \Phi(b, V_b)\delta'(0) + \frac{\partial V_b}{\partial \xi}\left[\left.\frac{\partial V_a}{\partial \xi}\right|_{\xi = 0} - \Phi(a, V_a)\delta'(0)\right].$$

### *Proof*

We have

$$V_b(\delta, \Delta) - V_b(0, 0)$$
$$= f[a, V + \Delta](b + \delta) - f[a, V](b)$$
$$= f_\Delta(b + \delta) - f_0(b)$$
$$= [f_\Delta(b + \delta) - f_\Delta(b)] + [f_\Delta(b) - f_0(b)]$$
$$= \int_b^{b+\delta} \Phi(x, f_\Delta(x))dx + [V_b(0, \Delta) - V_b(0, 0)]$$
$$= \int_b^{b+\delta} \Phi(x, f_\Delta(x))dx + \int_0^\Delta \frac{\partial V_b}{\partial V_a}(0, \sigma)d\sigma$$

and in similar fashion

$$V_a(\delta, \Delta) - V_a(0, 0)$$
$$= [f_\Delta(a + \delta) - f_\Delta(a)] + [f_\Delta(a) - f_0(a)]$$
$$= \int_a^{a+\delta} \Phi(x, f_\Delta(x))dx + \Delta.$$

By combining these two equations we have

$$\frac{V_b(\delta, \Delta) - V_b(0, 0)}{\xi} = \frac{1}{\xi} \int_b^{b+\delta} \Phi(x, f_\Delta(x)) dx$$

$$+ \frac{1}{\Delta} \int_0^\Delta \frac{\partial V_b}{\partial V_a}(0, \sigma) d\sigma \left[ \frac{V_a(\delta, \Delta) - V_a(0, 0)}{\xi} - \frac{1}{\xi} \int_a^{a+\delta} \Phi(x, f_\Delta(x)) dx \right].$$

The desired result follows by taking the limit as $\xi \to 0$.

$\square$

### 13.3.4  Calculation of some Useful Derivatives

We use the standard notation for the train control problem. To begin we observe that
the constraint $V_{n+1} = 0$ determines $\xi_{n+1}$ implicitly as a function of $\xi_1, \xi_2, ..., \xi_n$.
In the case where $V_{n+1}(\xi_1, \xi_2, ..., \xi_{n+1}) = \varepsilon > 0$ we note that for each $h \le n$ the
equation

$$\frac{\partial V_{n+1}}{\partial \xi_h} + \frac{\partial V_{n+1}}{\partial \xi_{n+1}} \frac{\partial \xi_{n+1}}{\partial \xi_h} = 0$$

implies

$$\frac{\partial \xi_{n+1}}{\partial \xi_h} = (-1) \frac{\dfrac{\partial V_{n+1}}{\partial \xi_h}}{\dfrac{\partial V_{n+1}}{\partial \xi_{n+1}}}.$$

We also have

$$\frac{\partial V_{n+1}}{\partial \xi_{n+1}} = \frac{u_n(x_{n+1}, \varepsilon)}{\varepsilon}.$$

In the notation of the previous section we write

$$V_{n+1} = v[x_n, V_n](x_{n+1})$$
$$= V_{x_{n+1}}(0, 0).$$

If we replace $\xi_h$ by $\xi_h + \xi$ then $V_{n+1}$ is replaced by

$$V_{n+1} + \Delta V_{n+1} = v[x_n + \xi, V_n + \Delta V_n](x_{n+1})$$
$$= V_{x_{n+1}}(\xi, \Delta V_n).$$

This is illustrated in Figure 13-1.

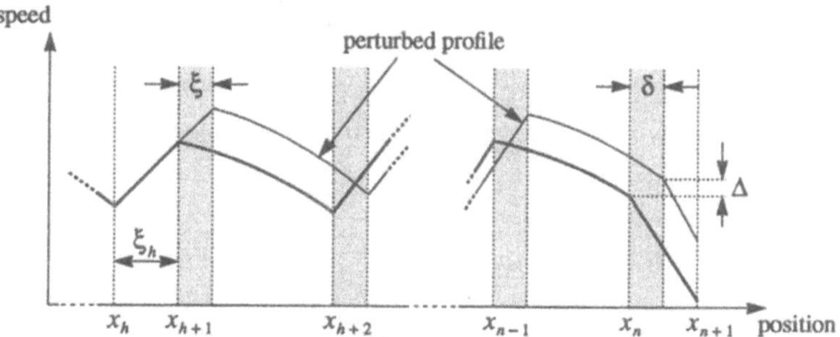


Figure 13-1: Change in speed profile cause by a change in $\xi_h$

Thus we have $\delta(\xi) = \xi$ and $\Delta(\xi) = \Delta V_n(\xi)$. By applying Corollary 13.4 we obtain

$$\frac{\partial V_{n+1}}{\partial \xi_h} = \frac{\partial}{\partial \xi} V_{x_{n+1}}[\xi, \Delta V_n]\Big|_{\xi=0}$$
$$= \frac{u_n(x_{n+1}, \varepsilon)}{\varepsilon} \delta'(0) + \frac{\partial V_{n+1}}{\partial V_n}\left[\frac{\partial V_n}{\partial \xi}\Big|_{\xi=0} - \frac{u_n(x_n, V_n)}{V_n}\delta'(0)\right]$$

$$(13.16)$$

and using (13.15) gives

$$\frac{\partial V_{n+1}}{\partial \xi_h} = \frac{u_n(x_{n+1}, \varepsilon)}{\varepsilon}\left\{1 + \left[-1 + \frac{V_n}{u_n(x_n, V_n)}\frac{\partial V_n}{\partial \xi_h}\right]e^{-\mathcal{D}_n}\right\}. \qquad (13.17)$$

It follows that

$$\frac{\partial \xi_{n+1}}{\partial \xi_h} = (-1)\left\{1 + \left[-1 + \frac{V_n}{u_n(x_n, V_n)}\frac{\partial V_n}{\partial \xi_h}\right]e^{-\mathcal{D}_n}\right\}. \qquad (13.18)$$

Since (13.18) is independent of $\varepsilon$, it will also be true when $V_{n+1} = 0$, with $\mathcal{D}_n$ well defined by the formula

$$\mathcal{D}_n = \int_{x_n}^{x_{n+1}} \frac{1}{u_n(x, v(x))} \, dg(x).$$

In general, for $h \le n$ we have

$$\frac{\partial V_h}{\partial \xi_h} = \frac{u_{h-1}(x_h, V_h)}{V_h}$$

and for $h < k+1 \le n$ we use an argument similar to that used to establish (13.17) to deduce that

$$\frac{\partial V_{k+1}}{\partial \xi_h} = \frac{u_k(x_{k+1}, V_{k+1})}{V_{k+1}} \left\{ 1 + \left[ -1 + \frac{V_k}{u_k(x_k, V_k)} \frac{\partial V_k}{\partial \xi_h} \right] e^{-\mathcal{D}_k} \right\}. \qquad (13.19)$$

By an inductive process applied to (13.18) and (13.19) we obtain

$$\frac{\partial \xi_{n+1}}{\partial \xi_h} = (-1) \left\{ 1 + \left[ (-1 + Q_n) + \sum_{i=h}^{n-1} (-1 + Q_i) \prod_{j=i}^{n-1} Q_{j+1} e^{-\mathcal{D}_j} \right] e^{-\mathcal{D}_n} \right\}$$

and

$$\frac{\partial V_{k+1}}{\partial \xi_h} = \frac{u_k(x_{k+1}, V_{k+1})}{V_{k+1}}$$

$$\times \left\{ 1 + \left[ (-1 + Q_k) + \sum_{i=h}^{k-1} (-1 + Q_i) \prod_{j=i}^{k-1} Q_{j+1} e^{-\mathcal{D}_j} \right] e^{-\mathcal{D}_k} \right\}.$$

For $h \le n$ and $V_{n+1} = \varepsilon > 0$ we use the formula

$$\tau_{n+1} = \int_0^{\xi_{n+1}} \frac{1}{v(x_n + \xi)} d\xi$$

to deduce that

$$\frac{\partial \tau_{n+1}}{\partial \xi_h} = \frac{1}{V_{n+1}} \frac{\partial \xi_{n+1}}{\partial \xi_h} + \int_0^{\xi_{n+1}} \frac{-1}{[v(x_n + \xi)]^2} \frac{u_n(x_n + \xi, v(x_n + \xi))}{v(x_n + \xi)}$$

$$\times \left\{ 1 + \left[ -1 + \frac{V_n}{u_n(x_n, V_n)} \frac{\partial V_n}{\partial \xi_h} \right] e^{-\mathcal{D}_n(x_n, x_n + \xi)} \right\} d\xi$$

$$= \frac{-1}{V_{n+1}} \left\{ 1 + \left[ -1 + \frac{V_n}{u_n(x_n, V_n)} \frac{\partial V_n}{\partial \xi_h} \right] e^{-\mathcal{D}_n} \right\} + \left\{ \frac{1}{V_{n+1}} - \frac{1}{V_n} \right\}$$

$$+ \left[ -1 + \frac{V_n}{u_n(x_n, V_n)} \frac{\partial V_n}{\partial \xi_h} \right] \int_0^{\xi_{n+1}} e^{-\mathcal{D}_n(x_n, x_n + \xi)} d\left[ \frac{1}{v(x_n + \xi)} \right].$$

Using integration by parts and (13.4) we find that

$$\int_0^{\xi_{n+1}} e^{-\mathcal{D}_n(x_n, x_n + \xi)} d\left[ \frac{1}{v(x_n + \xi)} \right] = \left\{ \frac{1}{V_{n+1}} - \frac{1}{V_n} \right\} e^{-\mathcal{D}_n}$$

from which it follows that

$$\frac{\partial \tau_{n+1}}{\partial \xi_h} = (-1) \left\{ \frac{1}{V_n} + \left[ -1 + \frac{V_n}{u_n(x_n, V_n)} \frac{\partial V_n}{\partial \xi_h} \right] \frac{e^{-\mathcal{D}_n}}{V_n} \right\}. \tag{13.20}$$

Since (13.20) is independent of $\varepsilon$ it will also be true when $V_{n+1} = 1$. By an inductive process applied to (13.20) it follows that

$$\frac{\partial \tau_{n+1}}{\partial \xi_h} = \left[ (1 - Q_n) + \sum_{i=h}^{n-1} (1 - Q_i) \prod_{j=i}^{n-1} Q_{j+1} e^{-\mathcal{D}_j} \right] \frac{e^{-\mathcal{D}_n}}{V_n} - \frac{1}{V_n}. \tag{13.21}$$

In general for $k < n$ we use a similar argument to show that

$$\frac{\partial \tau_{k+1}}{\partial \xi_{k+1}} = \frac{1}{V_{k+1}} \tag{13.22}$$

and for $h < k + 1 \leq n$ that

$$\frac{\partial \tau_{k+1}}{\partial \xi_h} = \frac{1}{V_{k+1}} - \frac{1}{V_k} + \left[ -1 + \frac{V_k}{u_k(x_k, V_k)} \frac{\partial V_k}{\partial \xi_h} \right] \left[ \frac{1}{V_{k+1}} - \frac{1}{V_k} \right] e^{-\mathcal{D}_k}.$$

Once again, we use an inductive argument to show that

$$\frac{\partial \tau_{k+1}}{\partial \xi_h} = \left[ (1 - Q_k) + \sum_{i=h}^{n-1} (1 - Q_i) \prod_{j=i}^{n-1} Q_{j+1} e^{-\mathcal{D}_j} \right] \left[ \frac{1}{\mathcal{V}_k} - \frac{1}{V_{k+1}} \right] e^{-\mathcal{D}_k}$$

$$+ \frac{1}{V_{k+1}} - \frac{1}{V_k}. \tag{13.23}$$

The formulae (13.16), (13.21), (13.22) and (13.23) of this section have essentially the same form as the corresponding formulae (12.12)–(12.15) in Section 12.7.1.


## 13.4  An Algorithm for Solving the Key Equations

The algorithm for solving the key equations is similar in principle to the algorithm of Section 12.10. However, it is now possible to describe the process more effectively.

We consider a strategy of optimal type from the subset $S(\{j(k+1)\}_{k=0, 1, \dots, n})$ where we assume that $n$ is even and that

$$j(k+1) = \begin{cases} m & k \text{ is even} \\ 0 & k \text{ is odd} \end{cases}$$

for $k < n$. We assume for the moment that $\lambda$ and $\mu$ are known, and calculate the positions of the switching points in reverse order. We assume that the track is not steep. That is, we assume that as $x$ increases, $v(x)$ increases during each power phase and decreases during each coast phase. We also assume that $g(x) = 0$ for $x \notin [0, X]$.

During the final phase, $v = v(x)$ decreases as $x$ increases and we can see from the formula

$$\frac{d}{dx} \left[ \frac{1}{\mathcal{V}_n(x)} \right] = \frac{-1}{v^2} \frac{dv}{dx} e^{\mathcal{D}_n(x)}$$

that the expression $1/\mathcal{V}_n(x)$ increases as $x$ increases. There is therefore at most one point $x = x_n$ where (13.5) is satisfied. We use an appropriate numerical scheme to find $v = v(x)$ in the region $x < X$ from the differential equation

$$v \frac{dv}{dx} = u_n(x, v)$$

with $V_{n+1} = v(X) = 0$. We can now calculate $\mathcal{V}_n(x)$ and hence find $x = x_n < X$ and $V_n = v(x_n)$.

During the semi-final phase $v = v(x)$ also decreases as $x$ increases. Thus from the formula

$$\frac{d}{dx}\Big[\{\mathcal{E}_\mu\}_{n-1}(x)\Big] = E_\mu'(v)\,\frac{dv}{dx}\,e^{\mathcal{D}_{n-1}(x)}$$

and from Lemma 11.1 we see that the expression $\{\mathcal{E}_\mu\}_{n-1}(x)$ increases as $x$ increases when $v < w_\mu$ and decreases as $x$ increases when $v > w_\mu$. Since $r(v) > g(x)$ during this phase,

$$\{\mathcal{E}_\mu\}_{n-1}(x_n) = E_\mu(V_n) \;>\; \frac{\mu}{V_n} + g(x_n)$$

and there is either no solution to (13.6) or else one solution with $v < w_\mu$ and one solution with $v > w_\mu$ in the region $x < x_n$. The former solution does not allow solutions to the remaining key equations. Thus we choose the solution $x = x_{n-1} < x_n$ with $V_{n-1} = v(x_{n-1}) > w_\mu$. We make the required calculations by solving the differential equation

$$v\frac{dv}{dx} = u_{n-1}(x, v)$$

with $v(x_n) = V_n$ in the region $x < x_n$.

In general, we note that

$$\frac{d}{dx}\Big[\{\mathcal{E}_\mu\}_k(x)\Big] = E_\mu'(v)\,\frac{dv}{dx}\,e^{\mathcal{D}_k(x)}$$

and from Lemma 11.1 we see that for a power phase the expression $\{\mathcal{E}_\mu\}_k(x)$ increases as $x$ increases when $v > w_\mu$ and decreases as $x$ increases when $v < w_\mu$. Since

$$\{\mathcal{E}_\mu\}_k(x) = E_\mu(V_{k+1}),$$

where $V_{k+1} = v(x_{k+1}) > w_\mu$, there is at most one solution $x - x_k$ with $v < w_\mu$ in the region $x < x_{k+1}$. In this case,

$$V_k = v(x_k) \;<\; w_\mu \;<\; v(x_{k+1}) = V_{k+1}.$$

We make the required calculations by solving the differential equation

$$v\frac{dv}{dx} = u_k(x, v)$$

with $v(x_{k+1}) = V_{k+1}$ in the region $x < x_{k+1}$. A similar argument applies for a coast phase, but in this case

$$V_k = v(x_k) > w_\mu > v(x_{k+1}) = V_{k+1}.$$

Finally we note that the task of determining $\lambda$ and $\mu$ is an iterative process. For level track, the key equation (13.3) reduces to the form

$$E_\mu(v) = \lambda.$$

For $\lambda > E_\mu(w_\mu)$ this equation has precisely two solutions $V$ and $W$ with $V < w_\mu < W$. Thus we obtain an approximate speed-holding strategy with the speed oscillating between $V$ and $W$. Since $w_\mu^2 \, r'(w_\mu) = \mu$ and since $w_\mu > X/T$ we can begin the iterative process with $\mu = \mu_{est}$ where

$$\mu_{est} = \left(\frac{X}{T}\right)^2 r'\left(\frac{X}{T}\right) + \varepsilon$$

and $\varepsilon$ is a small positive number. Since the value $\lambda$ must be greater than $E_\mu(w_\mu)$ we can begin with $\lambda = \lambda_{est}$ where

$$\lambda_{est} = E_{\mu_{est}}(w_{\mu_{est}}) + \delta$$

and $\delta$ is a small positive number. As $\delta$ decreases, the values of $V$ and $W$ move closer together and the length of each phase decreases. Thus we adjust $\delta$ until we obtain $x_0 = 0$. This means that the distance constraint is satisfied. It is also necessary to satisfy the time constraint. We calculate

$$T_\mu = \int_0^X \frac{1}{v(x)} dx.$$

If $T_\mu > T$ then $\mu$ must be increased. If $T_\mu < T$ then $\mu$ must be decreased. The whole process is now repeated. On level track it is easy to see that $\delta$ and $\varepsilon$ are both positive. On non-level track, this may not be the case.

## CHAPTER 14

# PRACTICAL STRATEGY OPTIMISATION

For a prescribed sequence of fuel supply rates it has been shown that a strategy of optimal type depends on two real number parameters, $\lambda$ and $\mu$. The parameter $\mu$ determines the hold speed for the journey, and the parameter $\lambda$ determines the size of the *coast–power* pairs used to approximate speed-holding. These parameters also determine the switching points and ultimately determine the distance travelled by the train and the time taken for the journey. As the magnitude of the *coast–power* pairs decreases the strategy approaches an idealised strategy that is determined by the hold speed alone. By adjusting the value of the hold speed it is possible to construct a feasible strategy.

The Scheduling and Control Group have developed a prototype computer program that calculates an energy-efficient strategy for any given journey. These strategies are based on the strategies of optimal type. The program has been used with realistic data supplied by Australian National to calculate strategies for typical long-haul journeys. The calculations can be performed in real time. In this chapter we discuss the development of the algorithm.

## 14.1 A Simple Journey

To begin, we consider a simple journey on flat track with no speed limits, and with the train starting and finishing at rest. The optimal strategy has four phases: maximum power, speed-hold, coast and brake. The hold speed $Z$ and brake speed $U$ are related by the formula

$$U = Z - \frac{\varphi(Z)}{\varphi'(Z)}$$

where $\varphi(v) = vr(v)$. Thus the hold speed determines the duration of the coast phase. For now, assume that the duration of the hold phase can be adjusted so that the journey covers the required distance. An optimal strategy can therefore be constructed given only the hold speed. The idealised strategy for Example 8.5 is a typical optimal speed profile, and is shown in Figure 14-1.

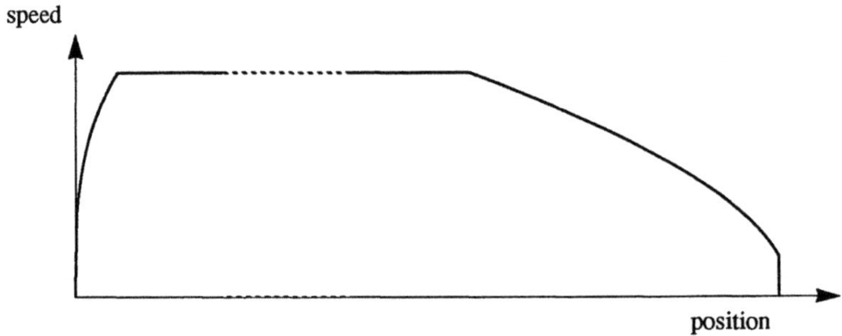


Figure 14-1: Optimal speed profile for flat track

The time taken to traverse the required distance can be adjusted by increasing the hold speed to give a reduced journey time, or by decreasing the hold speed to give an increased journey time.

A numerical root finder can be used to find the correct hold speed $Z$ for which $t(Z) - T = 0$, where $t(v_h)$ is the journey time corresponding to the strategy with hold speed $v_h$ and $T$ is the desired journey time. The optimal strategy for a given hold speed $v_h$ can be calculated as follows:

1.  calculate the corresponding brake speed, $v_b$;

2.  for the brake phase, use a differential equation solver to calculate the brake speed profile from the final position and speed $(X, 0)$ back to the brake point $(x_b, v_b)$, and the braking time $\tau_b$;

3.  for the coast phase, use a differential equation solver to calculate the coast speed profile from the brake point $(x_b, v_b)$ back to the coast point $(x_c, v_h)$, and the coasting time $\tau_c$;

4.  for the power phase, use a differential equation solver to calculate the power speed profile from the initial position and speed $(0, 0)$ to the hold point $(x_h, v_h)$, and the power time $\tau_p$;

5.  calculate the holding time $\tau_h = [x_c - x_h] / v_h$;

6.  calculate the journey time $t(v_h) = \tau_p + \tau_h + \tau_c + \tau_b$.

As noted earlier, this algorithm assumes that $x_h \leq x_c$.

## 14.2 Undulating Track

The *power–hold–coast–brake* strategy is also optimal for undulating track, provided that the track does not contain steep inclines or steep declines.

An incline is steep if the power required to maintain the desired hold speed is greater than the maximum train power. A decline is steep if braking is required to maintain the desired hold speed. Note that steepness depends on both the gradient of the track and the desired hold speed.

The algorithm for calculating an optimal strategy has the same structure as for flat track, but uses a more complicated iterative procedure to determine the switching points for the coast and brake phases. This procedure uses the $\mathcal{E}$-curves of Section 14.3. Figure 14-2 shows idealised optimal speed profiles for a small valley and for a small hill.

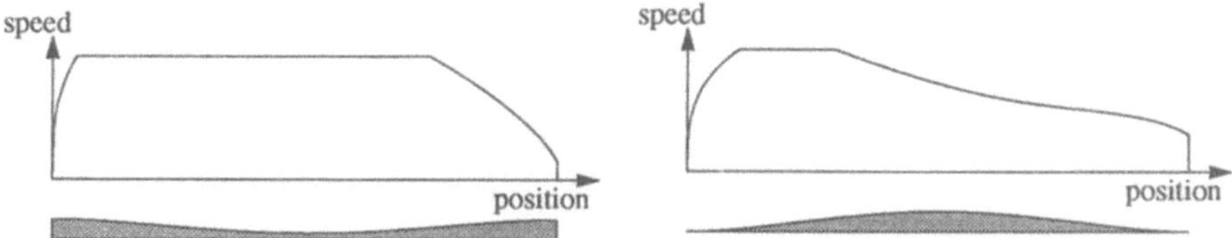


Figure 14-2: Optimal speed profiles for non-steep undulating track

## 14.3 Speed-Holding on Steep Track

If a steep incline is encountered during the proposed speed-hold phase, full power must be applied before the incline and maintained until the train returns to the hold speed beyond the incline. The speed of the train will increase until the base of the steep incline, decrease to below the hold speed while on the incline, and then return to the hold speed after the incline. Figure 14-3 shows an optimal speed profile over a steep incline.

The interval $[x_1, x_2]$ for which power is required is determined by equations of the form

$$\mathcal{E}(x_1) = \mathcal{E}(x_2) \tag{14.1}$$

and

$$\mathcal{E}'(x_1) = \mathcal{E}'(x_2) = 0 \tag{14.2}$$

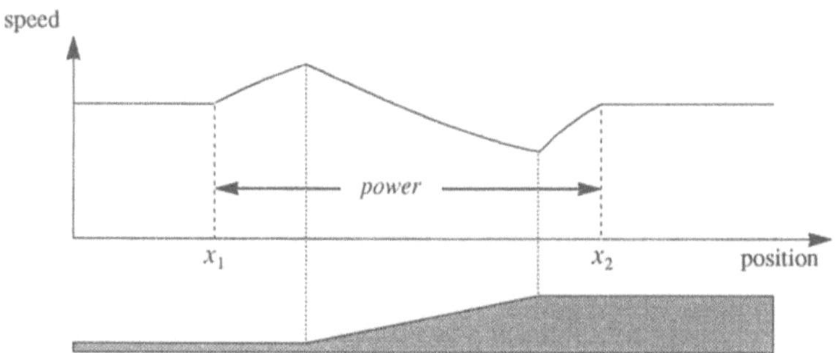


Figure 14-3: Optimal speed profile on a steep incline

where $\mathcal{E}(x)$ depends on the train characteristics and on the track gradient, and is the value of the function $\mathcal{E} = \{\mathcal{E}_\mu\}_k$ of Section 12.8. Note that (14.1) is the familiar key equation (12.22), and (14.2) is an additional condition obtained by recognising that in the idealised case, $v = w_\mu$ throughout the speed-hold phase. The curves $y = \mathcal{E}(x)$, which we shall call $\mathcal{E}$-curves are obtained by solving a set of differential equations, starting from a point $x_2$ beyond the steep track section and working back to a point $x_1$ before the steep track section. The end point $x_2$ is adjusted until the above equations are satisfied. Figure 14-4 shows candidate speed profiles and $\mathcal{E}$-curves for a steep incline. The middle curve on each graph is optimal.

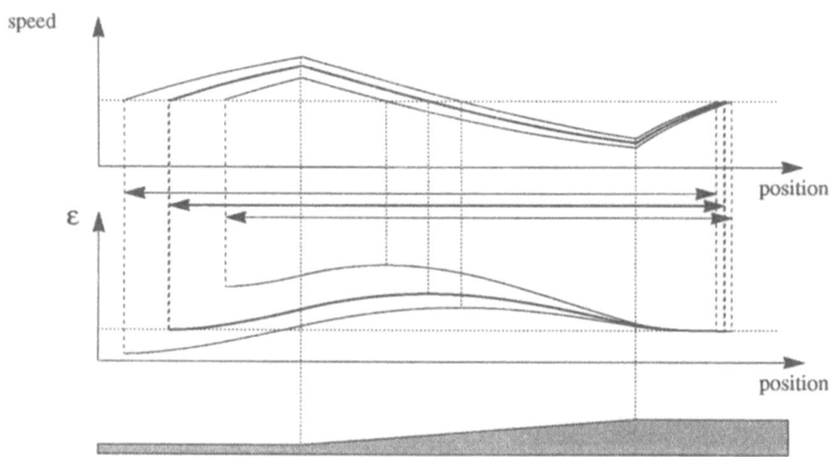


Figure 14-4: Candidate power intervals for a steep incline

Note that the $\mathcal{E}$-curves have turning points whenever the hold speed is reached.

To find $x_2$ we search in a prescribed interval, which is obtained as follows. The procedure is depicted in Figure 14-5. We wish to calculate the optimal power interval for the central incline. The desired hold speed is $v_h$. At the top of the incline, the speed of the train cannot exceed

$$v_M = \min \{ v_h, v_1(b) \},$$

where $v_1(x)$ is the speed profile if power is applied at the end of the previous steep gradient. Furthermore, the speed of the train at the top of the incline cannot be less than

$$v_m = \max \{ v_2(b), v_3(b) \},$$

where $v_2(x)$ is the speed profile if power is applied at the base of the incline and $v_3(x)$ is the speed profile that reaches the hold speed at the beginning of the next steep gradient. If $v_M(x)$ and $v_m(x)$ are the speed profiles with $v_M(b) = v_M$ and $v_m(b) = v_m$ then the shaded area in Figure 14-5 represents the search interval $[x_M, x_m]$ given by $v_M(x_M) = v_h$ and $v_m(x_m) = v_h$.

If an appropriate $x_2$ cannot be found by this procedure then the control interval overlaps with the control interval for an adjacent steep section, and an alternative procedure must be used to find the combined control intervals.

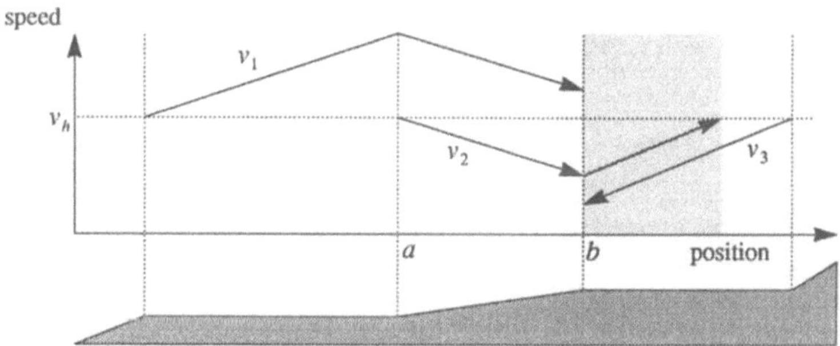


Figure 14-5: Search interval over a steep incline

The coast interval over a steep decline can be found in a similar way. Figure 14-6 shows the construction of the search interval for a steep decline. At the bottom of the decline, the speed of the train cannot exceed

$$v_M = \min \{ v_3(b), v_4(b) \}$$

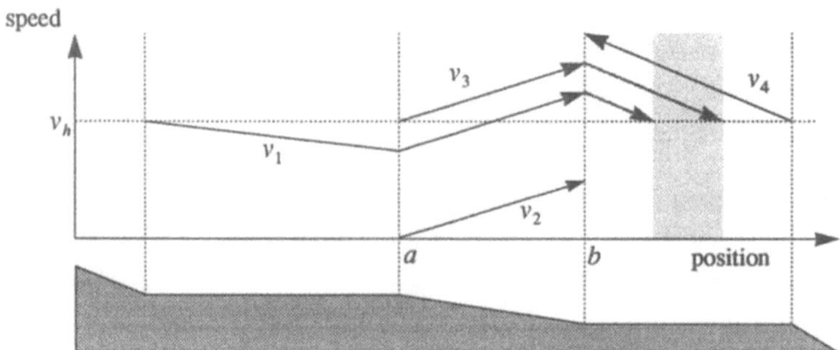


Figure 14-6: Search interval over a steep decline

and cannot be less than

$$v_m = \max \{ v_h , v_1(b), v_2(b) \} .$$

The interval $[x_M , x_m]$ can now be searched for an $\mathcal{E}$-curve with $x_2 \in [x_M , x_m]$ and $x_1 < a$ satisfying (14.1) and (14.2). Note, however, that the interval may be empty or may not contain an $\mathcal{E}$-curve satisfying these equations. If an appropriate $x_2$ cannot be found by this procedure then the control interval overlaps with the control interval for an adjacent steep section. Even if an $\mathcal{E}$-curve can be calculated for a steep track section by the above procedure, it is still possible that it will overlap the $\mathcal{E}$-curve from another section. In such cases it is necessary to find combined control intervals.

## 14.4 Overlapping Control Intervals

So far, we have seen that an optimal strategy may contain the following control phases:

- an initial power phase;
- a speed-hold phase, which may be interrupted by a power phase for each steep incline and by a coast phase for each steep decline;
- a coast phase; and
- a brake phase.

For a given hold speed, some of these control intervals may overlap. In particular:

- the initial power interval may overlap with the final coast interval, or even the final brake interval;
- the initial power interval may overlap with a steep control interval;

- control intervals for steep sections may overlap each other; and
- a control interval for a steep section may overlap with the final coast interval, or even the final brake interval.

In each of these cases we must calculate a modified control interval.

### 14.4.1 Overlapping Control Intervals for Steep Sections

Two steep track sections are said to be *close* if the corresponding control intervals overlap.

Figure 14-7 shows speed profiles and $\varepsilon$-curves for a double incline with overlapping control intervals. The dark curves correspond to a single optimal power interval. Notice that this single control interval sits inside the interval containing the two individual, overlapping control intervals. Double declines also exhibit this behaviour.

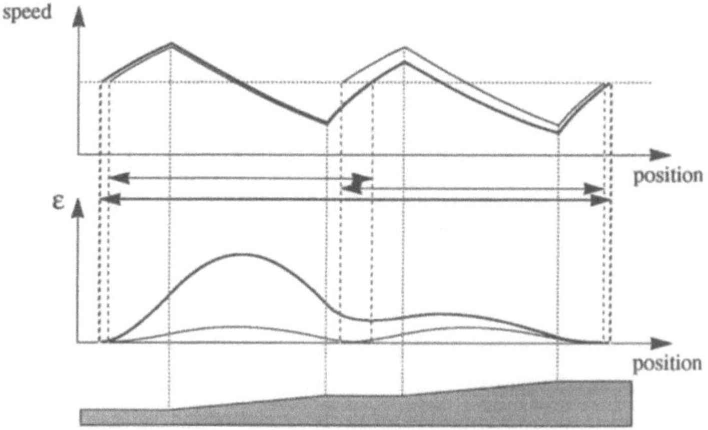



Figure 14-7: Power intervals for a double incline

Figure 14-8 shows speed profiles and $\varepsilon$-curves for a steep incline followed by a steep decline. The dark curves correspond to the optimal control interval. The vertical scale for the optimal $\varepsilon$-curve has been exaggerated for clarity. The control is changed from power to coast on the incline, when $\varepsilon = 0$. Notice that each of the modified control intervals sits inside the corresponding original control interval.

We make the following general observations:

- when the control intervals for two steep inclines or two steep declines overlap the two speed profiles do not intersect, and the modified control interval contains the two original control intervals;

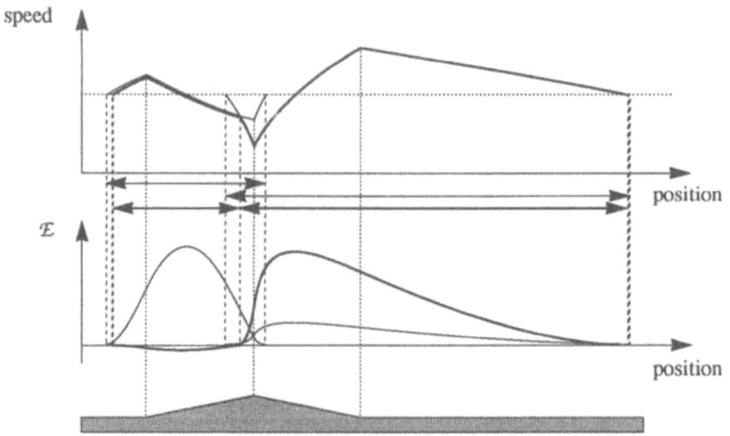

Figure 14-8: Control intervals for a steep hill

- when the control intervals for a steep incline and a steep decline overlap
  the two speed profiles intersect, and each modified control interval lies
  within the corresponding original control interval.

For more complicated combinations of steep sections, it is not necessarily possible
to make general observations about the optimal control intervals.

### 14.4.2 Other Overlaps

In practice, we suggest that an overlap between the initial power phase and a steep
control interval can be determined by first calculating an extended speed-hold profile
with $v(-\infty) = v(\infty) = v_h$, and then finding the intersection of this speed profile with
the speed profile for the initial power phase. The final coast and brake phases can be
determined in a similar manner.

## 14.5 Initial and Final Speeds

If the initial speed of the train is above the initial speed on the extended speed-hold
profile, the train should coast until the speed-hold profile is encountered; otherwise
power should be applied until the speed-hold profile is encountered.

If the final speed of the train is above the final speed on the extended speed-hold
profile, the train should power to the final speed. If the final speed of the train is

between the final speed on the extended speed-hold profile and the brake speed, the train should coast to the final speed. Otherwise the train should coast to the brake speed and then brake to the final speed.

## 14.6 Speed Limits

For a journey with speed limits, the strategy is essentially unchanged. Where the desired hold speed is above the speed limit, the speed of the train should be held at the limit. Full power should be used to change from a low speed to a higher speed. On the approach to a more restrictive speed limit, the train should coast so that it arrives at the new speed limit at the correct speed. If the new speed limit is below the brake speed, the train should coast to the brake speed and then brake so that it arrives at the new speed limit at the correct speed. In practice, the brake speed varies with the speed limits and with the gradient, and is difficult to determine.

## 14.7 A Practical Algorithm for Energy-Efficient Strategies

### 14.7.1 Overview

An energy-efficient strategy can be constructed as follows:

1.  Choose an initial hold speed, $v_h$.

2.  Construct an extended speed-hold profile, ignoring speed limits, as shown in Figure 14-9.

3.  Calculate a speed-limited speed-hold profile, as shown in Figure 14-10.

4.  Construct initial and final phases, as shown in Figure 14-11. The initial phase takes the train from its initial state onto the speed-limited speed-hold profile. The final phase takes the train from the speed-limited speed-hold profile to its desired final state.

5.  Calculate the journey time, $t$.

6.  If $t > T$ increase $v_h$ and return to step 2; if $t < T$ decrease $v_h$ and return to step 2.

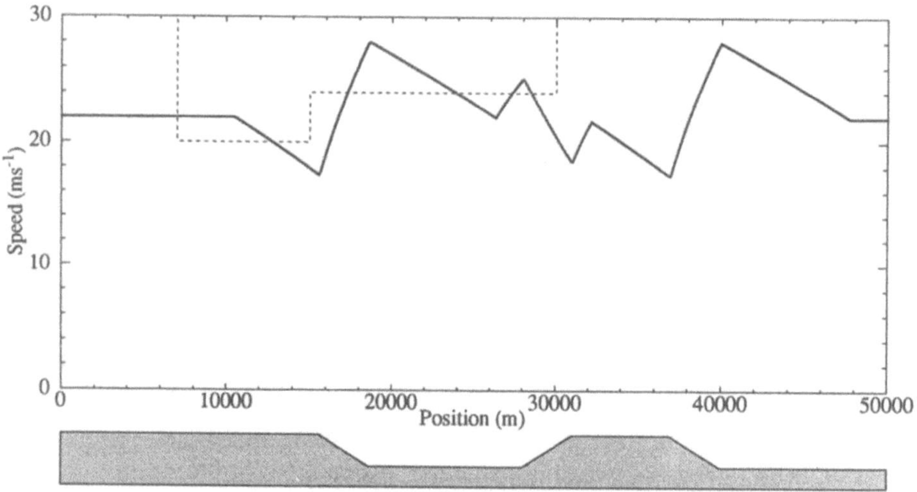

Figure 14-9: Extended speed-hold profile

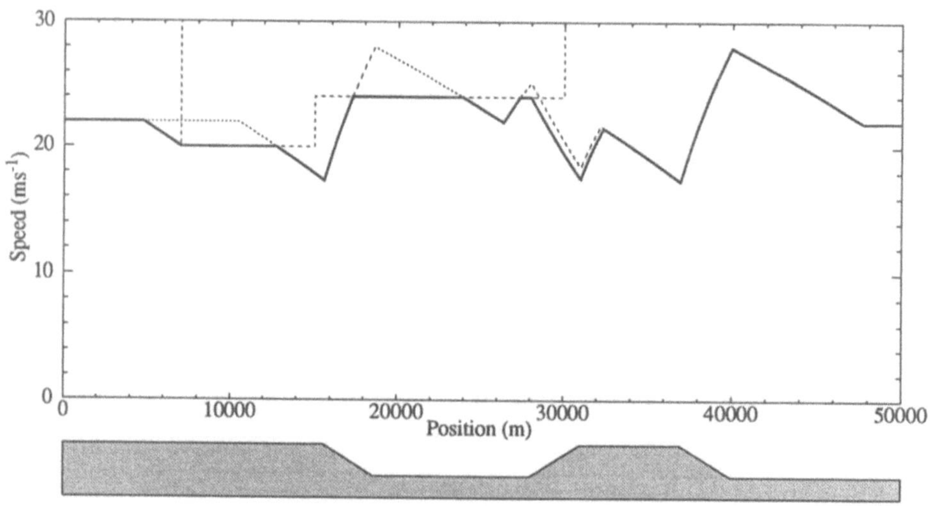

Figure 14-10: Speed-hold with speed limits

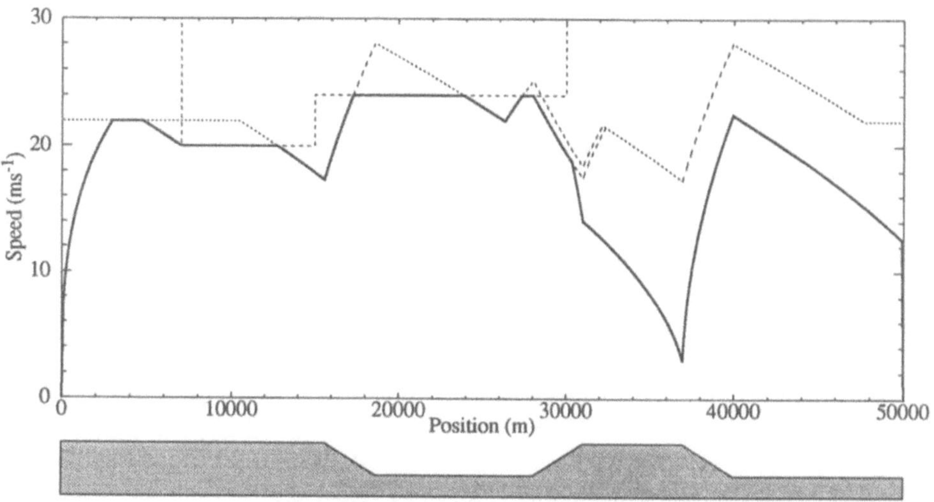


Figure 14-11: Initial and final phases

## 14.7.2 Speed-Holding

During a speed-hold phase, optimal control intervals can be calculated for each steep track interval in isolation. Overlaps can then be resolved as follows:

- For a sequence of close inclines or a sequence of close declines, the individual speed profiles do not intersect. A single optimal control interval must be calculated, as shown in Figure 14-7.

- For a close incline and decline, the individual speed profiles intersect. Instead of calculating a single optimal control interval, the individual speed profiles can be followed and the control changed at the intersection point.

## 14.7.3 Speed-Holding with Speed Limits

To incorporate speed limits into the speed-hold profile, the original profile from step 2 is processed in the direction of travel as follows:

- where a drop in speed limit causes the hold profile to exceed the speed limit, the new profile must coast (or coast and brake) into the speed limit;

- where the hold profile accelerates through a speed limit, the new profile must switch to speed-hold at the limit;
- if the new profile is below the speed limit and the original speed-hold profile, full power should be applied.

If the train does not have enough power to speed-hold, full power should be applied. If speed-holding is required on a steep decline, the brakes must be partially applied to maintain the desired speed.

A drop in speed limit may require the train to coast or to coast and brake. For flat track, the optimal brake speed is given by

$$v_b = v_h - \frac{\varphi(v_h)}{\varphi'(v_h)}.$$

The approximate algorithm uses this brake speed throughout the journey. When a drop in speed limit requires the train to reduce speed, the new profile is calculated as follows. Start at the lower speed and calculate a brake profile back until the brake speed is reached; then calculate a coast profile back from the brake point until the speed profile from step 2 is encountered. If the target speed limit is greater than the brake speed, the brake phase is omitted.

Note that a steep decline may cause the coast speed to drop to zero while calculating back to a speed profile. If this occurs, a higher brake speed is required for this particular *coast–brake* phase. The new brake speed can be calculated by calculating the coast profile forward from rest at the top of the steep decline and then brake when required.

### 14.7.4 Initial and Final Phases

The initial phase takes the train from the initial state to the desired speed-limited speed-hold profile. If the initial speed of the train is above the profile then the train should coast; if it is below the profile then maximum power should be applied.

The final phase takes the train from the desired profile to the required final state. If the final speed is above the profile then full power should be applied; if it is below the profile then the train must either coast, or coast and brake, to the final speed.

### 14.7.5  Calculating Journey Time

Once the complete speed profile has been calculated, the time for the journey is given by

$$t = \int_{0}^{X} \frac{1}{v(x)}\,dx.$$

### 14.7.6  Finding the Correct Hold Speed

For any given hold speed $v_h$, steps 2–5 of the algorithm give a corresponding journey time $t$. The correct hold speed can be found using a numerical root-finder.

# REFERENCES


[1]   Milroy IP. Aspects of Automatic Train Control. PhD thesis, Loughborough University, 1980.

[2]   Milroy IP. Minimum-Energy Control of Rail Vehicles. Proceedings of the Railway Engineering Conference, Sydney, Institution of Engineers Australia, 1981, pp 103–114.

[3]   Lee DH, Milroy IP, Tyler K. Application of Pontryagin's Maximum Principle to the Semi-automatic Control of Rail Vehicles. Proceedings of the Second Conference on Control Engineering, Newcastle, Institution of Engineers Australia, 1982, pp 233–236.

[4]   Benjamin BR, Long AM, Milroy IP, Payne RL, Pudney PJ. Control of Railway Vehicles for Energy Conservation and Improved Timekeeping. Proceedings of the Conference on Railway Engineering, Perth, Institution of Engineers Australia, 1987, pp 41–47.

[5]   Howlett PG. The Optimal Control of a Train. Study Leave Report, University of South Australia, 1984.

[6]   Howlett PG. An Optimal Strategy for the Control of a Train. Journal of the Australian Mathematical Society, Series B, 1990, vol 31, pp 454–471.

[7]   Asnis IA, Dmitruk AV, Osmolovskii NP. Solution of the Problem of the Energetically Optimal Control of the Motion of a Train by the Maximum Principle. USSR Computational Mathematics and Mathematical Physics, 1985, vol 25, no 6, pp 37–44.

[8]   Benjamin BR, Milroy IP, Pudney PJ. Energy-Efficient Operation of Long-Haul Trains. Proceedings of the Fourth International Heavy Haul Railway Conference, Brisbane, Institution of Engineers Australia, 1989, pp 369–372.

[9]   Mills RGJ, Perkins SE, Pudney PJ. Dynamic Rescheduling of Long-Haul Trains for Improved Timekeeping and Energy Conservation. Asia-Pacific Journal of Operational Research, 1991, vol 8, no 2, pp146–165.

[10]  Howlett PG, Milroy IP, Pudney PJ. Energy-Efficient Train Control. Control Engineering Practice, 1994, vol 2, no 2, pp 193–200.

[11]  Pudney P, Howlett, P. Optimal Driving Strategies for a Train Journey with Speed Limits. Journal of the Australian Mathematical Society, Series B, 1994, vol 36, pp 38–49.

[12] Howlett PG, Cheng J. Optimal Driving Strategies for a Train on a Track with Continuously Varying Gradient. Journal of the Australian Mathematical Society, Series B, (to appear).

[13] Howlett P, Pudney P, Benjamin B. Determination of Optimal Driving Strategies for the Control of a Train. In: Noye BJ, Benjamin BR, Colgan LH (eds) Computational Techniques and Applications: CTAC-91. Computational Mathematics Group, Division of Applied Mathematics, Australian Mathematical Society, 1992, pp 241–248.

[14] Cheng J, Howlett PG. Application of Critical Velocities to the Minimisation of Fuel Consumption in the Control of Trains. Automatica, 1992, vol 28, pp 165–169.

[15] Cheng J, Howlett PG. A Note on the Calculation of Optimal Strategies for the Minimization of Fuel Consumption in the Control of Trains. IEEE Transactions on Automatic Control, 1993, vol 38, no 11, pp 1730–1734.

[16] Davis WJ Jr. The Tractive Resistance of Electric Locomotives and Cars. General Electric Review, October 1926.

[17] Bazaraa MS, Shetty CM. Nonlinear Programming—Theory and Algorithms. John Wiley & Sons, New York, 1979.

[18] Craven BD. Mathematical Programming and Control Theory. Chapman and Hall, London, 1978. (Chapman and Hall Mathematics Series).

[19] Curtain RF, Pritchard AJ. Functional Analysis in Modern Applied Mathematics. Academic Press, London, 1977. (Mathematics in Science and Engineering, vol 132).

[20] Dunford N, Schwartz JT. Linear Operators, Parts I & II, Interscience, New York, 1958.

[21] Girsanov IV. Lectures on Mathematical Theory of Extremum Problems. Springer-Verlag, Berlin, 1972. (Lecture Notes in Economics and Mathematical Systems, vol 67).

[22] Hermes H, LaSalle JP. Functional Analysis and Time Optimal Control. Academic Press, New York, 1969. (Mathematics in Science and Engineering, vol 56).

[23] Lee EB, Markus L. Foundations of Optimal Control Theory. John Wiley & Sons, New York, 1967. (SIAM Series in Applied Mathematics).

[24] Luenberger DG. Optimization by Vector Space Methods. John Wiley & Sons, New York, 1969.

[25] Yosida K. Functional Analysis (5th Edition). Springer-Verlag, Berlin, 1978.

[26] Kreyszig E. Introductory Functional Analysis with Applications. Wiley, 1978.

[27] Milne RD. Applied Functional Analysis. Pitman, 1980. (Pitman Advanced Publishing Program).

[28] Naylor AW, Sell GR. Linear Operator Theory in Engineering and Science. Springer-Verlag, New York, 1982.

[29] Oden JT. Applied Functional Analysis. Prentice-Hall, 1979. (Civil Engineering and Engineering Mechanics Series).

[30] Hocking LM. Optimal Control—An Introduction to the Theory with Applications. Oxford University Press, Oxford, 1991. (Oxford Applied Mathematics and Computing Science Series).

[31] Howlett PG. Existence of an Optimal Strategy for the Control of a Train. School of Mathematics Report #3, University of South Australia, 1988.

[32] Kraft KH, Schnieder E. Optimale Trajektorien im spurgebundenen Schnelklverkehr. Regelungstechnik 29 Jahrgang Heft 4, 1981.

[33] English GW, Moynihan TW, Boumeester H. Reassessment of the Economics of a Coast Advice System for GO Transit. Report for the Canadian Institute of Guided Ground Transport, Queen's University at Kingston, Ontario, 1989.

[34] Howlett PG. Necessary Conditions on an Optimal Strategy for the Control of a Train. School of Mathematics Report #4, University of South Australia, 1988.

[35] Howlett PG. An Optimal Strategy for the Control of a Train. School of Mathematics Report #5, University of South Australia, 1988.

[36] Howlett PG. Optimal Strategies for the Control of a Train on a Track with Non-constant Gradient. University of South Australia School of Mathematics Report #6, 1992.

[37] Birkhoff G, Rota G. Ordinary Differential Equations, 3rd Edition. John Wiley & Sons Inc, 1978.

# INDEX

MIX
Papier aus verantwortungsvollen Quellen
Paper from responsible sources
FSC® C105338

If you have any concerns about our products,
you can contact us on
ProductSafety@springernature.com

In case Publisher is established outside the EU,
the EU authorized representative is:
Springer Nature Customer Service Center GmbH
Europaplatz 3, 69115 Heidelberg, Germany

Printed by Libri Plureos GmbH
in Hamburg, Germany

# Contents

# Acknowledgments

The authors are indebted to many individuals without whom this book could not have been written.

The following people kindly agreed to be interviewed for this project, generously sharing their time, their memories and their opinions: Brian Barnett, the late Peter Bredl, Alf Brogan, Neville Burns, John Cann, Jeff and Mare Carter, Greg Churchill, Hal Cogger, Marjorie Compton, Lynda Croxford, Dorothy Dale, Jon Dekkers, Bev Drake, David Doherty, Jim Edwards, John Edwards, Allan Fox, Jan Forbes, Arnold Gaunt, Tim Hawkes, Paul Horner, Grant Husband, Raija Krauss, Roy MacKay, Steve McEwan, David Millar, Greg Miles, Peter Mirtschin, Michael O'Brien, Mary Rayner, Wilma Pond, Trent Russell, the late Vincent Serventy, Rick Shine, Irwynne Symington, Brian Starkey, Pam Tinsley, Joyce Vereyt, Richard Wells, Wally Wickers, Ray Williams, Bob Withey and Dina de Wolff. In addition, we corresponded through telephone calls, letters and/or email with Terry Boylan, Gary Chivers, Jeanette Covacevich, David Farrimond, Rosemary Fleay-Thomson, Neville Goddard, Sue Hawkins, Lyall Naylor, Geoff Pickering, Meg Slater, Susi Sutherland, Karl Switak and David Williams.

A very special thank you to Lyn Abra, John Dwyer, Jack Green, Peter Krauss and John McLoughlin, who made themselves available to us through ongoing discussions, whether face to face or via email or letter, throughout the life of this project.

We also thank the officers at the following institutions for making available their collections of documents and other archived material: Ross Sadlier, Collections Manager, Herpetology, and the staff of the library, Australian Museum; Australian Reptile Park, Somersby; Australian Science and Technology Heritage Centre, and in particular, Dr Gavan McCarthy, Melbourne; Dr Ken Winkel, Director, Australian Venom Research Unit, University of Melbourne; Cairns City Library; City of Adelaide Archives; Donald Thomson Collection, National Museum of Victoria; Geoff Potter, Local Studies Librarian, Gosford City Library; Mitchell Library, Sydney; National Film and Sound Archive, Canberra; National Library of Australia, Canberra; Dr Dianne Bray, National Museum of Victoria; State Library of South Australia; State Library of Western Australia; Taronga Conservation Society, Australia.

Funding for this project was generously provided by the School of Humanities and Social Science and the Faculty of Business and Law, University of Newcastle; and the School of Tourism and Hospitality Management and Division of Research, Southern Cross University. For excellent research assistance, we thank Ms Jennifer Debenham, Dr Annona Pearse and Ms Lena Rodriguez. At UNSW Press, we thank Elspeth Menzies, publisher, Dr Heather Cam, managing editor, and Chantal Gibbs, project editor, for their ongoing advice, support, enthusiasm and encouragement, and Vanessa Mickan, our copy editor, who patiently worked through our manuscript with an eagle eye for detail and whose constructive refinements to the text have added clarity to the narrative and greatly enhanced the book's readability.

Kevin Markwell would like to thank his parents, Bert and Elaine, who encouraged his interests in wildlife, permitting him, as a boy, to occupy much of the backyard with reptile pits; colleagues and friends at the University of Newcastle and Southern Cross University for their support and interest in the project; and his partner, Steve Harrison, for putting up with ongoing reptilian intrusions into our life together.

Nancy Cushing would like to thank her family in Canada and Australia for moral and material support and encouragement, and all of the colleagues, conference organisers and journal editors who expressed interest in the project over its five-year gestation.

Finally, we would like to express our deep gratitude to John and Robyn Weigel and gratefully acknowledge the ongoing support and encouragement that they have given to us from the very beginning of this project in 2005. The Weigels graciously provided us with all manner of assistance whenever it was requested and without their enduring interest in this book it could not have been realised.

# A note on sources

Writing a history of the Australian Reptile Park has been a challenging task. Our research has taken us across the continent from Cairns and Darwin to Perth and Melbourne to conduct interviews with former and current staff of the park; collaborators and competitors; and friends, relatives and associates of Eric Worrell. We have spent many hours sifting through all manner of documents in archives and libraries, including newspaper coverage, park records and Worrell's own extensive writings. We have viewed television and film footage of and about the park, toured other wildlife parks and examined preserved snakes and live spiders.

We were generously assisted by many of the park's and Eric Worrell's employees and associates, who spent hours answering our questions. In some cases, an initial formal interview led to additional contacts by telephone, in person and through email. Their memories have been placed in the context of the documentary record of the park. We travelled to Melbourne, Canberra, Cairns, Gosford and Sydney to access newspapers, books, articles and papers held by a range of institutions including the National Library of Australia, the State Library of New South Wales, Gosford City

Library and the Commonwealth Serum Laboratory (CSL). We also accessed Worrell's own extensive collection of published and unpublished writings, works of other contemporary naturalists, studies of tourism and captive animal displays, histories of antivenom production, and unpublished journals and reports of scientists and artists. Both still photography and moving images were important to Worrell and the park and these were examined at the State Library of NSW and the Australian Film and Sound Archives. Although many of the park's operational files were destroyed in a fire in 2000, hundreds of them survive and they have been carefully read for detailed accounts of day-to-day operations and correspondence with other bodies, including the CSL and the NSW National Parks and Wildlife Service. We regret that we are unable to acknowledge each of these sources individually.

While we have immersed ourselves in the written and spoken word, we were not witnesses to any of the events that give life to the history of the park and we know Eric Worrell only vicariously, through the memories of others and through his own works. This detachment brings with it a level of objectivity that helped to guide our way as we tried to understand and make sense of all the individual stories that have shaped our narrative. We have drawn from many sources, but the interpretation of the information collected is our own.

Any history is, to a certain extent at least, a product of the authors' own critical imagination, and our interpretations of certain events may not necessarily be the same as those of others. There is also a multitude of other stories and pieces of information that we have not been able to access, stories that have been hidden or that remain untold. This book is our interpretation of the often disparate and sometimes contradictory information that we have been able to obtain, and its analysis is underpinned by our training and experience in the fields of cultural geography and history. We hope that it brings a balanced and respectful approach to the subject matter and that it communicates what we found to be a fascinating tale of the triumphs and frustrations of pursuing an unconventional dream.

That the Australian Reptile Park continues to flourish is a tribute to the pioneering vision of Eric Worrell as its founder and to the creativity and unwavering commitment of Robyn and John Weigel, the park's owners. But as both John and Robyn Weigel would freely admit, the park could not have thrived without the commitment of staff. There have been many dedicated staff members throughout the years, without whom the park wouldn't be what it is today, and indeed, over its lifetime the park has employed almost 200 animal keepers. They have all played their part in ensuring that the Australian Reptile Park continued to innovate, educate and entertain. We have featured some of their stories in this book. We acknowledge that there were many more whose contributions were significant.

# Introduction:
# The only good snake
# is a dead snake?

A young girl stares in quiet fascination as a small green python slides effortlessly out of its dry wafery skin to reveal its vivid, flawless body. People gather around a large case, mesmerised by a 3-metre-long king cobra as it methodically pokes its fist-sized head into nooks and crannies within its exhibit, searching for something to eat. The two snakes ignore the onlookers but the humans are transfixed by their sinuous grace. Since the Australian Reptile Park first opened its doors to the public, millions of visitors just like these have come to gaze at one of the largest collections of deadly snakes on public display.

Snakes seem to us almost other-worldly creatures. They are so unlike us. They have no limbs and slide on their bellies, gazing about with lidless eyes. They are silent except for their hiss, emitted through a mouth that can swallow prey many times greater in circumference and weight than the snake itself. These mouths often conceal fangs that can deliver a deadly toxin. Snakes hide under rocks or in foliage and often seem to appear from nowhere. When they feel under threat, most can strike with considerable speed, agility and accuracy. As a result, snakes both fascinate and repulse; and they have been stigmatised in Western cultures as far back as the story of the Garden of Eden and the snake's temptation of Eve.

A deep-seated anxiety about snakes was carried to Australia, stowed in the cultural baggage of British colonists, and passed down for generations. Some people believed snakes could turn themselves into hoops to roll faster than a man could run; suck the milk right out of a cow's udder; and swallow their young to protect them. It was thought that when a snake was dealt a lethal blow, if its body twitched, it would not die until after sundown. Snakes were supposed to have the power to hypnotise their prey into a trancelike state before covering them in slime in readiness for ingestion. All snakes, regardless of size or species, were assumed to be capable of a lethal bite and to be eagerly awaiting the opportunity to inflict one on any unwary man, woman or child. Snakes, cold blooded and scaly, were not animals with which most people could easily empathise. The fact that some of them were potential killers of humans and livestock made them outsider animals that had no value and deserved persecution.

Indeed, Australians prided themselves on their killing of snakes. They displayed their mangled, lifeless bodies along fences and in photographs; they held picnic days for group snake hunts in which hundreds of snakes were killed; and they read stories in which snakes were nasty, malevolent beings. Having grown up on lonely selectors' blocks with his mother Louisa managing as best she could, Henry Lawson drew a compelling picture of the solitary life of a colonial woman in *The Drover's Wife*, the air of desperation to her life made all the more desperate by her discovery of a black snake within the walls of her house. Her husband far away droving, without neighbours and fearful for her children, she remained vigilant all night, until the following morning when their loyal dog crushed the snake's head and broke its back. This snake remains a metaphor for the seeming hostility and danger of the Australian bush. In May Gibbs's children's classics, Mrs Snake is the wicked force seeking to hypnotise and terrorise the innocent gumnut babies. Steeped in tales of threatening snakes from childhood, Australians embraced the folk saying 'the only good snake is a dead snake'.

Captain Watkin Tench, an officer who arrived with the First Fleet in 1788, actually thought that the snakes common along rivers and sandstone cliffs were all nonvenomous.[1] Soon, the colonists were learning that the situation wasn't quite so straightforward. Convict William Noah wrote, in 1798, that the Sydney region had 'a Vast Number of Snakes of Various Colour & very Large & none Venimous [sic] But the Black Ones they are'.[2] By 1809, the colonists had gained more experience with Australia's snakes and fear was mounting. Naturalist and surgeon on the *Hindostan*, Joseph Arnold, reported that what frightened him most of all about the new land were the dangerous serpents, 'the bites of which are certain death'.[3]

Snakebite was not an especially common occurrence in the new colony, though, even as the first European settlers set about clearing and working the land. Fatalities occurred in perhaps less than 10 per cent of bites, but the lack of recorded deaths did not prevent a growing hysteria. Surveyor William Govett, who was working in the NSW Blue Mountains in the 1820s, viewed the black snake as an inveterate enemy and confessed that he wasted many hours on the government payroll searching the bush for snakes to kill. He thought it amusing to cut off the head of a black snake, wind the body around his arm and hold the head in his hand, waving his arm about to frighten people who thought he was holding a live snake. In 1858 in Victoria's Western District, at least one local newspaper carried weekly tallies of snakes sighted and killed.[4] Killing snakes using a stick, a rock, a gun or whip was an ordinary part of life for many boys in rural Australia. Organised snake hunts in which hundreds of tiger snakes were slaughtered were not uncommon even into the middle of last century, and no doubt still occur on a smaller scale today.

Author Alan Moorehead captured the prevalent attitude towards snakes:

> In the part of Australia where I grew up we used to come across snakes quite often when we were walking in the bush, and our fear and loathing of them was something more than the usual thing, mainly I suppose because very occasionally some of us really did get bitten … I never saw a snake – that furtive sliminess, that mad, hating eye – without a sudden instinctive constriction of the heart and after the first moment of panic was over we children had just one thing on our minds: 'Kill it. Do not let it get away.' And so we would grab a stick and in a spasm of furious terror we would beat at the hideous twisting thing until at last it lay inert in the dust.[5]

Into this world was born one boy who did not share the fear, Eric Worrell. He would grow up to be Australia's original snake man. His sustained and enthusiastic involvement in venom production would help save lives and he would do more than anyone before him to change Australian attitudes to snakes and other reptiles. Beneath the successes, he would face personal challenges and financial hardships that would bring him to the brink of ruin – yet the Australian Reptile Park stands to this day as his enduring legacy. This is his story, and it is the story of those who worked with him and who have expanded his vision to create today's Australian Reptile Park.

# PART ONE

# THE EARLY YEARS

1

# A childhood with snakes

Before retiring to bed in her Lilyfield, Sydney, home late one evening in the 1940s, a middle-aged woman walked the familiar steps down the damp brick path to the outside toilet. When she reached up to pull the chain, she fumbled a little in the darkness. Her hand sought cool metal, but instead it found the silky scales of a large diamond python. The woman screamed and immediately knew who the culprit was. 'Eric Worrell!' she shrieked, 'Get this snake out of my toilet!'

Worrell was a rarity. Like many children, he developed a love of animals from a very early age but he gravitated towards those that were less appreciated by others, the 'unpopular ones', as he would later write.[1] Worrell became interested in reptiles, particularly snakes, because he was puzzled by the fear people generally had of them, when they were so obviously more fearful of human beings – and with good reason.

He had empathy for the outsider but he was not one himself, for Worrell grew up in a home in which he was warmly nurtured and supported. Eric's parents lived in what was then one of Sydney's working-class western suburbs. His father, Charles Percival (Percy) Worrell, a carpenter, was born in 1892. He married 23-year-old Rita Mary Ann Rochester in 1923 in Granville. Their marriage was to last 59 years, until Percy's death in 1982 at the age of 90. Eric Frederick Arthur, their first child, was born on 27 October 1924. The Worrells were proud of their son, entering the curly-haired two-year-old in a beautiful baby competition in which he won first prize. Daughter Joyce followed three years later. This rather small family for the period was enlarged by foster children who were welcomed into the household.

The Worrells moved to Norfolk Street, Paddington, in Sydney's eastern suburbs, when Eric was still a young lad. This was not the exclusive Paddington that emerged later in the century, but a down-at-heels terrace-house suburb. There, Eric engaged in the usual money-making schemes of a Depression boyhood – collecting bottles and helping in a grocery store. He also did some things that weren't quite so common and reflected his budding interest in animals, such as going door-to-door selling raffle tickets for budgerigars he bought at Paddy's Markets.

Worrell's fascination with reptiles began with Sunday visits to La Perouse as a child. From the 1920s, the land along the shores of Botany Bay had been dotted with a series of shantytowns for Sydney's unemployed and homeless, who shared the area with Aboriginal people from the La Perouse mission. Among the hardship and poverty, an entertainment precinct developed on the La Perouse headland at the Loop, so called because it was the end of the tramline, where the trams circled round for their return trip to the city. A variety of spectacles came and went at the Loop, including performing dogs, camel rides and demonstrations of knife and tomahawk throwing. The longest-lived offering, the snake show, started there in 1897. Young Eric, curly headed and often well dressed in shorts and a blazer, stood fascinated beside the enclosure in which Snake Man, George Cann, demonstrated his mastery of dangerous snakes each Sunday at 1 pm. This snake show continued after Cann's death in 1965, performed by his sons, George Jnr, who died in 2001, and John. John, who was made a Member of the Order of Australia in 1992 for services to the community and the environment, officially retired from the pit in April 2010. George Cann was a snake showman of the old school. He had entertained, educated and thrilled his audiences for decades in travelling shows and at La Perouse, showing his acquired skill and quick reflexes as he grabbed and toyed with his snakes. His prowess with reptiles was recognised by Taronga Zoo, which appointed him to manage their reptile collection in 1939. Cann stayed with the zoo for 23 years, until his retirement in 1962, and he maintained his Sunday snake show until shortly before his death in 1965. Cann noticed Worrell's intense interest in snakes and slowly an apprenticeship of sorts developed between the two. His gift of a yellow-faced whip snake was treasured by Worrell, as were his grisly stories about the effects of snakebite. Worrell attributed his own caution when handling snakes to his early training with Cann.

A second mentor was George Longley. The two met in the late 1930s at a bushland exhibition run by the Rangers' League of NSW, where Longley displayed some of his extensive collection of lizards. Longley was a frequent contributor of short but insightful articles on aspects of lizards to the *Proceedings of the Royal Zoological Society of NSW* and had accumulated a wealth of knowledge about Australian lizards and their care in captivity, which he shared willingly with Worrell. Longley's more scientific approach to reptiles added a new dimension to Worrell's understanding of them. In his mid-teens, Worrell often travelled by tram to view Longley's collection of lizards at his home in Bronte. As an active member of the Zoological Society, Longley was able to extend Worrell's circle of acquaintances to include published naturalists, perhaps firing his desire to carve out a reputation as a naturalist himself one day. Later, when Longley died suddenly in July 1947, he left his collection in the care of the 22-year-old Worrell.

The raconteur Eric would become in later life was emerging even while he was still a youngster. At the age of ten, he created a menagerie in his family's tiny backyard, and exercised his considerable charm and persuasive power in promoting it. Passers-by were waylaid by Worrell and his mates and enthusiastically invited to view the collection of a bantam, two rabbits, guinea pigs, a white mouse and a tortoise named Phar Lap for a penny a visit. Sometimes the boys would set up the display on the nearby Trumper Oval, hopeful of pulling in a bigger crowd. All of the animals were housed in cages specially designed and built by Percy Worrell. Later, goldfish, a variety of common lizards in glass cases and a stray dingo were added to the menagerie.

As he grew and as his knowledge of snakes increased, Worrell moved from just watching snake shows to keeping them himself. He began his collecting expeditions around Trumper Oval, on Glenmore Road, a few blocks from his Paddington home. When he was old enough to range further afield, he found his first large snake, a diamond python, near bustling Central Railway Station, opposite the Hotel Sydney. Twelve-year-old Eric rescued the snake, which was the unhappy target of rock-hurling boys, and took it back to his home, where he persuaded his reluctant parents to allow him to keep it. Oscar, as it was called, made a few attempts at escape, and it is likely that this was the snake that found its way into the outhouse of Worrell's agitated neighbour a year or two later, when they had moved to Lilyfield. Oscar was eventually released into Centennial Park, a wild oasis that yielded snakes, lizards, frogs and fish to Eric and his schoolmates.

Worrell's ability to catch snakes in central Sydney is a reminder of the ongoing presence and resilience of native fauna even in Australia's largest city. Indeed, newspaper stories from the 1930s through to the 1960s regularly featured unfortunate snakes that had dared to show themselves in Sydney suburbs. A '6-foot' (1.8 m) tiger snake, seen watching its reflection in a dressing-table mirror in Brookvale in 1935, was subsequently battered to death with a scrub hook, a sickle-shaped metal blade attached to a long wooden handle, generally used for clearing low bushes. A 5-foot (1.5 m) brown snake accidentally thrown into an Earlwood tram by the wheels of a passing truck in 1948 was feverishly killed by the conductor, using the iron point hook that he had ready to change the points to direct the tram onto another line of rails.[2]

Worrell's father, Percy, a taxi driver, was a deep-thinking man of few words. His support for his son's projects was practical – driving him in his taxi to see George Cann at La Perouse and building backyard cages and pits for his growing collection of animals. His mother, Rita, shared Eric's love of animals, having kept her own pets as a child, and she welcomed his specialisation in reptiles. While Joyce, his sister, shied away from snakes, their mother had no fear of them, seeing them as objects of great interest. The Worrells continued to support and assist Eric with his reptile collection despite occasional escapes and the complaints of their neighbours. Worrell returned this family loyalty, looking after his sister, including her in his circle of friends when she was a teenager and paying bills for his parents as he started earning his own money.


Worrell enjoyed his years at Glenmore Road Public School in Paddington. Nature study, which was well established in the curricula of Australian primary schools by the 1930s, held a particular attraction for him as an art taught through drawing and painting as well as a science. He was dux when he completed studies at Glenmore Road in 1937, having his name inscribed on the school's honour board, to his parents' great joy.

Worrell continued his education at Sydney Boys High. The school had joined Sydney Girls High nine years earlier on the site vacated by the Moore Park Zoo in 1916. The animals had been moved to the Taronga Zoo site on the north shore of the harbour, but there were continuing reminders, such as the bear pits and the aquarium that was then used as a gymnasium.

The formality of secondary education in the period did not suit Worrell's self-directed style of learning and he missed nature study. By the time he was 14, he had left school and embarked on a series of jobs, while continuing to develop his skills and knowledge in drawing, photography and first aid in his spare time. His sister recalled that 'his brain was going all the time, you know, he was never still'. He excelled in several areas, and Percy and Rita at first expected their son to become a doctor, then they thought he would be a minister of religion. Whatever was to come, they had great expectations and were there to support Eric in whatever course he chose.

Worrell remembered his youth not as many people do, as an aimless series of experiences, but as a steady movement towards a specific goal. He recalled a particular day in Centennial Park when, as he threw cake crumbs to a colony of skinks, he formulated his life's ambition. He would later write:

> I wanted to live somewhere away from the city, on a large piece
> of land where I could keep all the strange reptiles I would bring
> back from my travels to all corners of Australia. I wanted to keep
> them in as near to natural conditions as possible so I could study
> their life histories, and so that people could look at them and
> learn that there was no reason to fear them.[3]

At age 14, having moved with his family to Cecily Street, Lilyfield, Worrell was wrestling with mates in a paddock when he fell on some broken glass and badly cut his ankle. A passing motorist took him to Balmain Hospital, where he was given surgery to rejoin the tendon.

After developing septicaemia, he was transferred to Carrington Hospital in rural Camden to convalesce. The grounds of the hospital ran down to the Nepean River, giving Worrell an opportunity to watch the antics of water dragons climbing in the trees and plunging into the water. Sitting by the river one day, he was a little taken aback to be approached by a fortune teller who offered to read his palm. Much later, he wrote that she had given accurate personal insights and had predicted many of the major events of his professional life: that he would be involved with the sciences; that he would travel extensively with this work; and even that he would be awarded an MBE for his achievements.

Over the next decade, Worrell worked in an array of jobs, developing a range of skills, experiencing far-flung parts of Australia and observing the wildlife wherever he went. In the late 1930s, Percy Worrell was a travelling salesman and while he was away from home for long periods Rita and young Eric would scour the job ads in *The Sydney Morning Herald*, looking for work. One of these ads required a couple to work in a café and farmlet in Eden, in southern New South Wales. Rita answered the ad, but for reasons best known to herself decided to pose as Eric's sister, lowering her age to 29 and raising Eric's to 16. By this stage, Eric was of average build with dark wavy hair and could pass as a 16-year-old, while Rita was a good-looking woman with dark hair, Mediterranean looks and a slim, petite figure. Rita cooked and waited on tables – and as Worrell later recalled, did what she could to avoid the unwanted advances of an amorous lion tamer passing through town with a circus. Worrell combined chores such as chopping wood and washing up at the café with work on the associated farmlet: chipping Noogoora burrs and picking fruit.

When Percy found himself unemployed, he secured relief work in the Blue Mountains, ultimately gaining a permanent position on the railways as a ganger. Eric became the billy boy, providing the workers with tea during their breaks. He and his father also joined other Depression-era fossickers looking for gold in nearby Oberon. These early work experiences forged in Worrell resilience, a love of independence and a hunger for risk-taking and adventure which were traits of his character throughout the next decades of his life.

The Worrell family heard the announcement by Prime Minister Robert Menzies of the declaration of the World War II as they gathered around their radio in Lilyfield in September 1939, a month before Eric's fifteenth birthday. His family, as did many others, remembered this time as the end of the Great Depression, when the demand for workers once again matched the supply. Percy was taken on by a railway workshop and Eric found a position in a plastics factory in Glebe. In the queue of men and women applying for work there, Worrell met another young hopeful, Wally Wickers, and the pair soon struck up a friendship. Wickers was a keen recreational shooter and invited Worrell to accompany him on hunting trips out to the Campbelltown area. This entailed travelling by train and hiking long distances, often carrying heavy equipment and specimens. On Friday afternoons, the two lads took the mail train to Glenfield and then made their way to a farm where they were allowed to sleep in the barn in exchange for catching rabbits, which they would later sell as food. Of course, they also found time for reptiles. They filled sugar bags with tortoises and sold them to Sydney pet shops. They caught blue-tongue lizards and goannas together, but Wickers left the catching of venomous snakes to Worrell.

At the age of 16, Worrell shifted into work more directly related to the war effort. Infrastructure construction projects took him to diverse locations where he could explore the local bush and collect reptiles. His first position was at Tocumwal on the New South Wales side of the Murray River, digging sewer lines for the new military airport. He discovered that the area was a rich source of tiger snakes. Worrell was later reassigned to Clermont in central Queensland, where he was set to work as an axeman on a bridge-building gang.

His gang was part of a larger team building an inland road to Charters Towers as an escape route for residents should the Japanese invade. This first visit to northern Australia was enormously exciting for Worrell as it brought him into contact with a whole new tropical reptile fauna that included black-headed pythons, king brown snakes, frilled lizards and an assortment of monitors. By the age of 17, he had also worked as a timber hand and truck driver and lived in the bush from Queensland to South Australia.

His experiences in Australia's far north during the war years would only deepen his commitment to his unusual vocation among the deadly and dangerous.

# 2

# Adventures in Australia's Top End

We threw ourselves flat as the bomb ploughed into the ground and rocked our heads with the terrific explosions. We clenched sticks between our teeth and I bit mine through. Debris spattered on our tin hats as trees crashed on all sides and blasts of wind hit our faces. Bombs were *pwooshing* and exploding through the bush around us and chopped leaves billowed into the air … We had our fingers jammed in our ears and the ground seemed to be rocking.

Eric Worrell[1]

In Darwin in 1943, 18-year-old Eric Worrell found a town that was, in his words, a 'shattered mess', the harbour littered with the wrecks of bombed ships.[2] Only days after the fall of Singapore on 14 February 1942, Japanese bombers had been sent on their first raid over Darwin. The attack was unexpected, the city's defences were inadequate and although women and children had been evacuated, the losses were heavy. While the wounded and remaining noncombatants were shipped out, fighting men and civilian war workers began to arrive in large numbers to re-establish infrastructure, build up military installations and repel the enemy. Worrell was one of them.

Working for the Civil Constructional Corps (CCC), his first job was the installation of 9.2-inch guns at East Point at the northern end of Fannie Bay, about 5 kilometres from central Darwin. He experienced some of the dozens of bombing raids, which continued until November 1943, and was struck by a piece of shrapnel during one bomb blast but was not seriously injured. It is not known precisely why Worrell did not enlist in the armed services when he turned 18 in 1942, but his sister's recollection is that he was rejected by the army because of permanent damage done when he cut his leg as a 14-year-old.

But even in the midst of a war, Worrell found that the Northern Territory was an ideal place to be. In this wildlife-rich environment, Worrell's expertise with reptiles soon came to the fore. He was regularly called upon to remove snakes that the other men encountered at camp. Happy to rescue these snakes, releasing some and keeping some for his own collection, he made a point of reassuring the men that many of the reptiles they encountered, such as pythons and tree snakes, were harmless. He recalled a fellow CCC worker called Snowie:

> When I first arrived at East Point 'Snowie' was literally 'scared stiff' of snakes. He could not even bear to see one. When I left East Point 'Snowie' was catching snakes for me. He gradually learned there is nothing to fear from them, and that when exercising commonsense precautions they could be handled with safety. Snakes and other reptiles are killed needlessly and only through ignorance.[3]

The mangroves and mudflats that lined the glistening aquamarine waters of Fannie Bay, the scrub at East Point and the fringing coral reef were a herpetologist's paradise, which Worrell scoured for reptiles and other animals in his free time. Wally Wickers, his friend from the Glebe plastics factory, was also stationed in Darwin and invited him on pig-shooting trips in the lush tropical wetlands and vast grasslands of what is now Kakadu National Park, where Worrell collected snakes and lizards. For a young Sydney man with a passion for reptiles, the Darwin posting provided access to an entirely new range of herpetofauna, little known both to him and to the scientific world.

Worrell began corresponding with Roy Kinghorn, curator of birds, reptiles and amphibians at the Australian Museum, who, in 1929, had published the authoritative text *Snakes of Australia*. From his letters to Kinghorn about the reptiles he was encountering in the north, it is clear that Worrell already had a sophisticated understanding of the methods used to classify snakes:

> (2) Regarding Boiga irregularis: In your book it is described
> as having the preoculars contacting the frontal. I have only
> struck one specimen … showing that condition from hundreds
> examined from live to decomposed condition from Katherine to
> Darwin. Although the other scalation (allowing for abnormalities)
> is identical with the Darwin specimens, the specimens in the
> central north are of a distinctive and different colouration. I did
> not examine the dentition of any Darwin specimens, but in these
> the anterior palatines are enlarged.[4]

Worrell kept notebooks containing field notes of specimens he collected or observed during his time in the north, sometimes accompanied by small pencil diagrams. From 1944 onward, some of these animals, including snakes, bats and marsupial mice, were sent as preserved specimens to the Australian Museum. He also sent live lizards such as sand goannas, spiny-tailed monitors and legless lizards to his friend George Longley, who wrote notes on them for publication in the *Proceedings of the Royal Zoological Society*, some of them illustrated with Worrell's photographs.


After serving their time in Darwin, Worrell and his workmates were sent to Sydney on recreation leave early in 1944. On their return journey, they were sent on cattle trucks to Terowie, South Australia; then on the Ghan train to Alice Springs in the Northern Territory; on semitrailers to Larrimah; and finally, on the narrow-gauge steam train nicknamed *Leaping Lena* to Katherine which, they found, to their surprise, was to be their new posting. Compared with the significant military establishment at Darwin, Katherine was a mere tin-shack settlement, strung out along the banks of the Katherine River. Worrell's task was to operate a blacksmith shop, assisted by Italian and German enemy aliens. The internees grew vegetables and kept poultry, providing Worrell with welcome relief from the tedious rations of bully beef, beans and powdered eggs supplied to CCC workers. It also meant he had the occasional rooster to feed to some of his larger pythons.

Soon after arriving in Katherine, Worrell found a like-minded companion in Roland Robinson, a 32-year-old writer who had joined the CCC in order to experience the Northern Territory. Born in Ireland, Robinson was part of the Jindyworobak group of poets, formed in 1938, who were creating a uniquely Australian literary style communicating the spirit of the land and of Aboriginal culture. Frustrated with the general lack of intellectual stimulation in his work for the CCC, Robinson gathered a small group of people with inquiring minds who met as a cultural group for discussions on Thursday evenings. Worrell was one of the guest speakers and his talk on snakes attracted such a large audience of CCC men and soldiers – even the local commanding officer of the army – that it had to be moved to the open-air cinema. There, among the fighting dogs and the red dust, he demonstrated his own style of snakemanship, showing his mastery over his animals and incorporating factual information about the snakes' biology and habits, and first aid advice. It earned Worrell widespread respect among the men and women stationed at Katherine.

The rich tropical environment of the Katherine River and its spectacular gorge country captured Worrell's interest and imagination. Worrell, Robinson and their mates made weekend treks driving along rough tracks to Katherine Gorge, where they camped under the shade of the rivergums, figs and paperbarks that grew along the banks of Paradise Creek. From a boat fashioned roughly from petrol drums, which they christened the *Croc Chaser*, they marvelled at the abundant wildlife both by day and by torchlight at night. Worrell discovered a species of water monitor that he thought was new to science and published a description of it later in 1956, naming it *Varanus bullewallah*,[5] unaware that the species had already been described by Louis Glauert of the Western Australian Museum as *Varanus mertensi* in 1951. Along the Katherine River in July 1945, Worrell collected a large-headed freshwater turtle. Initially identified by Roy Kinghorn as Krefft's turtle, *Emydura krefftii*, it was later confirmed as a new species, *Emydura worrelli*. Worrell called it the Boof-headed tortoise, although this common name did not stick.[6]

Because of his reputation as a reptile expert, Katherine residents brought all manner of animals to Worrell, either for identification, disposal or simply to add to his growing collection. He eventually had so many snakes they needed their own tent, which they shared with frilled lizards and an assortment of other lizards in makeshift cages. One friend, Big Jack, used a box of snakes under Worrell's bed as a safety deposit box for his gambling winnings, confident that no one but Worrell would open it. Worrell's most impressive captive was a 10-foot (3 m) olive python, which he named Percy, after his father.

Worrell soon joined the long tradition of naturalists writing about native Australian reptiles. HW Wheelwright, in his book *Bush Wanderings of a Naturalist, or Notes on the Fauna of Australia Felix* (1861), wrote that the Australian bush was infested with snakes and that a confrontation with the 'enemy, the black snake' was inevitable, yet that the widespread fear of snakes was simply unjustified. He had not had a 'single instance of a snake-bite ever [come] under my actual observation'. Frederick Aflalo, in his *A Sketch of the Natural History of Australia* (1896), suggested that contrary to popular belief, snakes did not lie in wait to ambush passing humans but were defensive in nature and disinclined to attack unless provoked. Snakebites occurred, said Aflalo, because of the 'carelessness of colonials' in the bush. He wrote, 'Unfortunately the snake hardly gets fair play, so many writers having a weakness for making "sensational copy" out of their snakes . . .'

Although reptiles have never had the number of public champions that birds and mammals have enjoyed, new authors did take up the challenge of educating the public about them in the twentieth century. One of the earliest of the home-grown Australian naturalists to write regularly on snakes and other reptiles was the prolific Charles Barrett. In *Australian Wild Life* (1943), Barrett tells the reader that although Australia is a 'snaky land', deaths from snakebite rarely occur. Nevertheless, he warns of the dangers of the north Queensland taipan, which is 'so aggressive that it will attack man on sight'; the tiger snake, the most dreaded of Australia's serpents after the death adder; and the common brown snake, which 'can never be trusted for a moment'. However, he writes, the red-bellied black snake 'as a rule is fairly good tempered and tries to get out of the way when disturbed'. His message is clear in the chapter 'Snakes and snake lore' in his book *Australian Wildlife Illustrated* (1950), where he states that 'most of our snakes are moderately shy' and that 'fatal cases of snake bite are never numerous enough in Australia to cause deep concern', with a mortality rate of about twelve people per year.

However, captions of photographs depicting the tiger snake as the 'dealer of death' and the common brown snake 'a swift destroyer' perhaps work against his argument. Barrett's *Reptiles of Australia* was published in 1950 and this was the first book that focused specifically on Australia's reptiles, although it omitted the turtles and tortoises.

The naturalist David Fleay published some articles on reptiles in the *Proceedings of the Royal Zoological Society of NSW*, such as 'Black Snakes in Combat' (1936–37). Fleay also wrote numerous articles for newspapers and for the Australian National Travel Association's *Walkabout* magazine in the 1940s and 1950s. Fleay's writing was always rich in detail and based on his close observations of reptiles, either in the wild or in captivity. After publishing a newspaper nature column for many years, he reissued this material in a series of edited books. His treatment of reptiles was always respectful and often very sympathetic as he argued strongly for their place in the ecosystem.

*Walkabout* magazine was also a vehicle for Melbourne-based zoologist and anthropologist Donald Thomson, who wrote a number of articles on snakes. His approach was scientific, providing factual information about snakes and their habits, as well as helpful advice on correct identification. But it would not be until later on in the century – thanks to the work of Worrell and other naturalists, such as Vincent Serventy and Harry Butler – that articles and books were published that sought to change people's perceptions of, and attitudes towards, reptiles.

Worrell was a compulsive writer as a young man and had already been pencilling drafts of articles while stationed in Darwin, and a children's book titled *In the Land of the Lizards*. This was a whimsical story about a boy named Irwin, his gecko companion and their quest among the billabongs of the Top End to meet the King of the Lizards. Hundreds of pages of early drafts of magazine articles remain in the archives at the Australian Reptile Park, written in Worrell's distinctive cursive script on anything he could get his hands on – scraps

others allowed us to within a foot or so, then dived as we
grabbed, reappearing perhaps on the other side of the boat, or in
equally annoying positions, when I am sure they grinned wide
crocodile grins as they observed our clumsy efforts.[11]

So he invented the 'doover', a long-handled mechanism with padded sprung fingers, which would gently close around a young croc's body, holding it firmly until it could be caught.

Worrell's entertaining writing style combined with the exotic nature of his subjects ensured that his articles were eagerly accepted by the copy-hungry editors of nature and travel magazines over the next two decades. From the frequency with which his articles were published, it seems that readers also enjoyed them. That he could accompany his writing with dramatic photographs of rarely seen animals and landscapes made them even more appealing.

In January 1945, Worrell was presented with two Johnstone's crocodiles, which had been caught in the Katherine River. These crocodiles were only babies, about 30 centimetres long, but still quite snappy and pugnacious. Keeping them in captivity for the first time was not easy. 'Turning *Crocodylus johnsoni* into their tank without receiving nips from their hissing, snapping jaws almost developed into a problem within itself, but eventually the feat was accomplished without casualties on either side,' Worrell wrote.[12] He held them in an iron tank filled with water kept to 32 degrees Celsius, with a base of river sand and fine gravel and a large rock for basking and sheltering under. In the first of his published scientific journal articles, he reported on how they held their bodies almost vertical with just their nostrils sitting above the waterline while resting in the tank and used their tails only when swimming rapidly. Worrell found that they preferred to eat live tadpoles, frogs and shrimp, rather than the bits of meat and poultry that were also offered to them.

At the end of the war, Worrell travelled to Sydney – but not with the usual postwar dream of settling down to a suburban existence. After catching up with family and friends, he planned his return to the Territory. Worrell had the idea of writing a comprehensive book about Australian reptiles and wanted to conduct extensive field-work, financing his research by collecting animals for southern zoos and museums and writing articles. It would take him almost two decades to complete, being published in 1963 as *Reptiles of Australia*. Roland Robinson, at a loose end and, like Worrell, infatuated with the Northern Territory, agreed to accompany him.

Having arranged for his parents and friends to look after the reptile collection he had been accumulating since boyhood and had added to during his time in the Territory, Worrell prepared for a long journey. Setting aside initial plans to sail to Darwin, he and Robinson bought a second-hand panel van, which they christened the Bug Buggy. They set out in August 1946 to travel the 4000 or so kilometres between Sydney and the Territory, rarely getting above 50 kilometres per hour. Robinson had to drive them out of Sydney. Though Worrell had been driving for years, including at work, he did not yet have a driver's licence. He took the wheel once they reached country roads. In Brisbane, they sent the panel van on by train while they boarded another train, to Townsville. Collecting the van at Mount Isa, they drove to the Stuart Highway in the Northern Territory and headed north. The sweltering days on the road were punctuated by stops to catch snakes and lizards and to make run-ning repairs to the Bug Buggy.

Mataranka homestead on the sluggish Waterhouse River, 400 kilometres southwest of Darwin and 100 kilometres south of Kath-erine, was their base of operations for the next few months. Jeannie Gunn's autobiographical novel *We of the Never Never* (1908) told the story of life at a neighbouring property, Elsey Station. In Gunn's day, this was wild and isolated cattle country where people from widely differing backgrounds battled the elements and one another to earn a living. During World War II, it drew a new type of visitor, attracted by the soothing hot springs that rose near the Mataranka homestead. A serviceman, Herbert Victor (Vic) Smith, who had helped to enlarge the hot springs when stationed there during the war, secured a sublease on Mataranka Station in 1946. He was just beginning to build it up as a tourist resort based around the hot springs and the local scenery and wildlife when Worrell and Robinson happened along. Smith allowed the pair to set up their army tent by a billabong about a kilometre or so downstream from the homestead. Shaded by giant figs and Leichhardt trees, the campsite provided the men with a comfortable base for their collecting of fish and reptiles.[13] There, living among sun-blasted buffalo hunters, southern tourists and Aboriginal people, as well as a large colony of quarrelsome and pungent little red flying foxes, the men could pursue their interests, assisted by a letter of introduction from Arthur Walkom, the director of the Australian Museum.

Worrell and Robinson set nets for fish and turtles and caught lizards and snakes by hand. Archerfish, eel-tailed catfish, glassfish and rainbow fish were netted and held in ponds. One small fish Worrell collected was new to science and subsequently named *Craterocephalus worrelli*[14] in 1948 by Gilbert Whitley, curator of ichthyology at the Australian Museum. Not all of their collecting went smoothly, though. The men had been offered the substantial price of £5 each for live archerfish by a southern aquarium dealer. Although the fish were readily netted, they did not survive the long flight south. Unable to recoup their freight costs, Worrell and Robinson were under financial pressure for the rest of their stay.

They hunted young Johnstone's crocodiles at night using a dugout canoe or a 'boat' made of two truck engine covers that Worrell had welded together. A spotlight, powered by a car battery, made the eyes of the crocodiles glow orange-red, allowing Worrell to pluck the crocs from the water and tuck them into a sugar bag, while Robinson silently paddled.

Worrell tried to learn as much as he could during his stay, including the traditional knowledge of the Aboriginal people who lived and worked on the property, most of whom belonged to what was then called the Djauan language group and is now known as the Jawoyn nation. In order to earn their confidence, Worrell and Robinson lived as the Djauan people did, eating bush food such as crocodile eggs, waterlily bulbs and goanna, and going barefoot. Worrell eventually developed badly infected feet and most of his toenails dropped off. He learned the Djauan names for local fauna by meticulously writing them out on scrap pieces of paper and in notebooks. He listened to Djauan stories, which he noted went further than 'merely explaining the existence of things. Some songs were immense geography lessons describing areas hundreds of miles distant with the localities of water-holes and descriptions of game.'[15] Worrell took on the name Karliboodi, meaning 'black snake' in the Djauan language, as a pseudonym that he sometimes used for his magazine articles.

Living and working with Aboriginal people gave Worrell an insight into and a respect for their culture that few other southerners of the period had. He came to know a number of Aboriginal men and women well, observing their often uneasy negotiation between traditional and contemporary lifestyles, the callous way in which they were treated by the authorities and the subterfuges necessary for them to keep some control over their lives. He got to know Clara, whose elderly husband, Jack, was thought by a Department of Native Affairs officer to have developed leprosy and had been ordered to the leper island in Darwin Harbour. The couple disappeared before he could be taken. Clara returned a short time later, gashing her head in sorrow and declaring that Jack had died in the bush. Worrell noticed Clara regularly spending time away from Mataranka and drew out her story. Jack, blind but still very much alive, had been brought back to the camp so he could live out his final years with his wife and his people. They hid him under a blanket whenever strangers approached. Eventually, a cattleman revealed Jack's presence to the local policeman and he was taken to the leper colony. By sharing their experiences and their stories, Worrell felt at times that 'the barriers that are forced between the black and white men seemed to lift.'[16] Robinson made a career of publishing Aboriginal traditional stories, poetry on Aboriginal themes and books about the lives of Aboriginal Australians.

The traditional beliefs Aboriginal Australians held about snakes were very different to the phobic outlook of Western culture. For many Aboriginal cultures of northern Australia, the snake, in the form of the Rainbow Serpent, was the origin of life and shaper of the landscape, in contrast with its role in Christianity of separating man from God in the story of creation. Some groups totemised and identified with certain snakes, leading to the recording of snake dreamings in paintings and songs. The Walpiri people of Central Australia were mindful of the rain-snake who, if angered, would fail to cooperate in rain-making rituals. As well as having a spiritual role, snakes were good eating. The remains of meals of snakes have been found in the Devil's Lair cave in south-west Tasmania, dated at 30 000 years BP. Other reptiles – crocodiles, turtles and lizards – were considered totemic by various tribes and clans and were afforded some protection while also being enjoyed as a foodstuff from egg to adult.

In exchange for camping and collecting rights, Worrell and Robinson were expected to contribute to Vic Smith's efforts to establish Mataranka Station as a tourist destination. Even with the natural attractions of warm mineral springs, tropical vegetation and abundant wildlife, Smith knew he had to enliven the tourists' experience. He persuaded Worrell and Robinson to become the resident naturalists, showing tourists the specimens they were collecting, talking to them about the local ecology and Indigenous people and taking them on guided excursions.

Robinson had little tolerance for southern tourists and quickly became annoyed with their never-ending questions, lack of understanding of the natural world and negative assumptions about Aboriginal people and their culture. In contrast, Worrell took readily to the role of interpreter and tourist host. But he could also be disappointed by the outlook of the tourists, as expressed in this excerpt of a poem that he handwrote on a torn-out page of an exercise book in 1949, during a later stay at Mataranka homestead.

**Rest in Pieces**
Here lies the body of a poor quiet snake
Murdered for warped conversation's sake
Who lived where the nativewood and coolabah grows
And the fork-tailed kite-hawks fly with the crows
But one fatal day as he crossed the road
Came a Pioneer Coach with a tourist load
Then with sticks and stones, in sadistic glee,
They battered the snake they could not let be …

Mourned by Eric Worrell

He mourned but did not despair, applying a mix of the Snake Man's showmanship and his own scientific knowledge to attempt to educate and change attitudes. He had his own ways of putting his point across. One wealthy tourist insisted on obtaining a complete crocodile skin, including the head, legs and claws. Desperate for cash, Robinson and Worrell agreed to shoot one for him. The pair set out in a dugout canoe at night and encountered several freshwater crocs, which they tried unsuccessfully to shoot. As they were paddling back towards shore, they met up with a large freshie that had apparently been stunned earlier by one of Worrell's bullets. They managed to get the unconscious croc to shore and while Worrell straddled it and wrapped his hands around its head, Robinson ran back for more ammunition to finish the unlucky animal off. When he told the tourists relaxing back at the homestead that they had managed to catch a croc, the tourists followed him back down to the river, where Worrell was waiting.

> When I reached Eric, the croc was showing signs of recovering. I jammed a bullet in our single shot, but could see that Eric was going to put on a show. 'Oh! Oh!' they exclaimed. 'Is it still alive?' Then as the croc started to come to and thrash its tail … Eric 'wrestled' the monster and some of the women began to scream. When I reckoned they had their share of the spectacle, I put a bullet in its skull behind its eyes. The ladies, and the men, had had their thrill. No one could say that Eric and I did not entertain them.[17]

After this dramatic episode, a number of the men in the party badgered Worrell into taking them on another crocodile hunt. While Robinson refused to have anything to do with it – stating 'We were naturalists, not safari-conductors'[18] – Worrell agreed to take two out in a canoe. However, the canoe capsized just as the men were about to take a shot at a crocodile, and they all ended up in the water. Reading between the lines of Robinson's account, it seems that Worrell himself had tipped the boat over with the intention of terrifying his passengers. After that incident, 'no one pestered us to take them out croc shooting again'.[19]

Conscious of the power a healthy ecology has to attract tourists, Smith arranged to have Mataranka homestead formally gazetted as a flora and fauna sanctuary in February 1947. On Worrell's advice, harmless reptiles were protected from all but traditional Aboriginal hunting. Half a century before the term 'ecotourism' was coined, Worrell published an article about the sanctuary, discussing potential links between nature conservation and wildlife tourism.[20]

While at Mataranka, Worrell periodically travelled north to Darwin to meet up with mates Jim Edwards, Leo Hickey and George Haritos, who played an important role in the development of commercial crocodile shooting, being among the first to hunt crocodiles by night-spotting using torches. Worrell had met them at the end of the war, when they were running a coastal transport service using Edwards's cutter, the *Australia*. They made several trips along the Wildman River, northeast of Darwin, to hunt crocodiles, transport salt and other supplies to buffalo hunters and take the hunters' stiff, dried buffalo skins back to Darwin. Worrell recalled that on the Wildman River in 1945, he saw saltwater crocodiles (*Crocodylus porosus*) in their hundreds – more than he had seen anywhere else.

These expeditions with Edwards and his crew were tough going. During the day, the crocodiles were skinned and their hides treated with salt. There was little time for sleep. Sometimes the men would fail to beat the tide and the boat would be marooned overnight on a mudflat, forcing them to sleep among the putrefying bodies of the night's catch. The cold of the night would 'bite right through their bones' while clouds of mosquitoes and sand flies added to their misery.[21] Worrell used these trips to gain an understanding of the habits of the saltwater crocodiles and the impacts that hunting was having on their populations. He examined the smelly stomach contents of several hundred of the crocodiles killed by the hunters. He had X-rays taken of young crocodiles to show the pebbles that they ingested, either for reducing buoyancy or assisting with the digestion of food. He built up considerable knowledge of saltwater crocodiles in the wild and, at that time, probably possessed the most extensive field knowledge of them of any Australian naturalist. Whereas the various writings on crocodiles by naturalist Charles Barrett remained at a fairly general level, Worrell was able to provide detailed information about their diet and reproductive habits in *Song of the Snake* (1958) and *Reptiles of Australia* (1963).

After several months at Mataranka, Worrell and Robinson

returned to New South Wales. They were paid to drive back in war-surplus vehicles and deliver them to their new owner. This enabled Worrell to bring home four drums containing young crocodiles and pythons.

Worrell's next trip – to Darwin, in 1947 – was accomplished in greater comfort, by air. Robinson also returned to the north, pursuing his own interests in Aboriginal storytelling in other parts of the Territory as he prepared his first book on the topic, *Legend and Dreaming*, published in 1952. Worrell again spent some time at Mataranka and also took advantage of other opportunities that came his way as he collected specimens for the Australian Museum and other institutions. He accompanied Jim Edwards on his coastal transport barge to the mission stations and other Tiwi settlements on Melville and Bathurst islands, 80 kilometres northeast of Darwin. One of the missionaries he encountered, Brother Pye, suggested that a book on Australian's dangerous snakes would be useful for those engaged in mission work. This motivated Worrell to begin to research what would become his first book, *Dangerous Snakes of Australia* (1952), a catalogue that enabled ordinary readers to easily identify venomous snakes.[22]

After eight months in the Territory, Worrell returned to Sydney for Christmas, bringing with him 11 snakes and 30 lizards. The *Sun* newspaper reported his return, claiming: 'Everybody's got the cold shivers at the eastern end of Cecily Street, Lilyfield, because young Eric Worrell is back from Arnhem Land with some "stock".'[23] He was gaining quite a reputation as a reptile man. His renown grew even stronger in February 1948 when a small crocodile went missing from the University of Sydney's Zoology Department. After the police, university staff and students spent a day searching pools and drains, Worrell offered his services. Using his knowledge of the habitat preferred by crocodiles, and probably with a good dollop of luck, he managed to track the missing youngster to a patch of long grass within half an hour. The media were enthusiastic about his skills, referring to him as a professional crocodile hunter and reptile catcher from the Northern Territory who was unfazed by the task at hand. Worrell said the job was child's play. 'I'd sooner handle crocs than snakes any day,' he said. 'You know where you are with crocs.'[24]

3

# An aquarium at the seaside

When Tony Ormsby, a Sydney lawyer and amateur herpetologist, invited 22-year-old Worrell to a picnic in Woy Woy, on the NSW Central Coast, in 1947, it turned out to be an important day in his life, because there Worrell met the woman who would become his first wife. Carol Hawkins was one year his junior, living in the beachside Sydney suburb of Coogee and working as a shop assistant. She had a connection with the Woy Woy area as her parents ran a real estate agency at nearby Ocean Beach, now better known as Umina. Carol was a strikingly attractive young woman, tall and slender with stylishly coiffed hair and a stunning smile. When the picnic inevitably turned into a search for snakes, Carol had no qualms about joining in. Without her openness to snakes, a relationship may never have blossomed, for anyone who became close to Worrell had to get accustomed to seeing snakes everywhere: in his coat pockets, his shoes and even on his bed. The photogenic Carol began to appear holding reptiles in the photographs Worrell took to illustrate his articles. The *Sunday Sun* ran a story titled 'Snake Woman Finds a Mate,' in which Carol was described somewhat imaginatively as the 'snake woman of Woy Woy' who had found her 'soul mate', the snake man of Lilyfield. 'Eric and I were very lucky to find each other,' she said, 'There aren't many women who would care to marry a snake-catcher.'[1]

Under sunny skies, the couple married in July 1948 in the Presbyterian Church at Ocean Beach, only a few kilometres to the south of Woy Woy, where they met. With Eric's slightly greying dark wavy hair and chiselled features (he had not yet grown his distinctive beard), and Carol's movie-star smile, they made a handsome couple. Their partnership was both personal and professional. As newlyweds they worked together collecting diamond pythons to be sold to collectors and zoos overseas. Carol would sit at a sewing machine securing snakes into canvas bags for shipment. Eric and his father would then pack the bagged snakes into wooden boxes.

With a new wife and a growing collection of reptiles, Worrell left the family home in Lilyfield and relocated to Ocean Beach. In May 1949, he bought an average-sized block of land. Most other people would have built an ordinary family home on it, but Worrell had rather different plans. On this patch of land by the sea, he began to realise his childhood dream of opening a facility where he could conduct research and exhibit reptiles.

Worrell knew that reptiles had a tremendous drawing power. He had seen the crowds of day-trippers standing in awe watching George Cann demonstrate mastery over snakes and other reptiles on Sunday afternoons at his La Perouse snake pit. He would also have known about Adelaide Snake Park, the first known privately owned reptile park, opened in 1927 and still operating when Worrell opened his facility. There were more than 200 snakes on display when Adelaide Snake Park opened, including 'pythons, diamond snakes, tiger, the tiger copper headed with its burnished skin, the black, the black diamond, whip, carpet, brown and many more'.[2] A snake expert from South Africa, CJ French, accepted the position of curator at the park, arriving in January 1927 and bringing with him a collection that included puff adders, cobras and rinkhal cobras, as well as a couple of tiger snakes he had managed to catch while the ship was docked in Albany. An ill-tempered black tiger snake fatally bit French on the arm just a few months into the job. Worrell believed that it was French's death that triggered the development of a tiger snake antivenom.

Despite the untimely death of its curator, the Adelaide Snake Park proved successful. Visitors were entertained by snake-handling demonstrations and at four o'clock every afternoon could watch as 'the giant python [was] offered its substantial meal'.[3] Some local residents objected to sharing their refined North Adelaide address with several hundred dangerous snakes, which not only posed a menace should they escape but were also thought to 'foul the air with their stench'.[4] Likewise, there was some local opposition to Worrell establishing his facility. Some residents, not surprisingly, were less than happy about sharing the neighbourhood with a pit full of tiger snakes and a writhing assortment of other potential killers in the form of browns, blacks, death adders and, eventually, taipans.

Still, his building application was approved by the shire council – after Worrell had obtained a permit from the Department of Building Materials, a requirement in this period of postwar shortages of building materials. Construction began, and the Ocean Beach Aquarium opened to the public in 1950. We don't know whether Worrell's definition of the word 'aquarium' was a little broader than the Gosford Shire's aldermen realised – a hybrid reptile house and showcase for fish – but they no doubt saw how it would benefit the growing tourism economy.

Worrell had a keen interest in fish, seeing them as a more relaxing diversion from the main game of working with venomous snakes, and the aquarium provided him with an opportunity to exhibit some of the native species that he believed should find their way into the fish tanks of average Australians. As with his work to dispel myths about snakes, by advocating the keeping of native Australian fish he was championing a somewhat unfashionable cause. It would be a long time before keeping native species became popular among aquarists.

The aquarium building had a distinctively modern appearance.[5] A series of nine round windows suggesting portholes created a unique maritime look in a streetscape otherwise typical of an Australian beach-side village in the 1950s. A series of tanks was set into the walls of the interior in gallery style, modelled on the tanks at Taronga Zoo's aquarium. Visitors moved from exhibit to exhibit, gaining a close look at venomous snakes such as death adders and tropical species including king brown snakes; scrub and water pythons and the local diamond python; freshwater turtles; young crocodiles; frilled lizards; and smaller monitors and snakes. From 1954, taipans were periodically exhibited. Fish ranged from species from northern Australia such as red and silver scats, rainbow fish, catfish, glassfish, batfish and archerfish to local species such as bream, mullet, blackfish, anglerfish, seahorses, eels and crayfish. Then the visitors would go outside to view the temperate zone reptiles, which could be kept in open pits.

The snake pit housed large numbers of tiger snakes along with red-bellied black snakes, brown snakes and copperheads. Another outdoor pit housed a variety of local lizard species. In what was essentially an ordinary backyard edged by a picket fence set on a concrete wall, visitors gathered on Sunday afternoons in summer to watch Worrell milk tiger snakes of their venom for use in the production of antivenoms.

Local residents contributed to Worrell's collection by bringing in any reptiles they found or asking him to remove snakes from their properties. Worrell's national standing as a reptile expert was growing as more and more people read his *Dangerous Snakes of*

*Australia* or his column 'Don't You Believe It', which ran for a number of years in *Australian Country* magazine, in which he dealt with assorted myths and untruths about Australian wildlife. As a result, he received specimens – most often dead, but sometimes alive – from people throughout Australia. People sent their contributions by train and Worrell would collect them from Woy Woy train station or have them sent by taxi to the aquarium.

Just as Longley and Cann had influenced him, through the aquarium and his writings Worrell was now becoming influential in the world of natural history. A young Harold (Hal) Cogger, who would go on to become one of Australia's leading herpetologists and the deputy director of the Australian Museum, found the aquarium a fascinating place and visited from time to time. He and Worrell found a lot to talk about, as Cogger, too, had a strong interest in fish as well as reptiles.

Yet being in Ocean Beach made keeping up contacts with the herpetological community in Sydney somewhat more difficult for Worrell. George Cann's snake show at La Perouse was still an important meeting point for those with a passion for herpetology. Another focus was the Australian Museum, where Roy Kinghorn was the curator responsible for reptiles and amphibians. Kinghorn was a mentor to people such as Hal Cogger, who succeeded him as curator from 1960 to 1975; Roy Mackay, who worked at the Museum as a preparator; George Longley; and Tony Ormsby. Worrell and Mackay made the occasional collecting trip to Centennial Park and further afield to Leumeah near Campbelltown for freshwater turtles, frogs and fish, and traded the odd reptile with each other. Worrell met John Dwyer, who, with his brother Ken, was also a keen Sydney-based 'herp', at a snake show Worrell was conducting at Sydney Town Hall sometime in 1947 or 1948. Their shared interests sparked a life-long friendship, cemented by exchanges of animals. These men and others interested in herpetology began to meet to show their collections, discuss reptiles and conduct joint collecting expeditions.

Out of this loose affiliation, the Australian Reptile Club was formed in 1949, known since 1963 as the Australian Herpetological Society. Members of the club met initially at the home of Roy Mackay in Newtown, in Sydney. They made collecting trips, exchanged reptiles and shared information on keeping reptiles – but Worrell didn't get involved.

The next phase of his career was going to take him much further afield and give him the opportunity to save lives – while risking his own – by working with some of Australia's most deadly species.

4

# Getting started in
# the venom business

I picked up the biggest tiger snake in the collection to milk first.
Somehow it twisted and bit my thumb. Ken [Slater] whipped a tourniquet
round my arm above the elbow while I incised the thumb and sucked the
wound. Ken couldn't drive a car at the time, so with one arm I drove to a
nearby doctor's residence while Ken changed gears for me.

Eric Worrell[1]

Getting bitten by a deadly tiger snake (*Notechis scutatus*) was
not exactly the best way to begin a career in snake venom
production. Worrell started to feel better almost immediately after the doctor injected him with 3000 units of antivenom, and after an hour's observation he was allowed to go
home. Some days later, though, he suffered an allergic reaction to the
horse serum that formed the basis of the antivenom. He was nause-
ated, covered in hives, and his lips, eyes and feet had swollen, sending

him to his doctor's surgery looking like a modern-day Elephant Man. Fortunately, he responded to a series of injections of adrenaline and the next day he resumed milking, somewhat more cautiously.

Most of Australia's venomous snakes deliver their venom through fangs designed perfectly for their task, leaving nothing more than a couple of small holes and scratch marks at the site of the bite. Even so, the number of snakebite deaths in Australia per year was a dozen or so before antivenoms became available.[2] The venom being counteracted is needed to make antivenom. Venom also has a range of pharmaceutical uses. Some have been applied in drug therapies to treat high blood pressure while others have strong anti-bacterial effects. Venom is a powerful and valuable commodity. Of course, it can also be deadly.

Snake venom is a combination of proteins and enzymes dissolved in water. Depending on its composition, it can have varying effects on the body's nervous and circulatory systems. Neurotoxins block the signals sent from nerve cells to muscles, which can lead to the paralysis of respiratory muscles, while haemotoxins can destroy blood cells as well as cause clotting or haemorrhaging. Some species possess venoms that are primarily neurotoxins or haemotoxins, and some possess both forms. In addition, the venoms of some species, particularly the African vipers and American rattlesnakes, contain enzymes that can cause serious ulceration and necrotic damage, which results in the death of tissue.

In Australia, snakebite treatment was initially the standard technique known to all British colonists: applying a tourniquet to prevent the venom from moving through the body; attempting to remove the venom by sucking it out or cutting out the bitten tissue; and neutralising it with substances such as potassium permanganate, better known as Condy's crystals.

These measures were expanded upon by ever-inventive Australians. Snakebite treatments included the injection of horribly caustic or toxic substances such as ammonia and strychnine into the site of the bite, the inhalation of pure oxygen, the administration of spirits such as brandy or whisky, the eating of the venom gland of the same kind of snake that had been responsible for the bite, application of leeches to the bite wound and the use of exercise (or worse) to keep the patient from falling asleep. A Dr Macdonald wrote in 1891 of being called to attend to a victim of snakebite: 'On arrival I found the patient supported by two men, one under each arm, dragging him along and another man behind, flagellating him with switches.' Another poor soul, wrote Macdonald, 'had to contend with repeated injections of strychnine, frequent doses of spirits, forced exercise, injections of ether … the opening of a temporal vein and local cupping'. As Charles Campbell, an expert on snakebite treatments of this era, wryly remarked, 'To die from snake bite, at times, must have been a happy release from the treatment.'[3]

Drastic measures were not limited to the colonial period. 'We all knew what to do if we got bitten,' wrote author Alan Moorehead, reflecting on his youth in the 1920s, 'and most of us carried about with us a little tin box in which were one of father's razor blades and a phial of permanganate of potash crystals.'[4] Timber worker Ron Ward from the NSW Central Coast recalled the story of old Charlie Lauff of Yarramalong. Sometime between the wars, Lauff was out pigeon shooting when he was bitten by a death adder on the end of one finger. Lauff was decisive. He put his finger into the barrel of his shotgun and pulled the trigger, losing his finger but possibly saving his life.

The pioneering work in developing snake antivemons began in Melbourne after the death of CJ French in Adelaide in 1927. It was a collaborative effort between Charles Kellaway and his associates at the Walter and Eliza Hall Institute for Medical Research, and Frederick Morgan and John Graydon of the Commonwealth Serum Laboratories (CSL). Their efforts led to the commercial manufacture of tiger snake antivenenom, which became available from 1931 and was used as a treatment for all Australian snakebites. However, while tiger snake antivenom is effective in neutralising the venom of some other species such as the black snake and the copperhead, it is not particularly effective against brown, king brown or taipan venom. During the 1950s and 1960s, tiger snakes were the main cause of dangerous snakebites in Australia and prior to the development of antivenom, 50 per cent of bites were fatal.

The CSL took over the development and production of antivenoms, and continues today (as CSL Limited) to produce them for snakes, spiders and ticks that are harmful to people or domestic animals. Antivenom production was a relatively minor component of the laboratory's activities, however. Most of its resources were directed towards activities such as the production of insulin, the poliomyelitis vaccine and penicillin; some scientists working in these areas resented the focus on antivenoms. However, the fascination with all things venomous ensured that it was their work on venoms that captured the imagination of the Australian public.

The first stage of antivenom production is milking a snake for its venom. The snake bites through a membrane made of a substance such as latex or cellulose, stretched across the opening of a jar or beaker. The venom runs to the bottom of the vessel and is freeze-dried into crystals. In a process called hyper-immunisation, the crystals are diluted with water and injected into an animal – such as a rabbit, sheep, goat or horse – in quantities too small to cause harm. The animal's immune system responds, producing antibodies to neutralise the effects of the venom. Over a period of months, progressively higher doses of venom are given to the animal so that by the end of the process it is able to withstand very high doses that, prior to the immunisation process, would easily have killed it. Blood is then taken from the animal and the antibodies that have been developed against the venom are separated out and purified. The antivenom is developed from this purified serum.

In 1934, the CSL employed Tom Eades to collect the tiger snakes required for the production of a tiger snake antivenom, as well as other species such as death adders and copperheads, for venom research. It was also his job to milk these snakes. Eades had previously worked in the snake show circuit under the name Pambo Eades, painting himself brown and presenting himself as an Indian snake charmer. He originally maintained the CSL snakes at Melbourne Zoo but later a purpose-built snake house was constructed on site at the CSL. The entire snake collection was given to Melbourne Zoo shortly after the outbreak of World War II, possibly because of a redirection of priorities during the war or because of a potential security risk, given that the snake house was adjacent to a public road that bisected the CSL site. Additional quantities of venom from tiger snakes and other species such as the copperhead were also supplied to the CSL by David Fleay during the time he worked at Melbourne Zoo in the mid-to-late 1920s and later while he was the director of the Sir Colin MacKenzie Sanctuary for Australian Flora and Fauna at Healesville in the 1930s and 1940s. After Eades retired from the CSL in the early 1940s, another employee, Charles Ricardo, took over as his replacement for several years, milking the tiger snakes that were kept at Melbourne Zoo.

Eades was a hardworking employee and had provided such quantities of tiger snake venom that the CSL had a reserve of some years' worth. Ricardo supplemented these stocks during the period that he was collecting and milking the snakes. Then the Mount Lamington volcano in the Australian Territory of Papua erupted in 1951, bringing snakes and humans into much greater contact than usual as snakes fled the eruption, just as they flee during floods. Tiger snake antivenom was used to treat the resulting snakebites in Papua, significantly diminishing the CSL's reserve supplies. More venom was going to be required if stocks were to be replenished and an adequate reserve maintained. Another Snake Man was needed to keep up the venom supply.

Eric Worrell was on one of several trips to Arnhem Land in 1951, this time accompanying photographer Axel Poignant and another photo-journalist, Fritz Goro, who were on assignment for *Life Magazine*, when he received the message from the CSL in rather unusual circumstances:

> One day I was watching a spectacular corroboree depicting the moods of kangaroos. The Aborigines formed a large circle holding each other's waists as they hopped grunting like old Boomers. As they stirred up the dust it was almost hard to believe that they weren't kangaroos. An Aborigine came running from the nearby Oenpelli Mission Station and handed me a radio message. It was from the late Doctor Morgan, then the Director of the Commonwealth Serum Laboratories, in Melbourne. The message requested that I call into the Commonwealth Serum Laboratories at my earliest convenience.[5]

Some months later, Worrell visited the CSL and was offered the position of supplying them with venom. A regular nine-to-five type of job working for a large organisation did not really appeal to him, so he negotiated a compromise: he would supply venoms from his base at Ocean Beach Aquarium, in much the same way as David Fleay had supplied venoms while at Healesville sanctuary. Fleay was now otherwise engaged, preparing to pack up his family and animal collection and move to West Burleigh in south-eastern Queensland, where he would establish his own park, Fleay's Fauna Reserve. Early in 1951, during research for his first book, *Dangerous Snakes of Australia*, Worrell had visited the CSL to gather statistical information on snakebite and to better understand the effects of snake venoms. Worrell's knowledge of snakes and his ability to handle dangerous snakes had impressed the scientists and this led to the offer of a contract to supply venom.

Now Worrell needed large numbers of tiger snakes. In the

spring of 1951, he began the first of many annual trips to collect tiger snakes from the floodplain swamps of the Murray and Lachlan rivers, accompanied by his mentor, George Cann, and a friend and fellow reptile enthusiast, Ken Slater. The trips, coinciding with floods of a magnitude rarely, if ever, seen in much of that river system today, yielded hundreds of snakes. Worrell released them into his pit back at Ocean Beach, where they were milked and kept on display for the benefit of visitors. Surplus snakes were held in pits at Cann's house at Phillip Bay, in Sydney. Worrell visited periodically to milk these snakes and paid Cann five shillings a snake for looking after them.

Worrell needed more than just the snakes if he was to go into the venom supply business and the CSL provided him with the following equipment:

Desiccator 8" vacuum  1

Bottles, 2 litre with rubber stoppers  2

Gypsum as a drying agent $1/4 - 1\frac{1}{8}$ inch mesh

1 dracham vials and stoppers  12

Pasteur pipettes  4

Rubber dam

Rubber bands

Labels, glazed paper

Jar with parafined screw top [with paraffin wax added to the
   stopper to ensure an airtight seal]

Antivenene, 1500 units  6

Jars and medicine glasses for collection  18

Box to hold jars and glasses  1

Catching stick  1

No vacuum pump could be obtained to work the desiccator – the piece of equipment that freeze-dries and crystallises the venom – and so Worrell improvised. He did a test milking of a few snakes and then dried the venoms over oven-dried calcium sulphate, which acts as a desiccant, drawing the moisture out of the liquid venom and turning it into crystals. Then he scraped the dried venom crystals out of the jar and carefully poured them into a small glass vial. He did all of this inside a modified kerosene tin acting as a safety cabinet! Where would Australian ingenuity be without the versatile kerosene tin?

The CSL received the first two vials of dried venom from Worrell in November 1951. One vial contained 79 milligrams of dried venom from four 'small tiger snakes' and the second contained 133.7 milligrams of dried venom from one large king brown snake.[6] After the CSL had approved his procedure for drying the venom, Worrell began routine milking and venom processing. It was then that Worrell suffered the serious bite in the thumb from the large tiger snake mentioned at the beginning of this chapter.[7]

By 1952, Worrell had sent 17.6 grams of tiger snake venom but the CSL estimated that they would need between 25 and 35 grams per year, as well as requiring a reserve of several years' worth to be accumulated. In a memo to Worrell they noted: 'Venom we have received appears to have been handled well, and has been clean.'[8] The memo went on to generously suggest that Worrell retain the snake stick – or catching stick – that he had borrowed for as long as he was collecting for them.

The CSL officers and other experts seemed to take a strong interest in the improvised procedures Worrell was using to collect and process the venom. In a letter to the director of the CSL, the director of the School of Public Health and Tropical Medicine at the University of Sydney included this description of Worrell's unique approach:

The transfer is made in a box with glass roof and front wall.
As the box is enclosed, there is consequently little danger from
inhalation of the powder, whilst in addition he wears a mask of

some or other pattern. There is no exhaust fan but the floor of
the box is covered by black paper which is destroyed after each
operation. He works with ungloved hands but takes care to cover
any obvious cracks in his hands with adhesive tape. When filled
the bottles are wiped with a cloth which is subsequently burned.
At the end of each operation air is forced through the box with a
mechanical blower. [9]

The level of improvisation that Worrell employed in this early work
now seems staggering. A modified kerosene tin served as a protective cabinet, black paper was all that ensured that any spilled venom crystals could be easily seen and disposed of, adhesive tape over the scratches on Worrell's hands warded off the entry of venom into his bloodstream and a mask 'of some or other pattern' – in reality it might have been a handkerchief tied around his face – prevented him from breathing in the toxic dust. Strict occupational health and safety legislation was still several decades away.

Once he had perfected his process, Worrell became a prodigious supplier of tiger snake venom. In 1952, he supplied almost 50 grams of dried venom that had been obtained from over 2000 milkings. He did even better in 1953, with a total of 64 grams of dried venom from more than 3000 milkings, supplying the CSL with well over the 25 to 35 grams that it had requested.

To do this meant the annual acquisition of large numbers of tiger snakes. Worrell built up an extensive network of farmers and other landholders who provided him with intelligence on snake numbers as well as an open farm gate so he could come and remove them. It was a win-win situation: the farmers had their snake problem fixed and Worrell had his snakes. In the early days, Worrell would pack his panel van and drive the ten hours or so until he reached his destination somewhere along the Murray River or the Lachlan, and then spend several long, backbreaking days collecting the snakes, which he would drop into a canvas or hessian bag slung casually over his shoulder. At the end of each day, the snakes would be carefully transferred into specially made wooden boxes built by his father and, when he could fit in no more, he would pack up camp and make the long drive home again. Sometimes he went alone, other times he would have the company of George Cann and Cann's younger son, John. Jeff Carter, the photographer and editor of *Outdoors and Fishing*, who made Worrell the official consultant of the magazine on herpetological matters and published his articles in many issues, would sometimes also join Worrell on these journeys, along with his wife, Mare, who profiled the Worrell family for *Australian Country*. She recalled:

> My own approach to the exercise was tentative. I did occasionally
> whisper: 'There's one', indicating a body dozing on a fallen tree,
> its brown mosaic blending with the bark. Usually, Eric said,
> 'thanks, I saw him, I'll get him in a minute.' Then he would dart
> forward, with speed surprising in a stocky person, and grab up
> a whip-shaped animal instantly awake and writhing, trying to
> loop up and bite the fingers holding its tail. Sometimes he bent
> again and snatched a second reptile nipping into a hollow log. For
> an uncomfortable minute he would juggle the snakes, trying to
> manoeuvre them into a long calico bag he carried under his arm,
> while my heart stopped beating.[10]

The first few years after 1951 saw Worrell conduct on average more than 3000 milkings each year in order to satisfy the CSL's requirements for venom. He would milk some of these snakes for the public at Ocean Beach Aquarium and later at the Australian Reptile Park on Sunday afternoons during spring and summer. Dressed in his trademark safari jacket and slacks, he milked tiger snakes for audiences of 1950s Australian families: mums in floral sundresses, dads in white Chesty Bonds singlets, bespectacled Edna Everagesque grandmothers, and kids of all ages, some perched on their fathers' shoulders, some with mouths wide open, staring at Worrell as he massaged the venom glands of the tiger snakes to promote the flow of the toxic stream into the glass beaker. Faces were always tightly focused on the man with the snake and some strained to gain a better look. They gazed with a mix of fascination, disgust, concentration, admiration and awe. When Worrell worked with his snakes, milking them for the crowd, he melded modern science and medical research into his presentation along with great flair and showmanship. He made public snake milking into a form of the educational entertainment that is characteristic of zoos and wildlife parks today.

Maintaining the Ocean Beach Aquarium was costly. Feeding costs seem to have been the major expenditure, averaging about £400 per year, roughly $10 000 in today's money. Worrell required large quantities of mice, fowl, pigeons, chicks, eggs, fish, rabbits, guinea pigs, rice, fruit, horsemeat, liver, prawns, milk and cod liver oil to feed his collection. The aquarium did not pay its way as a stand-alone tourist attraction, with annual running costs of about £800 compared with income from admissions averaging only about £130 to £150. Things would have been grim had it not been for Worrell's venom work for the CSL and the income he derived from his writing. He also bred and sold laboratory animals to the CSL in the early 1950s.

A steady income was vital. No longer the single snake man living on a shoestring, Worrell had a growing family to support. Daughter Kim, his first child, was born in 1950, followed by a brother, Mark, in 1953. Worrell continued to travel for periods of several weeks or even months of fieldwork during autumn and winter, when the aquarium was closed. In the early 1950s, he was reported to be driving almost 50 000 kilometres a year on trips to the Riverina and Queensland. It was left to Carol to look after the reptile collection, remarkable in a decade that celebrated the virtues of domesticity for women. According to a local resident at the time:

> We're used to them [dangerous snakes] and the 'Doctor' as
> he's often called … But I still can't get over the way Mrs Worrell
> looks after the whole place when her husband goes snake
> hunting. You'd think she'd be scared, or worried about her
> two children. The window of their bedroom is right above the
> tiger snake pit![11]

Mare Carter described Carol as Eric's right hand: 'She did a terrific lot of work and did his books.' Carol attended to the needs of the reptile collection, dealt with her husband's correspondence, maintained the household and looked after the children during his absences. For Kim and Mark growing up at the aquarium, reptiles were always present and were sometimes literally embraced as playthings. As a preschooler, Kim was already a competent handler of non-venomous snakes, including Peter, a green tree snake:

> To this solidly built, fair-haired girl he is a playmate who likes to
> slither around her sun-tanned neck and chubby face … there is a
> touch of finesse in the way her hands carefully yet firmly guide
> the coils of the cool, pliable reptile.[12]

Worrell was proud of Kim's ability with snakes and often photographed her and her brother interacting with reptiles. A family Christmas card from this period showed Carol laughing as she held the head of a large python while Kim supported its tail. Between them sat Eric, pretending to milk the snake like a cow from a rubber glove attached to its belly, while young Mark looked on with interest. The caption read, 'So sorry we haven't written, we've been so busy milking.'[13]

In the late 1950s, Jack Green, a fish enthusiast living in Woy

Woy, kept an eye on the aquatic exhibits when Worrell was travelling. Green and Worrell had met while exhibiting animals at an event in Gosford. Worrell had asked if he could photograph Green's seahorses, and a friendship formed. Worrell gave Green his own key to the aquarium and asked him to take responsibility for the aquatic exhibits while he was away. Green was somewhat snake phobic, but Worrell assured him that Carol would maintain the reptiles and that he need not have anything to do with them. On his first evening of duty, he got more than he bargained for:

> Eric had shown me where the master switch was located to illuminate the whole aquarium section. It was situated behind a sliding panel above a couple of exhibition cases, and a small ladder was at hand to give you more height in order to reach the switch situated about 60cms behind the sliding panel. I positioned the ladder, climbed it, and slid the panel open. I reached in to the darkened interior, switched on the light, and there, looking me straight in the eye was this massive, coiled python, already positioned in a striking position. I let out an audible scream, stiffened and fell backwards off the stool. I could just see Eric at the time: sitting on a log beside his camp-fire, looking at his watch and figuring out the time when I would get the shock of my life from the vision of the DEAD [and mounted] python he had positioned there to introduce me to my new part-time job. [14]

While snakes would always be the primary focus of venom extraction at the park, in the latter part of the 1950s Worrell became involved in helping the CSL with its research into another type of venom – that of the Sydney funnel-web spider (*Atrax robustus*). These spiders are found in the sandstone region between Newcastle and Nowra and inland as far as Lithgow. Their large size, their habit of rearing up defensively when disturbed and their dangerous venom make them the most feared of Australian spiders. While the bite of the Sydney funnel-web is not toxic to cats, dogs, adult mice or guinea pigs, it can be fatal for humans and other primates. The bite of the male Sydney funnel-web is of greatest concern, being up to six times more toxic than that of the female.

The first determined effort to learn more about these spiders and their venom was begun by Charles Kellaway, Director of the Walter and Eliza Hall Institute of Medical Research in Melbourne, in the 1930s. This came as a result of several fatal bites between 1927 and 1935. Saul Wiener of the CSL began research on an antivenom in 1956. In order to secure the venom they needed for their research, they called for volunteers to bring spiders to collection depots in Sydney. There the spiders were carefully packed in plastic jars with perforated lids and damp cotton wool at their base and sent by air to the Melbourne laboratory to be milked of their venom. Not surprisingly, Eric Worrell responded to this call for spiders. Funnel-webs were left at Ocean Beach Aquarium by local residents and sent to the CSL.


In the 1950s, tiger snake bites were the most common types of snake-bite and the most medically significant, but common brown snakes (*Pseudonaja textilis*) also posed a considerable hazard because of the potency of their venom and the alarmingly large numbers that some-times occurred in the wheat-and-sheep country of New South Wales, Queensland and Victoria, especially after mouse plagues. However, the relatively small amount of venom that each brown snake yielded made the species a difficult prospect to milk and the CSL struggled to get enough supplies.

To overcome this problem, the CSL's Merv Hinton suggested that suppliers be paid more for the venom from this species. Worrell began supplying the CSL commercially with brown snake venom from December 1955. One of his most enthusiastic and long-lasting suppliers of brown snakes was Roy Reynolds from Yanco in the Murrumbidgee Irrigation Area of New South Wales, where the population of brown snakes had exploded following the excavation of drainage channels and a surge in the populations of prey such as frogs and mice. Worrell offered Reynolds 10 shillings per foot (30 cm) for snakes over 4 feet (1.5 m) long; £5 for snakes 6 feet (1.8 m) long; and £10 for snakes in excess of 6 feet 6 inches (2 m), which would have been very scarce. Reynolds sent up around 100 brown snakes in boxes on the train each spring for more than two decades, as well as many frogs that he collected in the drains and canals, which would be used to feed these and other snakes at the park. Throughout the 1950s, Worrell carefully built up a network of friends and associates who collected snakes on his behalf. There was John McLoughlin, Frank Little, Snr and Jnr, and Vince Edwards in Cairns (who all supplied taipans), Joe Bredl in Renmark, Silvester Smith at Lake George and Roy Reynolds in the Riverina, as well as George Cann, Snr and Jnr, and later, Brian Barnett in Melbourne. Worrell obtained death adders from the banana farmers of northern New South Wales.

The average yield of a common brown snake is around 2 milligrams of dried venom, but one monster Worrell milked in 1959 produced an astonishing 41.4 milligrams. The potential for this individual to produce a highly lethal bite so concerned a CSL medical officer that he sent Worrell six additional ampoules of brown snake antivenom in case he suffered its bite.

In order to maximise the venom output from these snakes, Worrell experimented with electro-stimulation: using electrodes to stimulate the snakes' venom glands with a low-voltage electrical current. Presumably because the yields were not substantially better and so it wasn't worth the extra expense and effort, he didn't persevere with this method, preferring the more conventional technique of applying pressure to the back of the snake's neck and forcing it to bite the membrane.

Worrell had become the primary supplier of snake venom to the CSL. Soon after he began supplying them with tiger snake venom, they had offered him a new challenge: to supply the venom of Australia's most dangerous snake, the coastal taipan.

5

# Quest for the taipan

ensing that the pressure around its neck had relaxed very slightly, the enraged snake suddenly squirmed free and snapped two or three times at the young man's hand before biting deep into his thumb. Another Australian was about to succumb to the powerful venom of the taipan. Kevin Budden, a 20-year-old reptile enthusiast from Randwick in Sydney, had travelled to far north Queensland in 1950 in search of snakes – in particular, the taipan – motivated by a genuine desire to contribute to the development of a life-saving antivenom.

The sound of desperate squeals had led him to a sheet of fibro under which he found the snake about to make a meal of a terrified rat. Budden secured the snake by placing his boot on its body near the neck, to which it reacted furiously, biting both the boot and Budden's trousers. He managed to grab hold of the writhing snake just behind its head and, realising that he would have little success in trying to bag the flailing reptile by himself, decided to walk out to the road in the hope of hitching a lift back to local naturalist Stephen

(Ern) Stephens's home. There he would be able to gain assistance in bagging the taipan more safely. The likelihood of a passing motorist giving a ride to a young man holding a long and angry taipan seemed slight to say the least, but so far luck was on his side and a local man stopped and offered Budden and the struggling serpent a lift into town. It was at Stephens's home that the taipan bit Budden while he was attempting to bag it. The snake collector's luck had run out.

Bravely, Budden recaptured the enraged snake. He told Stephens that regardless of whether he lived or died, the snake should not be killed. He wanted it to be sent to Melbourne, where it could be milked and its venom used in research to produce an antivenom. He declined to cut the bite site, then the standard first aid treatment, but a tourniquet was tied tightly around his upper arm and he was taken to Cairns Base Hospital. He died there the following day. He had paid the ultimate price and died far from family and friends.

In keeping with Budden's wishes, the taipan was flown to Melbourne, within two hessian bags inside a 'strongly nailed wooden box'.[1] It was milked a couple of days later by David Fleay, who was still the director of the Healesville Sanctuary at that time. This was an act of considerable bravery on Fleay's part, considering that he had no experience with the species and this individual was a proven killer. Fleay wrote shortly after the event that he had been in a 'blue funk' after being asked by the director of the CSL, Dr Frederick Morgan, to make the attempt. Milking the taipan was a big news story around the country. One newspaper likened the security precautions at the National Museum of Victoria, in Melbourne, as rivalling an 'atomic project'.

There are three species of taipans recognised in Australia. The one that bit Budden and that the CSL's antivenom efforts focused on was the coastal taipan, *Oxyuranus scutellatus*. It engendered widespread fear throughout its range in coastal Queensland from Cooktown south past Maryborough. That fear was fuelled by exaggerated media reports in the 1950s and 1960s, with headlines such as

'Australia's worst snake' and 'Naturally savage snake'[2] and accounts such as 'two taipans, caged, and so savage that they hurl themselves against the wire whenever anyone approaches'.[3] European migrants attracted to the region to cut cane were especially concerned that their already arduous work would put them in close proximity to the killer, the bite of which almost always proved fatal. The taipan's venom is a lethal cocktail of substances that clot blood, dissolve cell membranes, attack skeletal muscle and paralyse the muscles that control breathing.

Some six years before Budden met his untimely death, 20-year-old Worrell, still stationed at East Point in Darwin, had written to Roy Kinghorn at the Australian Museum:

> Dear Mr Kinghorn,
>
> I know you must be a very busy man but I would be very grateful if you could supply me with some information regarding the 'Tiapan' or 'Taipan' snake. I have been studying reptiles for quite a few years now but particulars pertaining to this snake seem to be very evasive.[4]

In a letter three months later, Worrell thanked Kinghorn for sending him the information and asked if he would 'kindly oblige by forwarding a description of the head scales which I omitted to request?'[5] These letters suggest that Worrell thought that the taipan might have occurred in the Darwin area (which in fact it was later shown to do) and he was equipping himself with the knowledge to make an accurate identification if he discovered this giant deadly snake in his travels in the Territory. His uncertainty over the proper spelling of its name is not surprising because little was known about the species at that time and indeed the common name, taipan, had only been suggested by zoologist and anthropologist Donald Thomson in a paper published in 1934,[6] which Worrell may not have seen at that stage. It would be eight years from the time that Worrell wrote these letters to Kinghorn until he saw a living taipan but it stayed on his mind, with plans to fly to the Cape York Peninsula with Carol in the winter of 1949 to search for it.

While the 1949 trip did not eventuate, Budden's death no doubt helped fan both Worrell's interest and the fire of anxiety about the taipan that swept through North Queensland. The CSL initiated research into the preparation of a taipan antivenom and an internal memo written early in 1952 put the question, 'If taipans were sent to Mr Worrell, would he be prepared to collect venom?' Ram Chandra, a snake handler in Mackay, had offered to collect taipans and forward either the snakes or the venom to the CSL. Worrell accepted the CSL's challenge and in 1952 he teamed up with Sydney men Wal Lorking and John Dwyer to search for the deadly snake. Dwyer had been in Cairns looking for taipans the previous year with his brother Ken and fellow reptile enthusiast Bill Hosmer. While they made plenty of valuable contacts with farmers in the area and had advertised a reward for information leading to the successful capture of a taipan in the *Cairns Post*, they returned to Sydney with only a couple of heads sold to them by a cane farmer. Dwyer was keen to return in 1952, as Sir Edward Hallstrom, president of the Taronga Zoological Park Trust, which governed Sydney's Taronga Zoo, had offered the hefty sum of £50 – the equivalent of almost a month's pay at the basic wage rate of the time – for a live taipan that he could exhibit. Hallstrom's own reptile curator, George Cann, sought agreement from the zoo's board of trustees to send him to North Queensland to also try and capture some taipans for display and study at the zoo, but his request came to nothing.

As Worrell told the story, he met up with Dwyer and Lorking coincidentally in Brisbane and they decided to join forces and head up to Cairns together. John Dwyer remembers differently: 'Ken and Wal and I were going to do the trip and Ken couldn't get time off from work. I got in touch with Eric and said, "We're going to go up and have a shot at taipans in Cairns. I've got some good locations," and he said, "I'd like to go with you" … [but] he didn't want to be associated with us. This is the real story. His idea was he wanted to be independent. He said, "I'll meet you in Brisbane." He got the same train [to Brisbane] as us!', Dwyer recounted with good humour.

Indeed, a story on the trio in a local newspaper reported that Worrell's 'presence with Mr. Lorking and Mr. Dwyer is accidental. He met them in Brisbane, having not known them previously, and the trio travelled north together.'[7] Perhaps wanting to be seen as an adventurer setting off by himself, just like Kevin Budden, Worrell made sure that this was the official story.

Once the three men arrived in Cairns, they headed out to the Freshwater and Redlynch districts on the outskirts of town, where Dwyer knew some of the cane farmers. In Freshwater they made camp and began the arduous task of searching for taipans, knowing all too well what would happen if they were unlucky enough to be bitten by one. The landscape here is one of abundant fertility – broad acres of sugar cane on the lowland flats overshadowed by the luxuriant rainforest-clad slopes of the Macalister Range. They would have had little time to appreciate this spectacular landscape, however, spending their days in the searing heat and high humidity searching for what Worrell called the 'killer of the cane fields'.[8] As it happened, several taipans had been killed by cane farmers in the area, the most recent just two days before they arrived. It wasn't long before they managed to bag their first snake. As they were searching a paddock, a canecutter came across a taipan and hurled a rock at it, intending to stone it to death. Worrell and his men rescued the snake, which measured 2 metres, capturing it without incident. Their next capture, however, was not to be so seamless.

The following day, they received word from the same canecutter that he had seen another taipan close to the spot where they had caught the previous snake. They hurried out to the site and began searching. Worrell was the first to spot the snake as it flashed past his boots in long grass and out towards a road. After much searching they failed to sight it again and so retreated to the shade of a tree by the roadside to take a break, keeping an eye on the road in case it reappeared there. In a matter of minutes, Worrell realised that what he had initially seen as a shadow cutting across the road some distance away was in fact a taipan. Dwyer and Worrell raced towards it, and it was Dwyer who got there first and grabbed hold of the great snake's tail. He then realised that the snake, terrifyingly angry at having being grabbed, was too large to control using the methods applied to smaller snakes. Dwyer recalled, 'It was about the second or third day that we spotted the big bloke. It took off and it was going into the bladey grass.' There was about 40 centimetres of its body still visible. 'I had done a 100-metre sprint ... grabbed it, and it's got its mouth open and it's heading straight at me. I thought, "Gee, it's got a big mouth, I better duck," so you know, it went straight over my shoulder. Then it came back at me and I stepped aside and I thought "Shit".' Dwyer spun around a few times. 'Funny what goes through your mind: "I wonder how strong you are, you bastard?" I slowed down a bit.' Then the snake started to double back on itself until its head was halfway up its body. 'So I started to swing it around. I started to swing it out a bit further.' Worrell and Lorking were standing about 3 metres away. They shouted, 'Fling it off!' Dwyer remembered: 'I didn't even know where the road was at this time. So I let it go ... unfortunately it went across the road, a cane field and then a strip of grass about six to eight feet wide and it went straight into the grass. Went around but couldn't find it.'

Lorking soon sighted the taipan again and this time Dwyer managed to pin its head. Worrell carefully picked it up and dropped it into one of their canvas snake bags.

While they were based at their campsite looking for more snakes, the two taipans they'd caught were left in the care of an acquaintance at Edge Hill, which is midway between Cairns and

Freshwater. Without the trio's permission, he had some photographs taken holding their taipans. This annoyed Worrell and he decided to have some professional photographs taken of him, Dwyer and Lorking holding and milking the larger of the snakes they had caught. Photographer Lionel Laws of Chargois Studios accepted the job. The three men spent much of the night prior to the photographic session planning how they would approach the deadly task and before they went to sleep they made a pact that everything would go to plan.

At the studio the next morning, Worrell systematically set out six ampoules of tiger snake antivenom and three syringes neatly on a table, together with a rubber tourniquet. He also placed on the table what John Dwyer referred to as 'Eric's tool of fear', a pair of sharp-pointed surgical scissors. In the case of a bite, Worrell suggested that someone should pinch up the lump of flesh and cut out the bite site with the scissors. Dwyer responded that the procedure was definitely not appealing and his choice would be to use a razor blade. Worrell instructed the photographer to vacate the room immediately if the snake got out of control, an instruction that he was apparently happy to follow without hesitation. Once all necessary precautions had been made, Worrell cautiously opened the lid of the wooden box containing the angry taipan and, after some struggling, managed to pin its head with his jigger – a Y-shaped metal frame at the end of a long handle with a narrow leather strap strung across it – before firmly taking hold of the snake's neck with his hand. Once the taipan was free of the box, Lorking held on to the back half to help steady it while John Dwyer held the glass beaker close to the snake's head. Worrell then brought the snake close enough to the beaker for it to bite. The taipan mauled the rubber membrane on the top of the beaker and one of its fangs overshot the edge, spilling venom down the outside of the beaker and onto Worrell's trousers. The men knew that the tiger snake antivenom they had brought with them wouldn't have effectively neutralised all the toxic agents within the taipan's venom and that had the snake managed to squirm free of Worrell's grip and bite one of them, the most likely outcome would have been death. Yet with their clean-shaven faces, slicked hair, neat shirts and good woollen trousers, the three young men look handsome and confident in the photograph. Their gaze is focused, concentrated and serious.

Three taipans were caught on this first expedition, but one escaped and the one rescued from the canecutter died soon after capture, presumably from the wounds inflicted by the rock that had been hurled at it. Only the biggest snake, the one in the photograph, remained in captivity. Because the expedition had been organised by Dwyer and Lorking, they took possession of it. When the men arrived in Brisbane they asked Paula Perry, a sideshow snake performer and friend of Worrell's who happened to be appearing at the Royal Brisbane Show, if they might team up and exhibit the snake as part of her act. The snake, exhibited in a glass-fronted wooden box within Paula's snake pit, was a big hit with the show-goers. Knowing very well how much of an audience puller snakes can be, 'snakeys' set up in parks and vacant allotments in the cities and travelled the agricultural show routes from the mid-19th to the mid-20th centuries.[9] They took on stage names such as 'Tambo the Reptile King', 'Cleopatra', 'Pambo' and 'Psycho', dressed in exotic costumes and draped themselves in snakes. They played up the aggressiveness of their snakes and sometimes allowed themselves to be bitten, then 'saved' themselves with patent remedies, which they then sold to audience members. Crowds gathered around the snake pits, frightened or disgusted, but at the same time drawn by the snakes and admiring of the bravery of their handlers. Female performers such as Melinda Lee, the Tennessee Snake Dancer, performed her 'exotic dance of the reptiles' to an admiring crowd at the Cairns Show in 1951, in competition with Paula Perry, known as the 'Queen of Snakes' and the 'Australian Bush Girl' who danced with a large collection of reptiles in the 'Pit of Death'.

Keen to take possession of the taipan, Worrell asked Paula to lend him money so that he could purchase the snake from Dwyer and Lorking; unhappily for Worrell, she declined. Once the show had ended, the taipan was bagged and the three men headed for the Roma Street train station for their trip back to Sydney. Transporting the creature down to Brisbane on the train from Cairns had proved to be problem free, but the same could not be said for the Brisbane–Sydney leg of the journey.

Dwyer had concealed the bag containing the taipan under his overcoat, carrying both over one arm. Unfortunately, just as his ticket was being checked, the bag containing the snake slipped off his arm and fell to the ground, right at the ticket inspector's feet. The inspector picked up the bag and realised it contained a snake, although he could not have known that he was holding one of the world's most dangerous species. Not unexpectedly, he ordered that it be carried in the luggage compartment. This option did not sit well with Dwyer, who feared that the snake might be crushed under heavy suitcases. Instead, he and Lorking quickly went back into the shops, where they purchased a stylish traveller's bag that would comfortably conceal the snake sack. A woman who had been working at the Brisbane show and who knew Dwyer agreed to take the bag containing the snake on the train with her, and so the taipan made its way to Sydney. Lorking and Dwyer handed their hard-won snake over to Hallstrom and, true to his word, he gave them their £50. The taipan was exhibited at Taronga under the care of George Cann for around three years.[10] Worrell periodically extracted its venom, which he sent to the CSL.

After this trip, Worrell rejected the prevailing belief that taipans were aggressive. He wrote: 'We found taipans to be extremely shy and surprisingly quick to vanish into heavy grass when approached.'[11] However, he confirmed that once disturbed or cornered, the taipan had a ferocity that was unequalled, behaviour that had previously been noted by David Fleay. His field observations and the dramatic photograph of him milking the taipan were featured in *Dangerous Snakes of Australia*, adding to the book's popularity.

Worrell also used the expeditions to north Queensland to hunt for the longest of Australia's snakes, the scrub or amethystine python, numbers of which congregated during the cooler months in the deep rainforest gorges. One of these trips ended in tragedy when his brother-in-law, Carol's brother, accidentally drowned in the Johnstone River.

Worrell and Dwyer headed north again in July 1955, with Worrell writing an account of the trip for *Modern Motor* magazine as a means of defraying the costs. On this trip, Worrell milked a taipan that John McLoughlin, a 19-year-old snake enthusiast Worrell had met in Cairns in 1952, had caught and given to Berkeley Cook, who owned the Brown's Bay Zoo. Dwyer remembered how incredibly difficult this snake was to handle, 'turning cartwheels' and narrowly missing biting Worrell.[12] This trip was particularly successful, with Worrell, aided by McLoughlin, catching a good-sized taipan in the Freshwater district, about 8 kilometres north of Cairns. Worrell and Dwyer left Cairns in mid-August for the trip home, stopping for a weekend in Townsville, where they collected native fish. Finally, at the end of August, having driven 6500 miles, they arrived back at Ocean Beach Aquarium with an extensive collection of animals: one taipan, one 2-metre crocodile, one python, 40 other snakes, 'sundry lizards and tortoises' and a variety of fish.[13]

The following year saw Worrell again back in Cairns, this time in October. He was accompanied by Eric West, a mate from Tocumwal, New South Wales, with whom he had caught tiger snakes in the Murray Riverland and on the islands of Bass Strait; and Charles Tanner, who had succeeded Fleay as the director of the Healesville Sanctuary. They had the help of John McLoughlin and another local man, Vince Reilly of Mount Molloy. The *Cairns Post* and local radio stations helped them in their quest by advertising that the men were looking for taipans and promoting the reward they were offering: £50 for the capture of a live taipan and £10 for any leads that assisted them in catching the snakes. The expedition was highly successful, bagging a total of five taipans. Reilly even managed the astonishing feat of catching two taipans at once, one in each hand.

After this trip, Worrell had an impressive total of nine taipans in captivity at the aquarium and he hoped that he would now be able to provide a regular supply of taipan venom to the CSL. Worrell had begun keeping taipans the year before, 1954, when John McLoughlin caught one and sent it to him. In 1950, someone had sent him a snake, supposedly a taipan, but it ended up getting lost in the post. The one sent down by McLoughlin made it safely to Ocean Beach and was the first that Worrell exhibited at the aquarium. Unfortunately, it bit itself in the spine when being fed and it died of its injuries.

Few taipans had been kept successfully in captivity prior to Worrell keeping them. Ram Chandra of Mackay – the man who had earlier offered to supply the CSL with taipans or their venom – claimed in 1955 to have been holding the first taipan in captivity in Australia for five months, according to the front page of the *Cairns Post*. Zoo owner Berkeley Cook was indignant and publicly claimed that he had been keeping one for the past nine months. Clearly, being the first person to successfully keep a taipan in captivity would be a record worth claiming. However, neither Worrell nor Cook nor Chandra could claim to have been the first to keep a taipan. That honour must go to Donald Thomson, an anthropologist at the University of Melbourne who also had a long-standing interest in birds and reptiles, particularly venomous snakes. Thomson had published the first scientific papers on the taipan and he kept several in the field for a couple of months in the 1930s. The first public display of a living taipan was of the Budden taipan at Melbourne Zoo. It had been transferred there from the National Museum of Victoria and put on display, albeit only for a week or two before it died in 1950. The second was the taipan brought back by Dwyer and Lorking from the 1952 trip; it lived for about three years in captivity at Sydney's Taronga Zoo. The first taipan kept for any period of time outside of the warm climate of northern Queensland, its survival was a credit to Cann as a keeper.

It took a few attempts before taipans were able to be kept successfully at Ocean Beach Aquarium. Initially, Worrell struggled to keep them alive at all. Some of the snakes may have been injured and stressed during capture and transportation, but probably the most significant factor was the difficulty in maintaining a suitably warm environment. Being tropical snakes, taipans need to be kept at temperatures in the high 20s to low 30s all year round, which would have been a challenge in the 1950s. Worrell kept the snake that he caught in Freshwater in 1955 alive for 18 months at the aquarium, an achievement of which he was quite proud. He claimed that such longevity demonstrated that regular milking didn't need to damage a snake's health. This was contrary to the wisdom of the day and a view he himself contradicted in later years.

Worrell's early taipan keeping suffered a serious setback at the end of 1956. One snake died and Worrell's friend Hilton Chalmers, a local pharmacist, performed an autopsy, determining that it had died of pneumonia. Treatment of the other snakes with antibiotics was unsuccessful and one by one they died within the next few days. The cause of the pneumonia was probably the stress of capture and captivity. As Worrell recounted in *Song of the Snake*, 'Outwardly they were in perfect condition, but the deadly bacteria worked rapidly on the lungs. I could have wept when I lost the original taipan that I had kept successfully for so long.'[14]

However, Worrell was stoic in his resolve to continue with the taipan project:

> It meant starting all over again. More expensive trips to the
> north, more expensive, air-conditioned cases to house them.
> However, it has become something I now accept. I must go north
> every year as I always have. Just as I must spend early spring
> along the banks of the Murray in search of tiger snakes.[15]

6

# The tigers
# of Tasmania

It was as if the mutton-birders were playing a game of Russian roulette as they thrust their arms deep within burrows to haul out protesting nestlings. Would their searching fingers touch a feathered chick – or a scaly snake? Here on Mount Chappell Island in the windswept, craggy Furneaux Group off the northeast coast of Tasmania lived tiger snakes unlike those seen on mainland Australia. These snakes were not striped but jet black, and sometimes grew to over 2 metres in length.

On one of the last few days of the hunt, luck ran out for mutton-birder Roy Goss. As he reached inside a burrow for one of the chicks – which were commercially hunted for their meat, oil and feathers – he groped a large black tiger snake instead of a soft, plump nestling, and was bitten. Worrell quickly tied a tourniquet then wielded a razor, opening up the bite wound, and then injected Goss's veiny arm with several ampoules of tiger snake antivenom. Goss looked to the ground, away from Worrell, grimacing. The pain was a small price to pay, for Worrell had saved his life. Goss was back birding the following day.

Worrell was hunting not mutton-birds but the snakes the birders so feared. Like many herps, he had heard of the legendary giant black tiger snakes on the island. Elsewhere, tiger snakes usually reach 1 to 1.5 metres but on Mount Chappell there are occasional giants over 2 metres in length. This geographically isolated population had first attracted the attention of venom researcher Charles Kellaway in the early 1930s. Now Worrell, along with his mate Eric West, had his sights set on the summit, in hopes of catching one of the largest of the large that were supposed to live on the higher parts of the cone-shaped 'mountain' which, although it only just scrapes 200 metres in height, dominates the island's topography. It is covered in a mix of tussock grasses and patches of scrubby vegetation. They set off to search the mountain, their route so steep that in places they had to haul themselves upwards. While they did not find the giant they were looking for, they bagged a couple of large specimens. Finally, they began descending to return to their hut. That was when West saw a big tiger snake disappear down a burrow. Worrell hooked the angrily coughing snake out of the hole and West grabbed it and dropped it in a bag.

Just when they thought they had him, the snake managed to bite West's hand through the bag. Both men knew immediately: West might well die. After treating Goss's bite, Worrell had only one ampoule of antivenom left, and it was back at the hut. He frantically applied a rubber tourniquet and cut the skin and flesh at the puncture site completely out with a razor blade. As they hobbled back to the hut, which fortunately was only about 100 metres away, Worrell had West's hand in his mouth, sucking at the bloody mess in the hope of removing as much venom as possible. By the time they reached the mutton-birders' hut, West's condition had deteriorated.

His face had lost its colour and he was clammy with sweat. Sitting him on a soapbox in the kitchen, Worrell injected the remaining ampoule of antivenom into the now blackening arm and slowly removed the tourniquet.

The only way off the island was by boat, but it wouldn't be able to be launched again until the next morning, so the two collectors, together with the Goss family, had to wait it out. Worrell detected a little improvement in West following the injection of antivenom, but his condition was still far from stable. Disturbed by the plight of the seriously ill man, the two Goss girls retired to their bed and wept, convinced that Eric West was going to die. Worrell spent the night worried about his friend's condition, staying awake as long as he could and checking on him periodically throughout the night. The next day, even though the weather was bad for sailing on the treacherous Bass Strait, Worrell, his snakes and West were taken to Flinders Island, where West was admitted to hospital. He recovered, but a partly paralysed hand would be a lasting reminder of how close he had come to death.

This was not the first time Worrell had travelled to the Furneaux Group. He had been studying and collecting the tiger snakes there to better understand the variations in venom toxicity of snakes from different islands and how they managed to thrive in such seemingly inhospitable conditions. The climate is harsh and icy strong winds frequently sweep across the islands. There is no permanent fresh water and apart from the mutton-bird chicks, which are available only during the annual nesting period, the snakes' other prey, skinks, are not abundant. Early in 1954, Worrell's friend and mentor George Cann had written to the head teacher of the Flinders Island Area School seeking information about the islands and their snake fauna. The teacher responded that snakes were present not just on Chappell Island but 'were also plentiful on Prime, Seal, Babel and

Sisters Islands, as well as parts of Flinders itself'.[1] He encouraged Cann and Worrell to come to Flinders Island.

In the late spring of 1954, Cann and Worrell flew down to Flinders Island from Melbourne. They checked into a guesthouse and must have been excited about what they might find on Chappell when the proprietor told them to expect tiger snakes that were 'longer than a man' and had 'heads bigger than match-boxes'.[2] After a few days' delay because of bad weather, they were ferried across to Mount Chappell Island by Roy Goss in his crayfishing boat. Goss left them there for a few days, with instructions that in the case of snakebite they should light a fire on the summit of Mount Chappell: 'plenty of smoke in daylight, plenty of flame at night'.

In a couple of days of easy hunting, Cann and Worrell bagged about 70 snakes, with some of the largest being in excess of 2 metres long. Observing the snakes in the wild and later in captivity, Worrell noted that the snakes took in enough water by drinking the mist that condensed on vegetation. This was an important observation because it helped him understand how the snakes could thrive on an island with no permanent water. The young snakes preyed upon small skinks that inhabited the island and adults fed on mutton-bird chicks for several months of the year. The snakes existed on stored fat for the remainder of the year, supplemented with occasional stranded fish and rodents. Although bleak and windswept, Mount Chappell Island, with its curious Cape Barren Geese, giant black tigers and mutton-bird rookery, captured Worrell's herpetological imagination.

Worrell collected almost 200 black tiger snakes from Chappell Island in just over six months. No doubt some of these animals would have gone to George Cann and Taronga Zoo; and Eric West, who had a small reptile exhibit, would also have kept some. Others went to the collection at Ocean Beach Aquarium. The cannibalistic habits of the species exasperated Worrell, who lost many from their propensity to eat one another. Worrell milked his captives and sent the venom to the CSL. He also studied their taxonomy, eventually designating two subspecies, *N. ater humphreysi* and *Notechis ater serventyi*. An early, undated typescript of Worrell's description of *Notechis ater humphreysi* shows that he originally intended to name it *N. ater yolla*, in honour of the hut they stayed in while on Fisher Island, which was called Yolla, the local Aboriginal word for mutton-bird. He ended up choosing the subspecies name *humphreysi* after fellow snake enthusiast and friend Bob Humphreys.[3]

When the ABC began its television broadcasts in 1956, Worrell's career went to another level. In the network's first week of broadcasting, Worrell was filmed milking a tiger snake and handling other snakes at Sydney's Royal Easter Show, making him, he believed, the first naturalist to appear on Australian television. Worrell already had some experience in front of the camera: he had been making and appearing in films since 1954, including documentaries such as *Our Australia: Dangerous Snakes* and short films about spiders, featuring him in his signature safari jacket searching under logs in the bush for funnel-webs.[4] He was well positioned to enter the new world of television, as the stations were hungry for local content to balance the many hours of American and British sitcoms, dramas and movies that dominated prime time. Not only could Worrell make films, he had the personality to step out from behind the camera to become a character himself. In February 1957, the ABC program *Australia Unlimited* featured 'Umina snake-catcher, Mr Eric Worrell, milking and force-feeding taipan snakes'. Capitalising on this early profile, he convinced the ABC to send him to film a story about the Bass Strait mutton-bird industry. It was the first of many short television documentaries he would make for the ABC and it gave him with an opportunity to visit other islands in the Furneaux Group than Chappell and to meet respected ornithologist Dominic (Dom) Serventy.

Serventy, the officer-in-charge of CSIRO's Wildlife Survey Division in Perth, had a passion for seabirds and was engaged in an extensive field study of the mutton-bird – the short-tailed shearwater, *Puffinus tenuirostris* – and the industry it supported. He was gathering important data on reproduction, diet and nesting behaviour, which would give biologists a better understanding of the species' ecology and be crucial in managing the effects of harvesting the birds. In 1957, Worrell met Serventy on Babel Island at the opening of the mutton-bird season, which runs from late March to the end of April. Worrell agreed to help Serventy band the mutton-bird chicks when weather conditions were too poor for filming and in return, Serventy provided assistance to the film crew during better weather, such as by taking them to the best areas to film the birds feeding and to their nests at the right time to film the birds returning or leaving. As well as obtaining excellent footage of the natural history of the birds, Worrell filmed the work of the mutton-birders, from catching the young birds in their burrows through to the processing of the carcasses. It says something of Worrell's charm and persuasiveness that he was able to win the confidence of the birders, who had initially refused to be involved in the project because of previous poor treatment by the media. Worrell presented a balanced view of the industry and included interviews he had done with some of the mutton-birders.

Worrell and his crew went to Serventy's base camp, on Fisher Island, where Serventy had banded every pair of breeding birds, making the entire rookery on this island the subject of intense field study. Their evenings were spent around a campfire at the researchers' cabin, 'discussing everything from Flinders's explorations and herpetology, to almost heated discourses on modern poetry'.[5] The two men must have hit it off because this trip was the beginning of a lasting friendship, which would be strengthened by Worrell's regular visits back to the Furneaux Group. Through Serventy, Worrell would also meet, and become firm friends with, Jock Marshall, foundation professor of zoology at Monash University, and Marshall's friend, the artist Russell Drysdale, both of whom played important roles in Worrell's developing career. Drysdale had an interest in the Australian bush and outback. He often accompanied Marshall on treks into the interior of Australia and probably went to Bass Strait on this trip for the adventure, to spend time in a wild environment and for the companionship. Worrell was to have an enduring and close friendship with Drysdale until the artist's death in 1981, naming a genus of snake *Drysdalia* in his honour.

Worrell spent the last couple of weeks of March 1960 on the Bass Strait islands in the company of Dom Serventy, Jock Marshall, Russell Drysdale and British zoologist Mary Gillham. Marshall shared with Serventy an interest in the breeding system of mutton-birds and they collaborated on a number of research projects. Gillham spent several years conducting field research on the ecological relationships between the mutton-bird industry, sheep grazing and the native flora and fauna. She affectionately called Serventy, Drysdale and Worrell 'The Terrible Trio'.[6]

Worrell collected about 40 tiger snakes from Mount Chappell Island on this particular trip and Gillham wrote, 'It was hot and his collecting bags were stuffed so full of reptiles that seven died from suffocation.'[7] Gillham did not pass judgment on Worrell for the death of almost 20 per cent of his catch; and he preserved the dead snakes and used them for his taxonomic and ecological work, such as inspecting their stomach contents to determine what they had been eating. Worrell's work on the Furneaux Islands has come in for some criticism, though, because of the number of snakes he did remove. In an article published some 30 years later, Tasmanian herpetologist Simon Fearn described Worrell's collecting expeditions as 'Viking-like raids to pillage the island of the cream of its snakes …'[8] Perhaps the abundance of tiger snakes on the islands coupled with the fact that they were frequently killed by the mutton-birders encouraged Worrell to take the numbers that he did. It is undeniable that he argued passionately for the conservation of the islands and their native species. Like Gillham, he was appalled by the ecological effects of overgrazing by sheep and the frequent burning of native vegetation by landholders. In a report she wrote to the Tasmanian Fauna Board soon after the 1960 visit, Gillham referred to the 'distressing condition of Chappell Island', caused by indiscriminate burning, overstocking and unchecked erosion.[9] The island's Cape Barren Goose population had also been decimated. She recommended a range of measures to improve the habitat quality of the island while at the same time permitting light grazing; and she urged the board to follow Worrell's suggestion to designate part of Mount Chappell Island a snake sanctuary because of the peculiarities of its subspecies of the tiger snake. Worrell wrote to the fauna board:

> It is a great pity that Chappell Island cannot be retained as an absolute sanctuary. The combination of shearwaters, Cape Barren geese and tiger snakes coupled with the insular vegetation and geographical isolation provide a remarkable ecological study area.[10]

He urged the board to 'afford legal protection to the snakes' on Cat and Chappell Islands and stressed the important role that the tiger snakes played in the ecology of the islands, being the apex predator and a natural control on mutton-bird numbers. Worrell clearly had a genuine concern for the conservation of the tiger snakes that inhabited this island, even though he had removed many from the islands for his own purposes.

Worrell only once came close to suffering a dangerous bite on Mount Chappell Island. During a trip with his friend, naturalist Vincent Serventy (Dom's brother), a captured tiger snake managed to bite Worrell through the bag. Serventy was concerned but Worrell brushed the bite aside, saying that it was only a scratch and that he felt fine. In reality, the reason for his show of bravado was that the tourniquet was inside the bag and he didn't want to lose the morning's catch by emptying out the bag to find it! Worrell did not get sick, as the snake must have failed to envenomate him. Still, Serventy remained worried for the rest of the day. He decided to get his own back at Worrell for causing him so much worry: just before dawn the following morning, Serventy found a church service on the radio, positioned the radio close to Worrell's ears and turned it up loud just as a hymn was being sung. Worrell awoke with a start, relieved when he realised that it was only his friend playing a practical joke and that he hadn't really passed over to the other side.

# PART TWO

# WYOMING

7

# Where oranges grew, snakes now slithered

Worrell's ambitions had outstripped his small aquarium in quiet Ocean Beach by the late 1950s. Now his aim was to build a much bigger facility, which would focus on medical research – that is, research into the venom of medically significant snakes – and zoological research, with tourism a lesser consideration. Indeed, his initial idea was to call it the Australian Reptile Research Institute and concentrate exclusively on research at first, and he did not intend opening the facility to the public for about three years.

He planned to build his new facility on a 4.5-hectare former citrus orchard at Wyoming, just north of Gosford, halfway between Sydney and Newcastle, on the Pacific Highway, some 18 kilometres from Ocean Beach. The site was reasonably flat and a small watercourse, Wingello Creek, complete with colonies of eastern water dragons, flowed across the northern end.

Just as he had when he first established the aquarium at Ocean Beach, Worrell encountered some strident local opposition to his plans, this time not because of what he intended to keep on the land, but rather because locals wanted the land dedicated as open space for recreation. The Wyoming Progress Association sent a letter of protest to Gosford Shire Council, but the councillors accepted Worrell's argument that the site was the best possible land in the shire for his purposes and that open-space alternatives were available.

Worrell's grand vision was for an institution of 'modern attractive design' set in parkland.[1] He would landscape the park to feature unusual native plants that would provide naturalistic settings for the animals. This put him ahead of contemporary attitudes. The native gardens movement was just beginning to stir and Australians were still generally reluctant to consider the use of Australian plants in gardens or public parks and reserves.

In September 1957, Worrell decided against the overly formal-sounding name of 'Australian Reptile Research Institute' and registered 'Australian Reptile Park' as a business name instead. The estimated building cost was £6000, more than $150 000 in today's terms. Worrell's mother, Rita, a slightly built woman with an exuberant personality, was always supportive of her son's unorthodox career and would do whatever she could to help him achieve his dreams. She remained his greatest supporter throughout her life. In February 1958, just as Worrell was seeking building approval, she took out a mortgage on the Ocean Beach Aquarium property, perhaps so she could help realise her son's boyhood ambitions of having his own park where he could conduct research on reptiles.

Around this time, Worrell also bolstered his finances by requesting and receiving from his publisher an advance of £200, to be drawn down from the expected royalties from his book *Song of the Snake*, which was to be published the following year. Since 1953, Worrell had been working on the book, a semi-autobiographical collection of some of his better stories from journals such as *Outdoors and Fishing* and *Wild Life*. When it was published in 1958, the book made Worrell a hero in the eyes of many for his extensive first-hand experience of the Australian bush and his in-depth knowledge of the reptiles, birds, mammals and people who lived in it. *Song of the Snake* is remembered fondly by many interviewed for this book and for some it provided a template on which to base their own adventures. For instance, Neville Burns, a professional reptile handler and someone who worked for the park in the early 1990s, was so inspired after reading it that 'at 16, with $10 in my pocket, I hitchhiked to Gargett in North Queensland in search of pythons'.[2] The American herpetologist D. Bruce Means, in his book *Stalking the Plumed Serpent* (2008), mentions Worrell's book as a source of inspiration to visit the black tiger snakes of the Bass Strait islands. The year following the book's publication, Worrell was approached by the ABC, who wanted to make a film adaptation for the series *The Land and Its People*. Worrell was to have written the script, but the proposed dramatisation never eventuated.

From mid-1958, Worrell maintained reptiles, mostly venomous snakes used for the venom extraction program, at the new site, while Percy and Rita managed the Ocean Beach Aquarium for another year or so. Worrell put on his first paid staff member at the reptile park, Jon Dekkers, a Dutch immigrant he had met at a pet shop in Gosford where Dekkers was working. Dekkers cleared the old citrus orchard, erected fencing and collected logs and rocks to landscape the reptile pits. A highly skilled timber craftsman, he later constructed a replica of an old settler's hut, which Worrell called Drysdale Hut, after his friend Russell. Worrell always planned on using it as a museum or a small theatrette for showing natural history films in, but it was rarely open to the public. Dekkers taught himself how to strip the bark from ironbark trees for the hut's roof. He also helped with the care of the first exhibits and captured bandicoots for Worrell to use as hosts for ticks, which he supplied to the CSL for their work developing a paralysis tick serum.

The Australian Reptile Park opened to the public on the October long weekend of 1959. The gleaming new turnstile was still partially in its cardboard wrapping. Above the entry, a painting of the Rainbow Serpent extended along the length of the building, Worrell's acknowledgment of Aboriginal spirituality. In the two sets of exhibits that became known by keeping staff as the 'north and south corridors', there were glass-fronted reptile tanks, their interiors painted white and decorated in a rather austere style – containing a feature plant and gravel but little else, against a plain white background. On display were 'brown snakes, mainland taipans, a New Guinea taipan, common black snake, shingle back lizard, water, diamond and carpet pythons. In outside segregated pits, surrounded by shoulder-high walls are lizards, goannas, and further snakes such as tiger and black snakes.'[3]

The park's star attraction was a 4.1-metre-long saltwater crocodile, supposedly the second-longest in captivity at the time. It generated enormous interest from visitors, whose access to such monsters of the deep north was very limited. This crocodile, which originated from the Weipa district, on Queensland's Cape York Peninsula, died a few months later, due to illness brought about by cold weather. Still, visitors streamed through the gates. Around 8000 visitors came during the summer school holidays of 1959–60, many drawn there by a 4.8-metre python at the park that had been featured on TV.

Another drawcard were 14 Papuan taipans and 10 Papuan black snakes Worrell had received in late 1959 courtesy of Ken Slater and the Division of Animal Industry, Territory of Papua and New Guinea. Slater, who had accompanied Worrell and their mutual friend George Cann on the first major tiger snake hunt to the Lachlan River in 1951, had moved to the Australian Territory of Papua (now part of Papua New Guinea) in 1952. A fitter and turner by trade, he initially worked there for an oil company but then became an animal ecologist with the Department of Agriculture Administration, based at Port Moresby. Like Worrell, he did not have any formal qualifications in ecology or zoology but was self-taught, learning from careful reading of the available literature as well as his own field observations. While in Papua, he described a subspecies of taipan – which he named *Oxyuranus scutellatus canni*, after George Cann – and began an association with the CSL, providing venoms from the taipan and other species such as the Papuan black snake (*Pseudechis papuanus*) and New Guinea small-eyed snake (*Micropechis ikaleka*). Using the snakes Slater had given him, and others he had collected on his behalf over the next decade or so, Worrell became the main supplier of venom from the Papuan black snake, which was considered responsible for the deaths of many people in the southern parts of the territory.

Worrell and Slater were good friends, even though Worrell was perhaps more focused on his work than keeping in contact with people. On a visit home from Papua in 1957, Slater had written to Worrell and Carol:

> Well old friend, I shall be seeing you soon, for Margie (my future wife) and I shall be in Sydney on the 12th November until the 17th when we will pass through Woy Woy on the way to Forster for a quiet fishing holiday. We would like to spend Sunday 17th with you and then drive on the following morning … Please, please Carol, get hold of him and twist his arm until he writes a reply to this letter, and don't let him put over that rubbish about being busy. I am busy too – damned busy – still I can always find a few moments to keep in touch with my friends.[4]

When Slater returned to Australia permanently in 1959, Worrell employed him as 'park ecologist', to undertake ecological studies of reptiles, a position well ahead of its time in the world of zoos.

This appointment was a means of making real one of Worrell's main intentions of the park – for it to be a place where research could be undertaken on reptiles. Worrell had already developed a simple mission statement for the park comprising three words: 'Research, Education, Conservation'. Slater's duties included the milking of snakes and he suffered a couple of bites from tiger snakes and another from a brown snake. In early 1960, he carried out a series of experiments on the venom toxicity of the black tiger snakes that Worrell had captured in Bass Strait. He made finely detailed and carefully crafted diagrams of the skull and dental characteristics of elapid snakes – the group of front-fanged venomous land snakes that accounts for the majority of Australia's snakes – that Worrell used in his taxonomic work. These diagrams were later reproduced in Worrell's most ambitious writing project, his book *Reptiles of Australia,* published in 1963. Slater also contributed to the Checklist of Australian Reptiles that appeared in the same book. Worrell and Slater were active members of the Central Coast Natural History Society, through which they established and co-edited a short-lived periodical they named *Cycad* in 1960.

The position at the reptile park should have suited Slater down to the ground, but he was soon dissatisfied. He had disagreements with Worrell about aspects of snake husbandry and he felt that Worrell's approach to scientific methodology was not as disciplined as it should be. Worrell sometimes had a cavalier approach to the truth; he might exaggerate a claim, or embellish or spice up a story to make it more compelling to a popular audience. These were usually very minor things, such as claiming to have bred a particularly attractive blue-tongue lizard in captivity when in fact it had been donated to the park, or claiming that he had set off on his own to capture taipans when in reality the expedition had been organised by John Dwyer. Such inaccuracies annoyed and irritated Slater, who was a stickler for absolute truth. Another matter that would have annoyed Slater was the lack of detailed notes accompanying the preserved reptile specimens that Worrell maintained. Worrell tended to rely on his own memory, which many said was encyclopaedic and usually reliable, but this was still an unconventional way of maintaining information about such a collection. This perhaps did not sit well with Slater's scientific stance.

Slater was also so concerned about a couple of aspects of venom production at the park that he wrote to Dr John Graydon, a research scientist involved in the production of antivenoms at the CSL. Worrell displayed dried venoms to visitors through a window into the venom laboratory. This exposed the venom to sunlight and Slater argued that this potentially degraded its potency, rendering it less suitable for antivenom production. The other issue was Worrell's reluctance to allow him to force-feed Papuan black snakes that had been refusing to eat since they had arrived at the park. When force-feeding was begun, 'venom yields jumped up amazingly after one week and now they are nearing the yield that can be expected for Papuan Blacks, allowing for their poor condition'.[5]

Slater left his job at the park less than a year after he started, and by then his friendship with Worrell was over. Slater worked for a short time in his original trade of fitting and turning before he, his wife and children sold their house in Wyoming and moved to South Australia, where he worked as a senior wildlife officer with the South Australian government, initially in the animal industry branch of the Department of Agriculture.

Slater's exact role would never be filled again at the park, but just as he was exiting, a new reptile enthusiast was beginning an association with Worrell and would soon be working for him on a casual basis. Towards the end of the 1950s, Peter Krauss had trained as a zookeeper at Stuttgart Zoo and had then worked for a short time for an animal dealer in Germany. His work exposed him to exotic animals from all over the world, giving him a healthy appetite for adventures in foreign places. Australia was offering assisted passage at the time and so in 1958, at the age of 20, he and a friend set sail.

After a stint working as an animal attendant at one of the CSL's primate facilities at Melbourne Zoo, Krauss took care of animals at the CSIRO's Animal Health Laboratory at Glebe, in Sydney. He had a passion for reptiles and often made the hour-or-so drive up to the Central Coast's Ourimbah district, as the sandstone ridges and gullies had a reputation as a hot spot for reptiles. It was during these trips that he got to know Worrell, remembering him as being supportive and generous in providing advice and information.

It was not long before Krauss began collecting for Worrell when he needed a particular species of reptile. Eventually, Krauss and his girlfriend Raija, a recently arrived Finnish immigrant who would later become his wife, began making trips down to the Riverina district of New South Wales looking for snakes needed by Worrell for venom. Finding brown snakes large enough for milking was not at all easy. Krauss recalled one snake they encountered on a trip to the Griffith area: 'I was trying to grab this brown that had got into a hole under a tussock, but it didn't go down the hole – it curled around and came back and gave me a quick bite – and it did bleed, it was a proper bite.' Tourniquets were still part of snakebite treatment, so one was applied – but Raija was still worried and wanted him to get medical help. Krauss wanted to keep collecting. They made a compromise, driving to the nearest hospital and camping outside on the lawn until several hours had passed without Krauss getting sick. 'We had a lot of magazines to read and there we were, sitting in the shade of a tree on the lawn and absolutely nothing happened,' he said. To their relief, the snake hadn't injected Krauss with venom and so they were able to get on with the work of collecting snakes.

Krauss recalled another trip that he and Raija made, this time into central Australia in search of Australia's largest lizard, the perentie, a species of goanna that can grow to more than 2 metres in length. The couple was dispatched to the arid interior of the continent, towing Worrell's battered old trailer behind their VW Beetle.

The trip was a success and they returned with three perenties, which were kept at the park. Krauss continued to do occasional jobs for Worrell, building a solid relationship with him.


Slater's resignation and the end of their friendship must have been a blow to Worrell, yet his output, and the achievements of his park, were immense in the first full year it was open to the public, 1960. In addition to overseeing the running of the park and the production and supply of the dried venom of nine species of snakes to the CSL, he made three major field trips to Bass Strait and to Lake Alexandrina in South Australia to collect reptiles. He compiled a checklist of the complete Australian reptile fauna, with Slater's assistance; delivered the final revision of the *Reptiles of Australia* manuscript to his publisher, Angus and Robertson; had two taxonomic papers published in scientific journals; and completely revised *Dangerous Snakes of Australia*, expanding it to include New Guinea. He also supplied live Sydney funnel-webs and paralysis ticks to the CSL for their research.

Meanwhile, he was never far from the public's view, giving numerous talks on ABC radio on topics such as snakebite first aid, taipans, and conservation, and in addition he made, produced or appeared on shows on topics spanning natural history, agriculture, forestry and tourism. He also worked on an educational film of *Dangerous Snakes of Australia* with Heinz Salkow, with whom he would go on to make several films.

After early one-off appearances on television, including an hour-long program with the curious title 'They don't shoot bathtubs, do they?', in which he built a strong but pragmatic case for wildlife conservation, and some 50 short films on a range of animals and rural industries, from the late 1950s to early 1960s, Worrell was a regular on an ABC TV show called *Animal Life and People's Pets*. Hosted by a young veterinary surgeon, Pam Tinsley, the program covered a wide variety of animals. Worrell was brought on as a native fauna expert. Tinsley recalled Worrell, attired in his best safari suit, conjuring up all manner of animals out of an assortment of sacks, bags and cages. The program went to air live at 8.45 pm, after a rehearsal around 5.00 or 6.00 pm. Part of the entertainment value of live shows such as this was the unpredictability of the animals.

'We had a lot of funny things happen,' Tinsley recalled. 'He brought a dingo on once, Diana, and she escaped from his ute and we couldn't find her, so they broadcast it on the ABC radio and she was found happily ensconced in a pub in Surry Hills and so Eric picked her up and we got her on the program then.

'We used to have Tasmanian devils on, with these enormous jaws and teeth, and Eric would just handle them with aplomb.' One night Worrell had borrowed a jumper from Tinsley's husband, Oscar. 'An enormous goanna sort of wrapped itself around him and Oscar said, "I'll never wear that sweater of mine again." It had relieved itself all over it, on camera! Lots of funny things happened – a wedge-tailed eagle swept across a studio once and landed on the sound boom and frightened the hell out of the cameraman.'

By now, the Ocean Beach Aquarium had closed and Worrell's parents, Percy and Rita, had moved into a house he'd built for them in the reptile park grounds. They lived there rent free and in return acted as caretakers. Rita would later run a kiosk in the park and then a small souvenir shop. She was a constant presence around the grounds, helping out. She invariably carried a whistle around her neck in case she needed to scold young visitors behaving badly. On frosty winter mornings before the gates opened she could be seen sweeping the paths – with newspapers wrapped around her legs below her skirt to keep them warm!

In the early 1960s, the park grounds continued to take shape. By 1962, a total of six pits had been made to house reptiles, and another for Tasmanian devils, while banks of reptile cases enabled Worrell to keep species from more tropical climates. There was a picnic area, where kangaroos grazed, and a small cage that housed foxes. Worrell experimented for a while with letting wombats roam free around the park, but their tendency to nip the ankles of unwary visitors and bulldoze young children meant they had to go into enclosures. Worrell had got the idea to keep free-ranging kangaroos, wallabies and waterfowl in the picnic area after observing the interactions between the residents of Morisset Mental Hospital near Lake Macquarie, on the NSW central coast, and native animals. He would often collect red-bellied black snakes in the wetlands near the hospital and had noticed how much enjoyment the free-ranging kangaroos gave to the residents, who fed the animals pieces of bread through the iron grates of their windows. He wrote, 'The moving experience of actually seeing … the pleasure that the patients derived from animal friendship led me to establish the tame kangaroo herd at the Australian Reptile Park.'[6]

A fairly substantial area was given over to the bandicoots used as hosts for the paralysis ticks supplied to the CSL. Adjacent to that was a display of baby animals, injured or orphaned joeys and baby wombats. Visitors could walk along a nature trail through bushland along the eastern end of the park, where wallabies were kept.

In the park's guidebook – first published in 1962 and a substantial publication in its own right, running to 50 pages and well illustrated with his own photographs – Worrell made it clear it was not a zoo, but rather a 'reptile research station set in natural parklands. Animals are confined only for special purposes.' He described the park's main focus as 'the collection and processing of various snake venoms' and listed a number of other activities including the breeding of laboratory animals such as rats and mice, tick-serum immunisation, bandicoot breeding, documentary film production and technical photography. 'All work is directed towards the building of a unique living museum in which the morphological exhibits stand side by side with the living reptiles for ready comparison,'[7] the guidebook said.

Although Worrell was very focused on research, he also quickly came to recognise the tourist appeal of the park and incorporated a kiosk, gift shop, train and camel rides, picnic area and barbecues. Visitation was more than just an end in itself – it provided an audience for the educational messages that were presented in signage and through his talks, which he hoped would equip Australians with factual information that would reduce the likelihood of death from snakebite and heighten national pride in Australia's fauna, and to a lesser extent, flora. His next major project was to create the iconic roadside marker that was to be emblematic of the Australian Reptile Park.

He would build a dinosaur.

Dear Father Christmas,
For Christmas I would like either a blotched blue tongue lizard, a blue-tongue with pink spots, or a fish aquarium with two tropical fish, or a rainbow lorrikeet, a young cock one, or a cinama projector, or a parrot's cage and a bag of canary seed.

I would like one of these items, more if there is enough money which mum hassent.

Yours Faithfully
Eric Worrell

P.S. or a steam engine. please
My address is 18½ Cecily Street
Lilyfield
if possible I would also ytact.



Eric Worrell's letter to Santa, circa 1934.

Eric Worrell holds a python, Northern Territory, 1940s.

Eric Worrell holds a large python, Northern Territory, 1940s.

ARP Collection

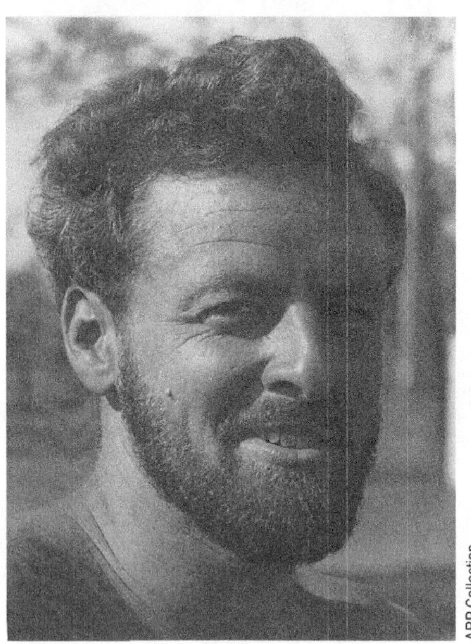

A newly bearded Eric Worrell during his time in the Northern Territory in the 1940s.

ARP Collection

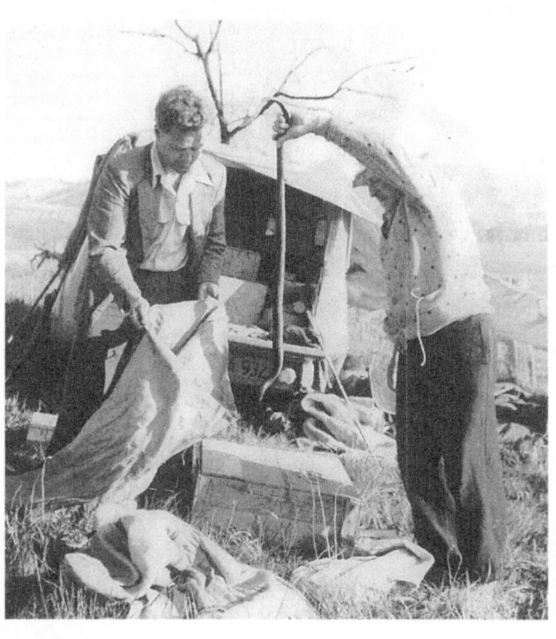

George Cann Snr manipulates an eastern brown snake into a bag held by Worrell, early 1950s.

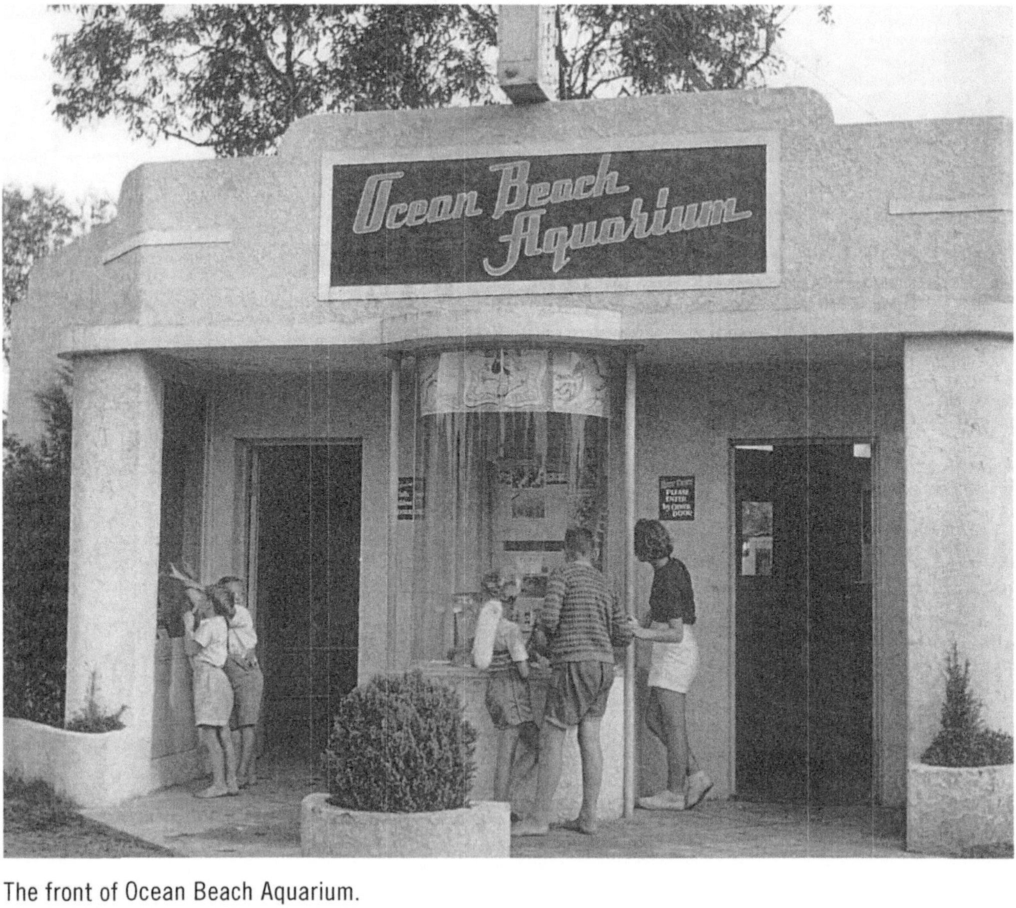


ARP Collection

The front of Ocean Beach Aquarium.

Eric Worrell with the freshwater crocodile that had escaped from its enclosure at the University of Sydney. He wrote on the photograph 'From your 'andsome 'erpetologist'.

Eric Worrell, the naturalist, Ocean Beach Aquarium, late 1950s.

Eric Worrell milks a tiger snake, Ocean Beach Aquarium, 1951.

The place where Worrell, Dwyer and Lorking found several taipans on their first expedition to Cairns in 1952. Worrell annotated the photograph to show where the taipans were caught.

ARP Collection

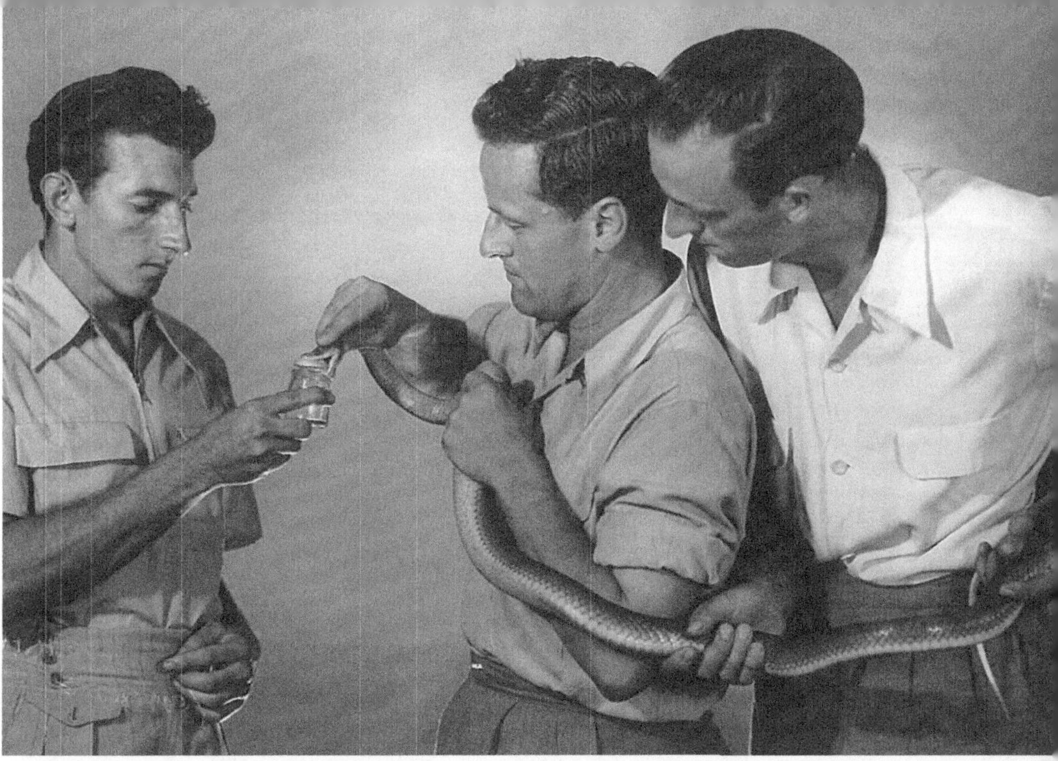

Milking the taipan caught on that first trip to Cairns in 1952. Worrell holds the snake's head while John Dwyer holds the glass jar and Wal Lorking peers over Worrell's shoulder. Had the snake bitten one of them, the most likely outcome would have been death.

ARP Collection

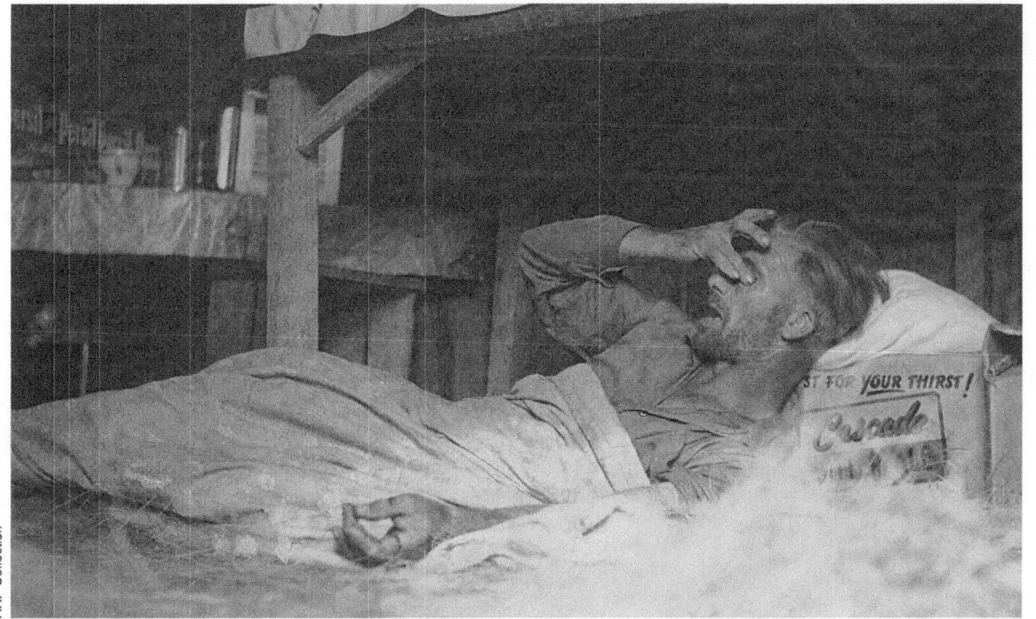

Worrell's photograph of his mate from the Murray Riverlands, Eric West, after a tiger snake bit him through the bag in which it was being carried on Chappell Island in 1955 captures both the pain West was experiencing and the rough conditions in which the mutton-birders lived during the season. West recovered.

Jeff Carter

Looking into the reptile park's central courtyard, where most of the venomous snakes were housed, in the early 1960s. The red-bellied black snake pit is on the left. Note the young free-ranging wombat near the kiosk.

ARP Collection

Eric Worrell with sculptor Ken Mayfield, in front of the Gosford Dinosaur, Wyoming, 1963.

A man and his dinosaur.
Jeff Carter

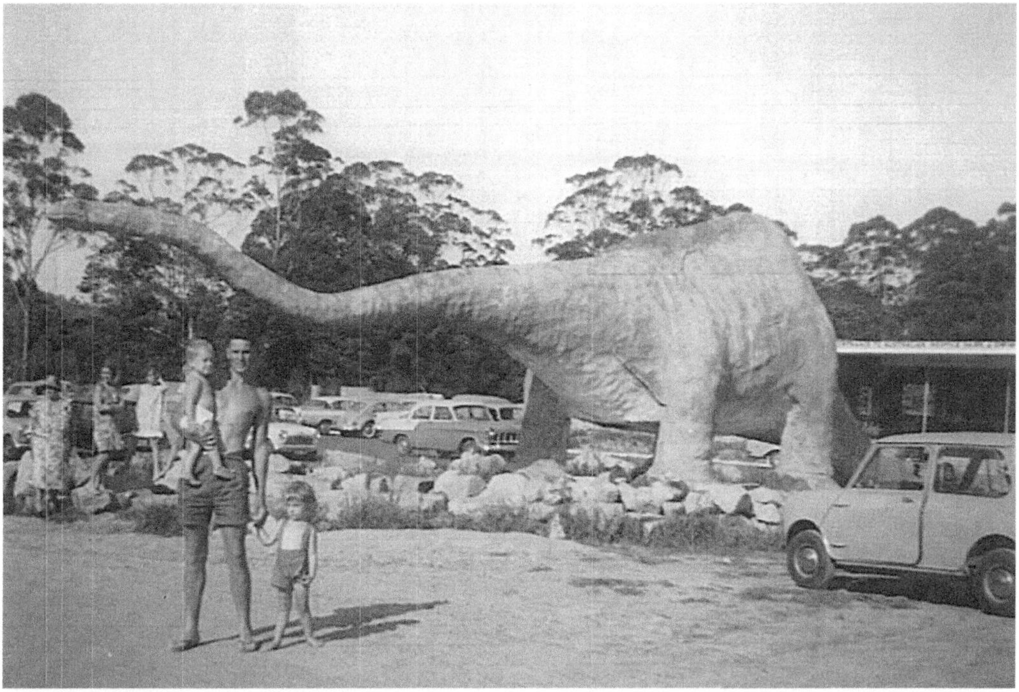

ARP Collection

Many hundreds of thousands of photographs similar to this would have been taken by tourists during the 33 years that the dinosaur stood sentinel in front of the reptile park, Wyoming.

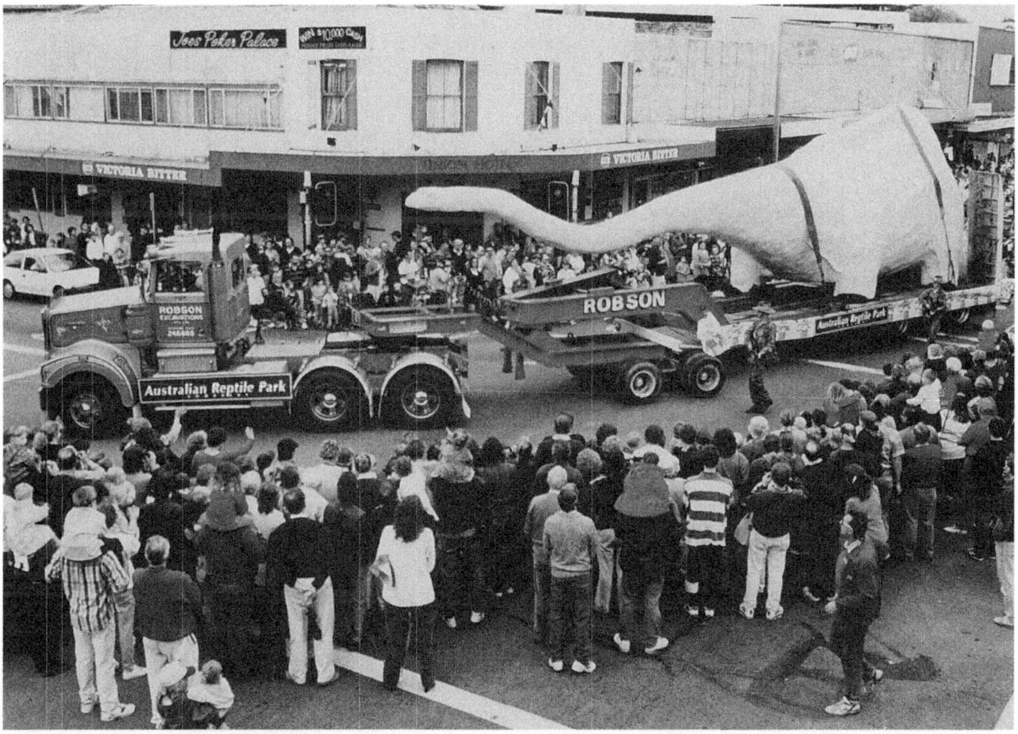


ARP Collection

The dinosaur receives superstar treatment as it makes its way on the back of a semi-trailer up Mann Street, Gosford, during the Ploddy Parade, 1996.

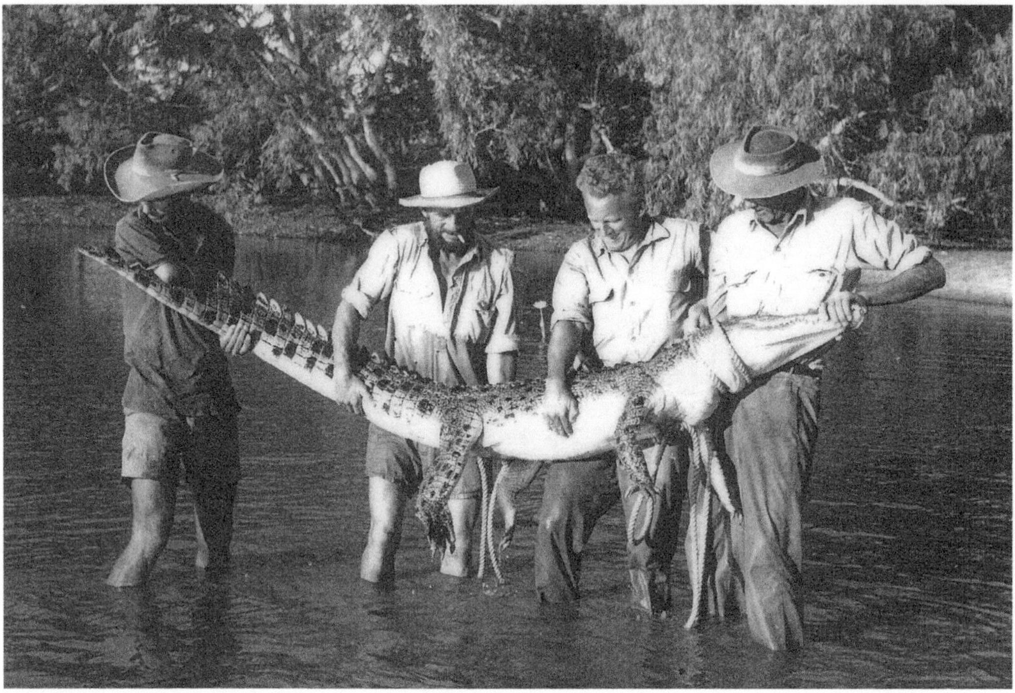

ARP Collection

Capturing a saltwater crocodile during 'Operation Crocodile' in Queensland, 1964. Local man, Terry Arnold, holds on to the animal's tail.

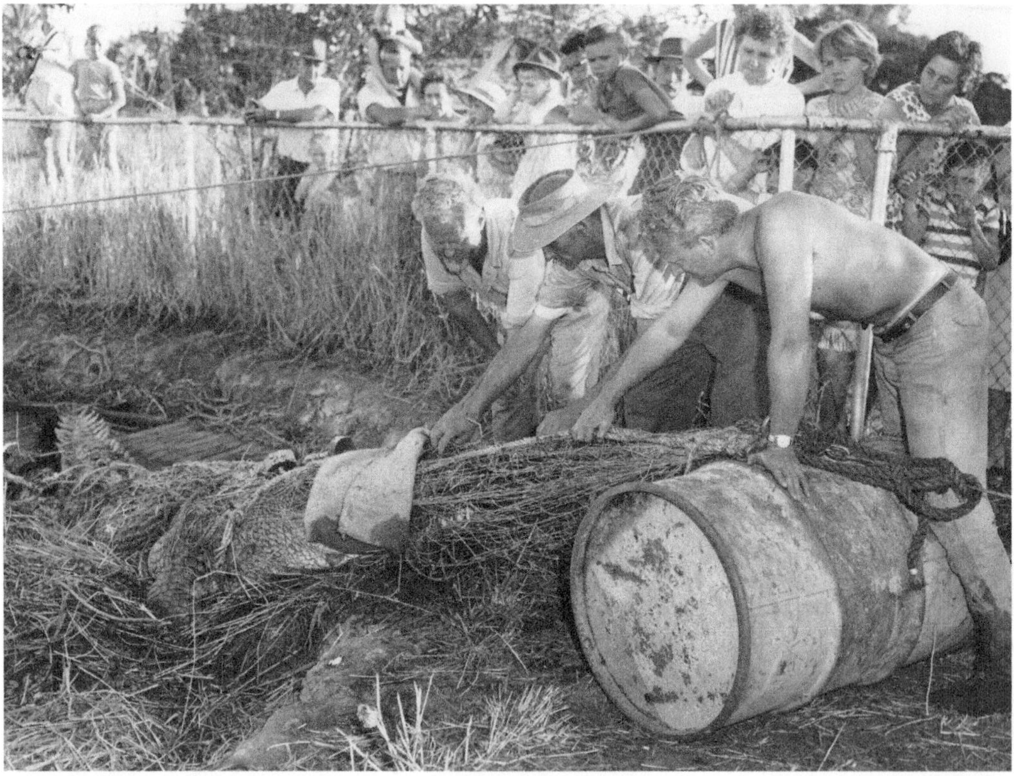

ARP Collection

Casanova the crocodile after being netted in Townsville in 1964.

A grinning Eric Worrell and Arnold Gaunt stand triumphantly in front of a now subdued Casanova, Townsville, 1964. Athol Compton is next to Eric.

Eric Worrell with Casanova at Wyoming, 1964. An American alligator can be seen in the enclosure behind Casanova.

Sometimes snakes need to be force-fed. Here Eric Worrell uses a large stainless steel syringe attached to more than a metre of rubber tubing to push a food mixture into the gullet of a king cobra while Jack Green and Robyn Innes hold the snake still.

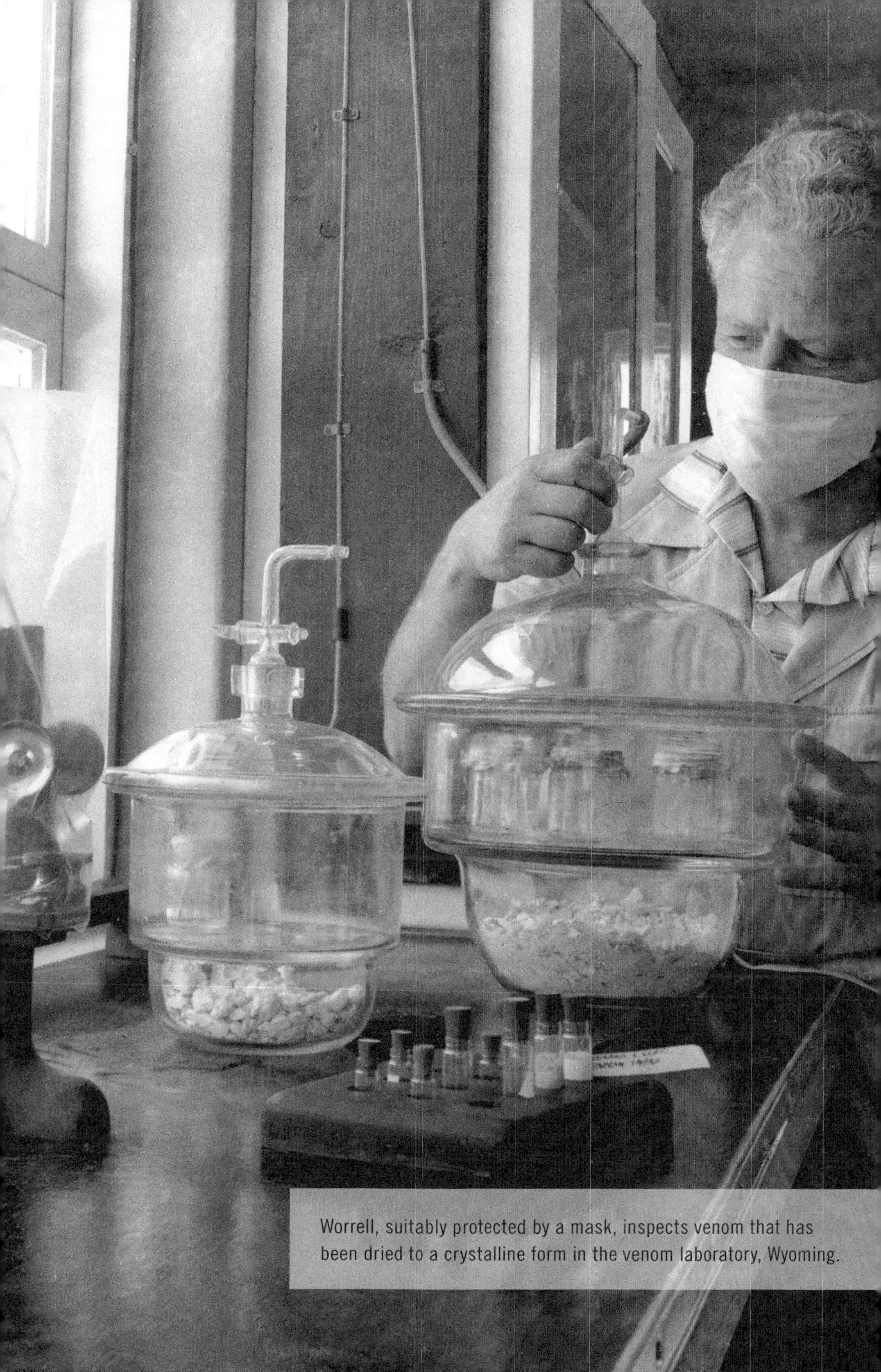

Worrell, suitably protected by a mask, inspects venom that has been dried to a crystalline form in the venom laboratory, Wyoming.

Worrell milks a red-bellied black snake for the benefit of visitors.

ARP Collection

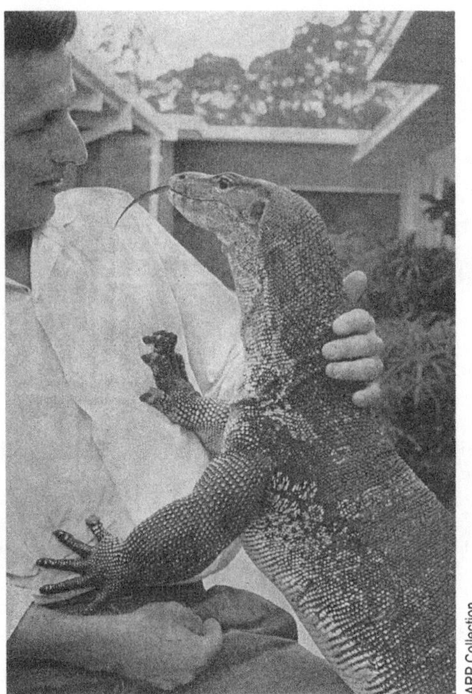

Australian Reptile Park curator in the 1960s, Peter Krauss, with a friendly monitor, Wyoming.

ARP Collection

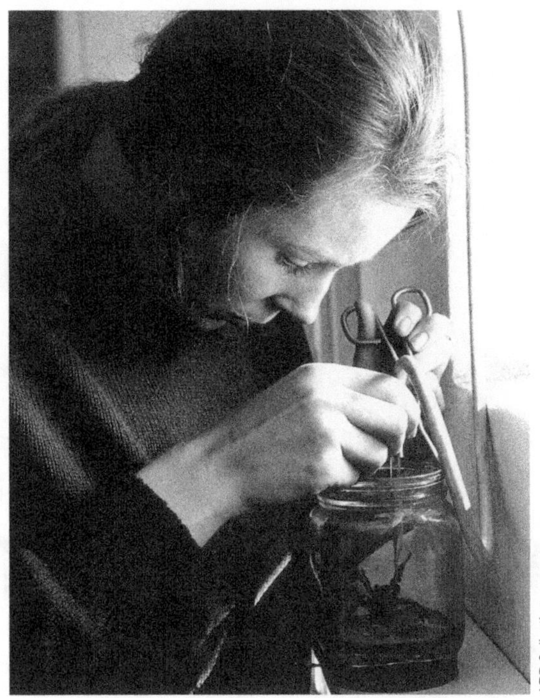

Robyn Worrell painstakingly extracts a few drops of venom from the fangs of a funnel-web spider to be used in the CSL's research into an antivenom.

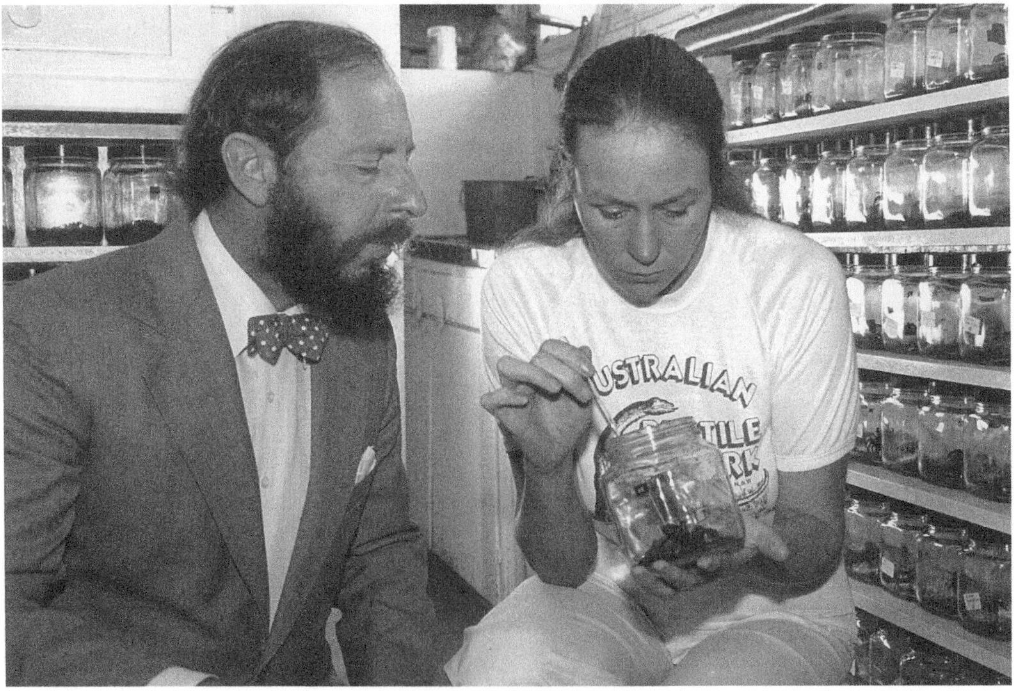


ARP Collection

Robyn Worrell and Dr Struan Sutherland look at a Sydney funnel-web spider in the park's spider laboratory, Wyoming, circa 1980.

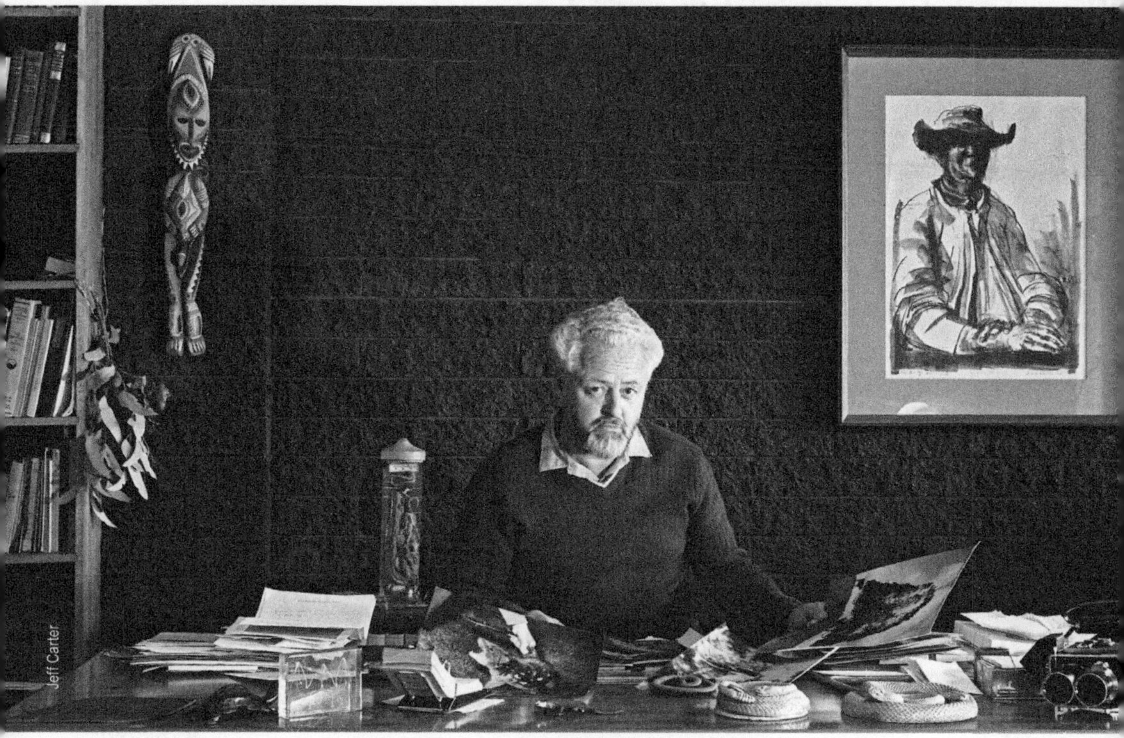

Jeff Carter

Eric Worrell at the desk in his study at the park in Wyoming, circa 1972.

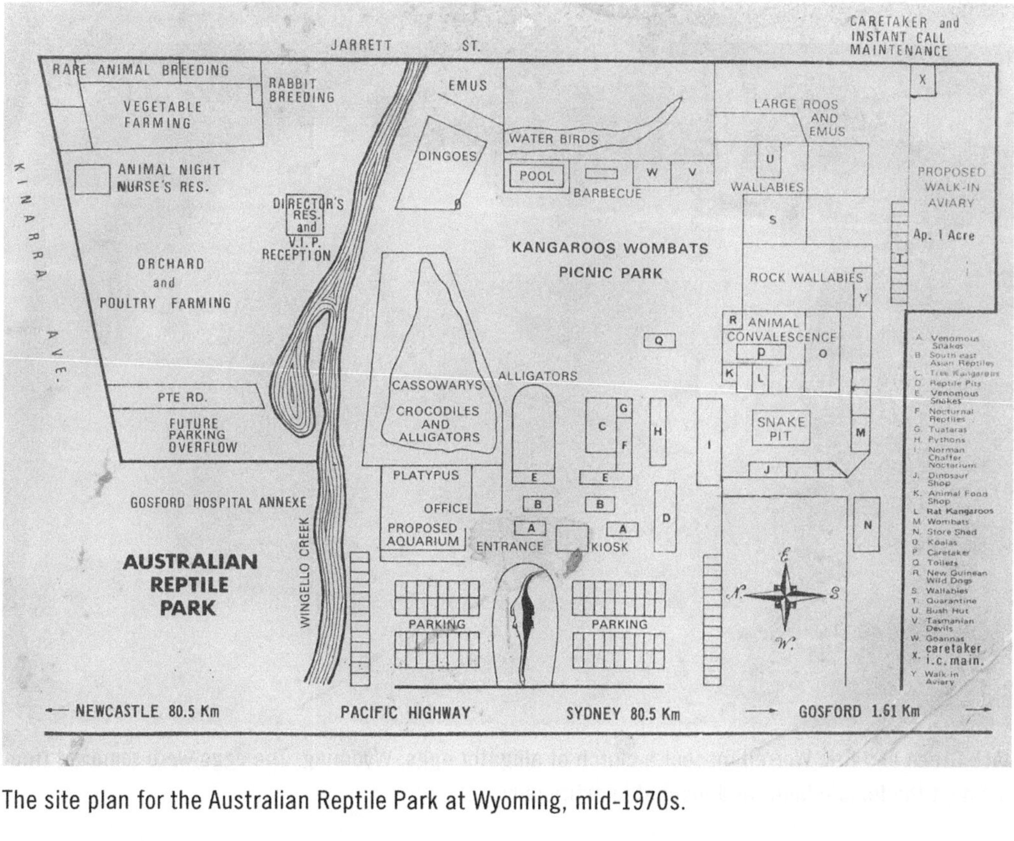



ARP Collection

The site plan for the Australian Reptile Park at Wyoming, mid-1970s.

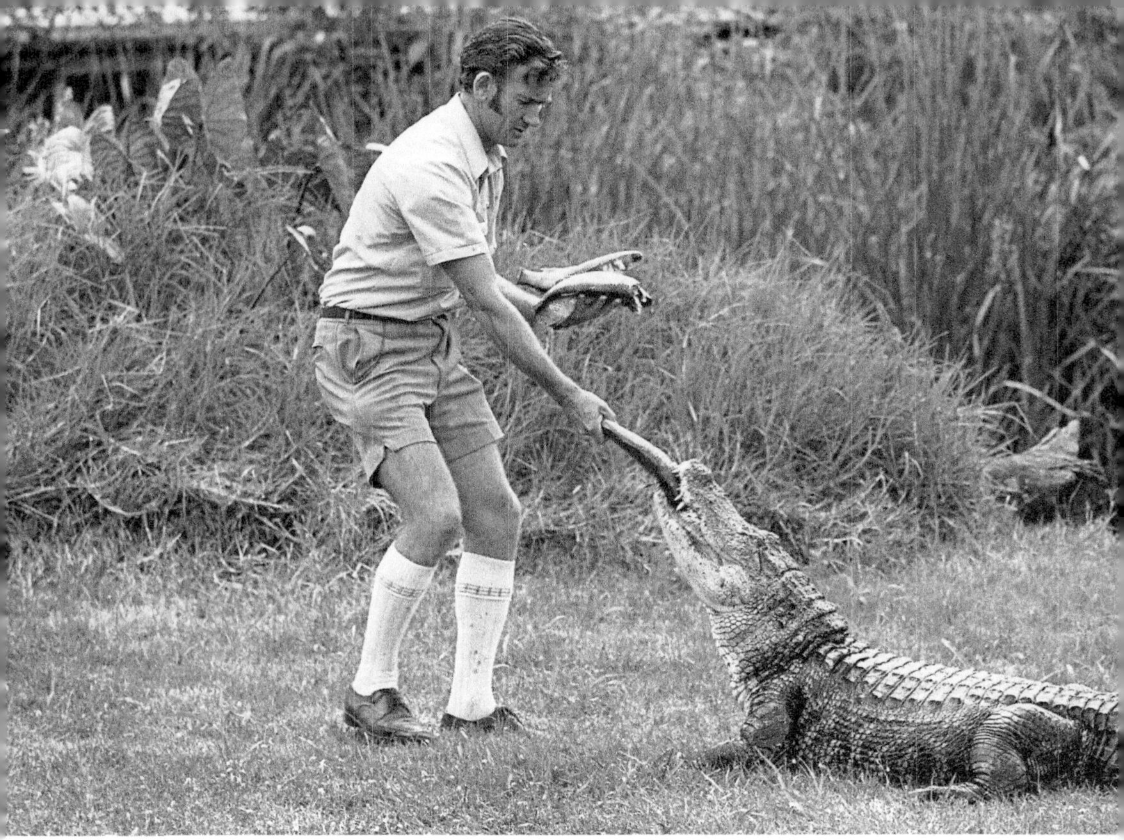

An agile Jack Green offers a mullet to one of the park's large American alligators, in 1972.

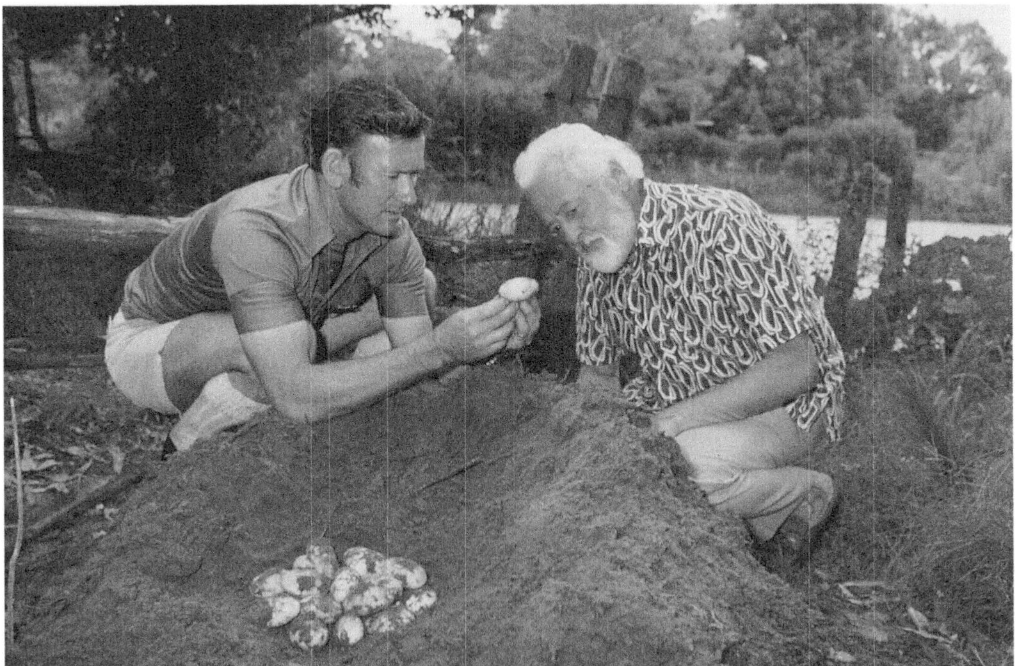

ARP Collection

Jack Green and Eric Worrell inspect a clutch of alligator eggs, Wyoming. The eggs were removed from the nest the female built, and incubated artificially.

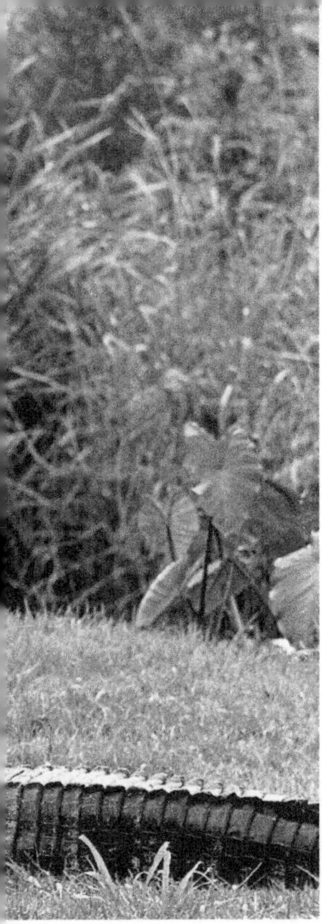

Jeff Carter

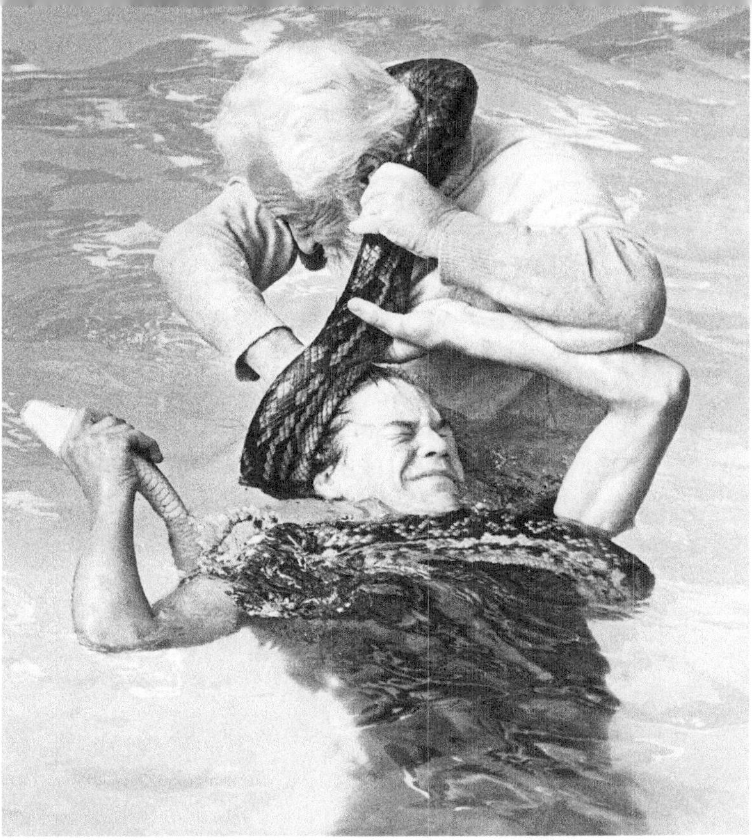

Sydney Morning Herald/Russell McPhedran

Young reptile keeper Steve McEwen, entangled in the coils of a scrub python while Eric Worrell grabs hold.

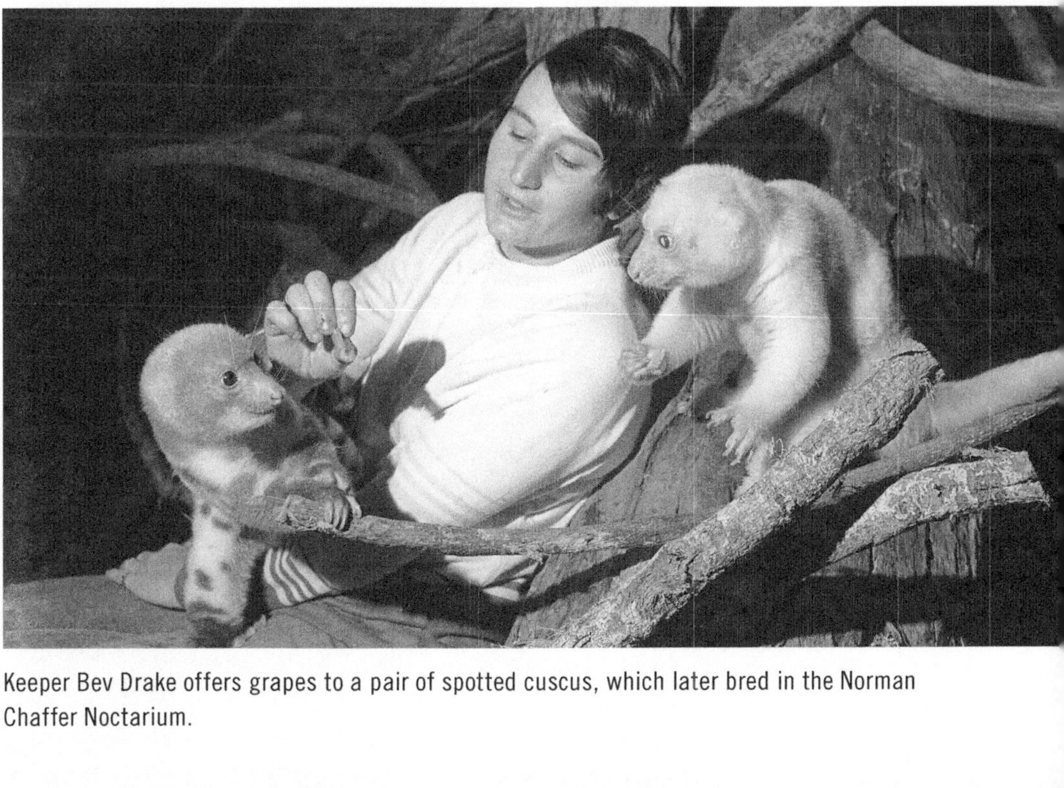

Jeff Carter

Keeper Bev Drake offers grapes to a pair of spotted cuscus, which later bred in the Norman Chaffer Noctarium.

Lyn Abra pins down an eastern tiger snake during a collecting expedition to the Lachlan River, filmed by Jeff and Mare Carter for their documentary, *Tiger Man*, 1972.

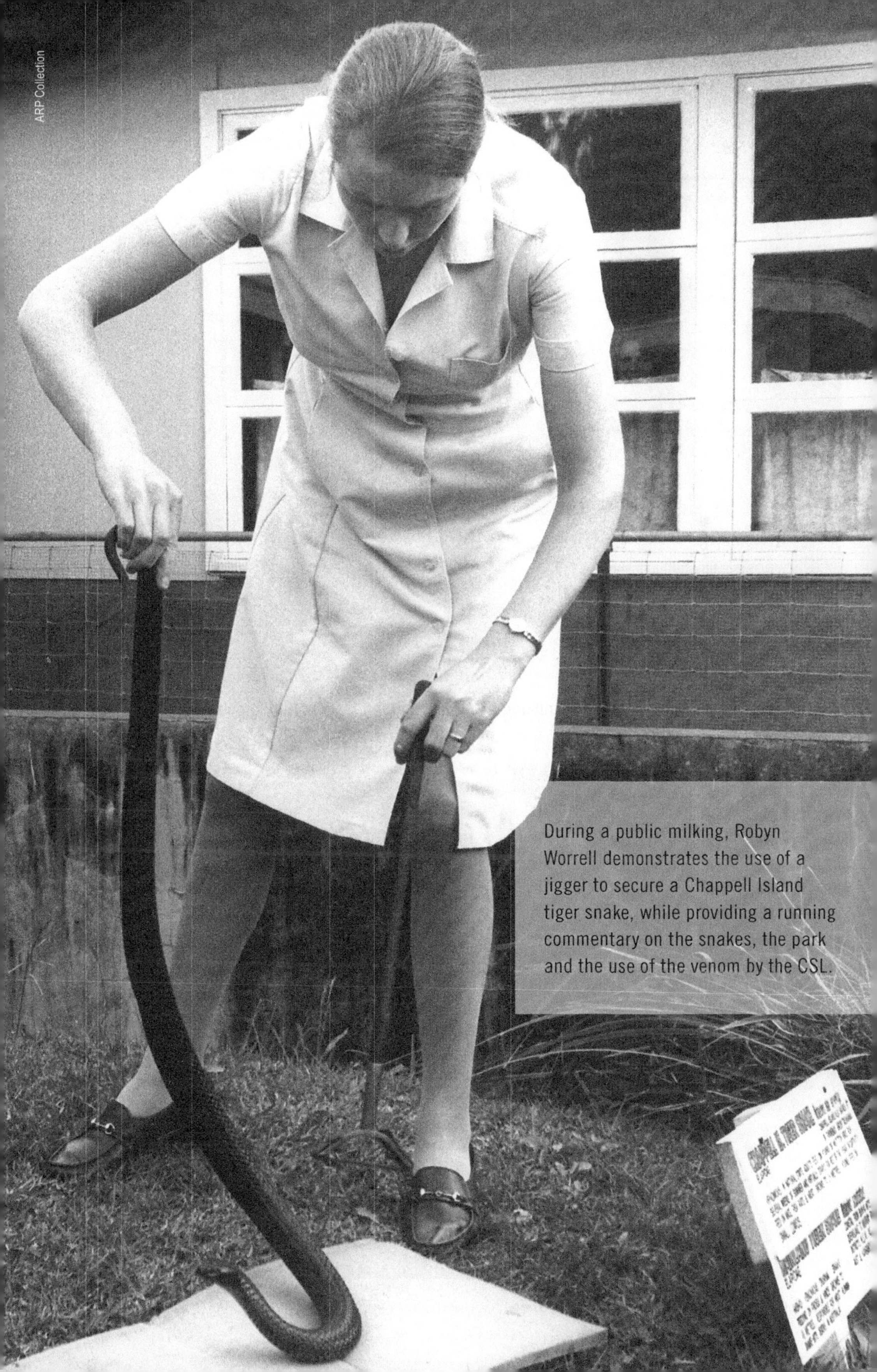



During a public milking, Robyn Worrell demonstrates the use of a jigger to secure a Chappell Island tiger snake, while providing a running commentary on the snakes, the park and the use of the venom by the CSL.

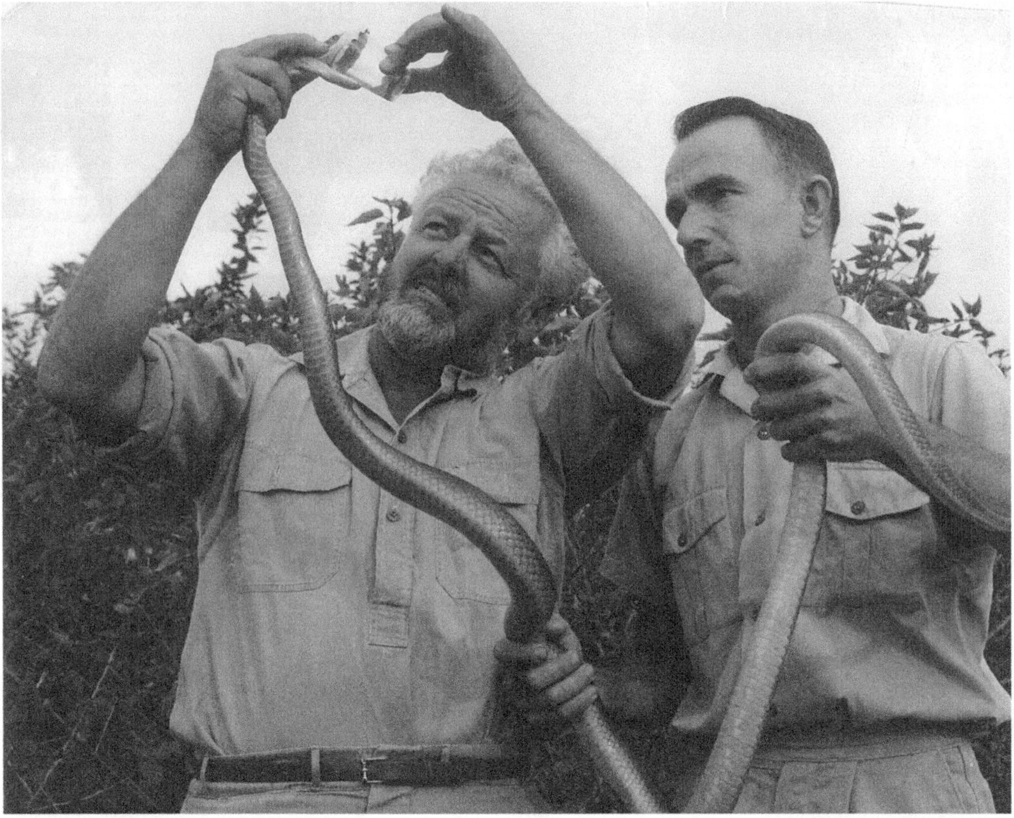

ARP Collection

Worrell exhibits his showmanship in milking a taipan, as John McLoughlin supports the impressively long snake.

ARP Collection

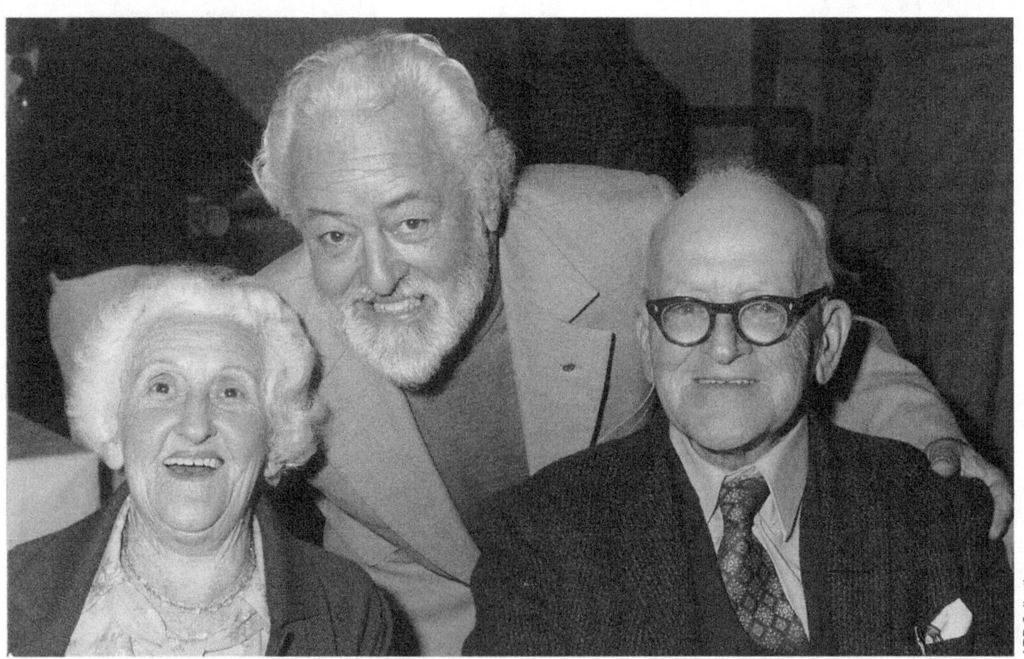

Rita, Eric and Percy Worrell, circa 1976.

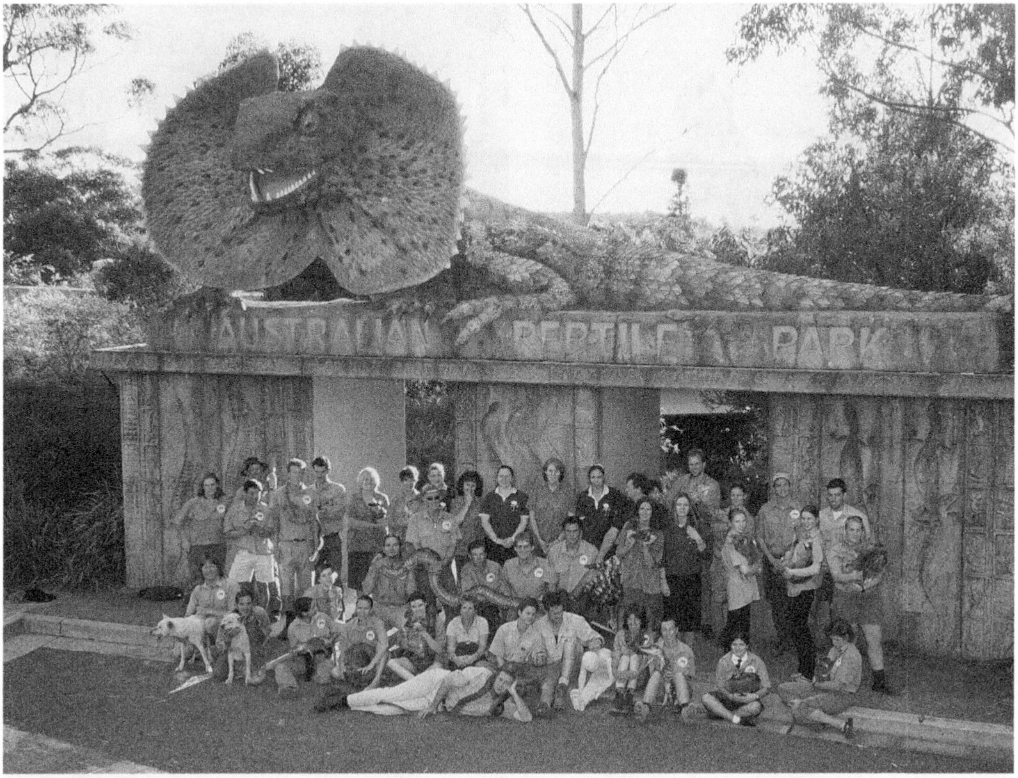


ARP Collection

Staff, volunteers and some animal friends at the front entry of the new park, Somersby.

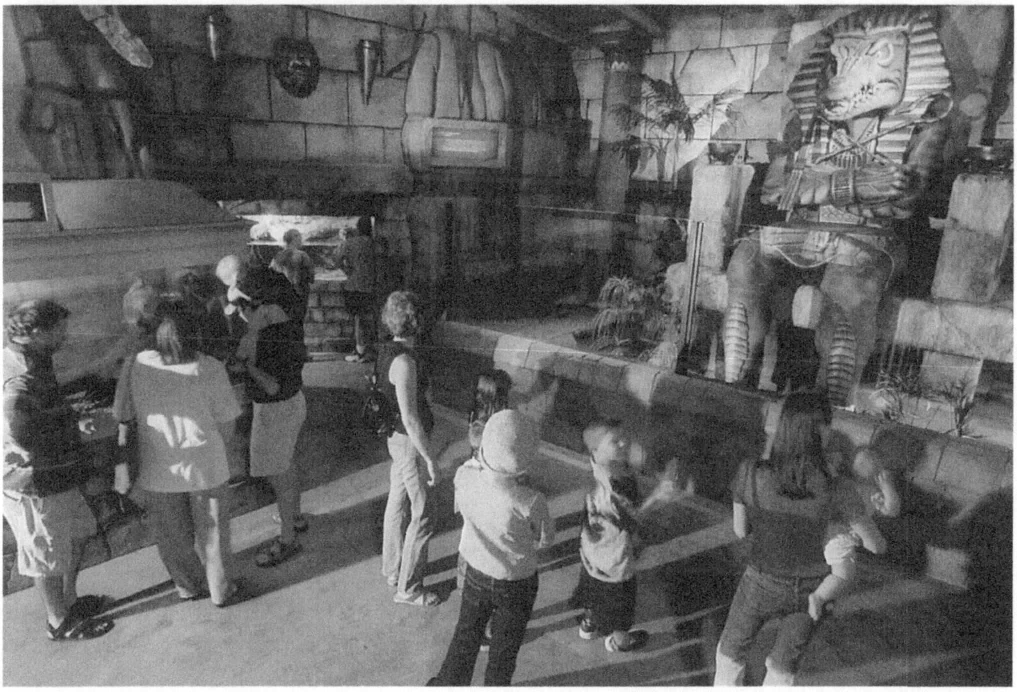

ARP Collection

Visitors listening to what the crocodile god, Sobek, has to say about the magnificence of reptiles, Lost World of Reptiles exhibit, Somersby.

Jeff Carter

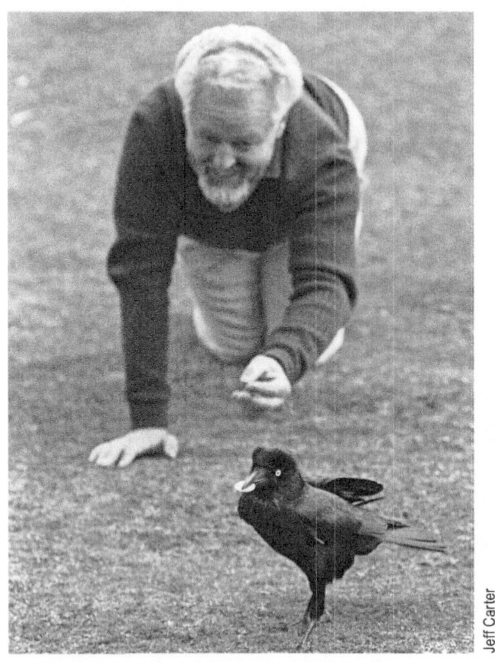

Athol the raven leaving the scene of the crime with a beakful of evidence.

ARP Collection

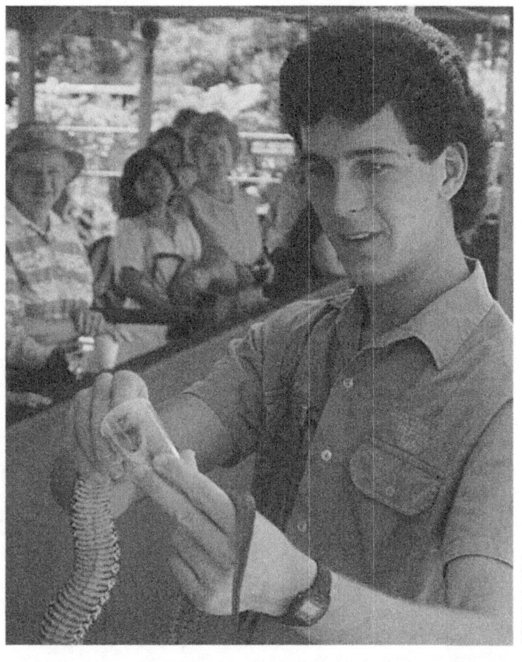

Reptile keeper Robert Montgomery milking a tiger snake, Somersby.

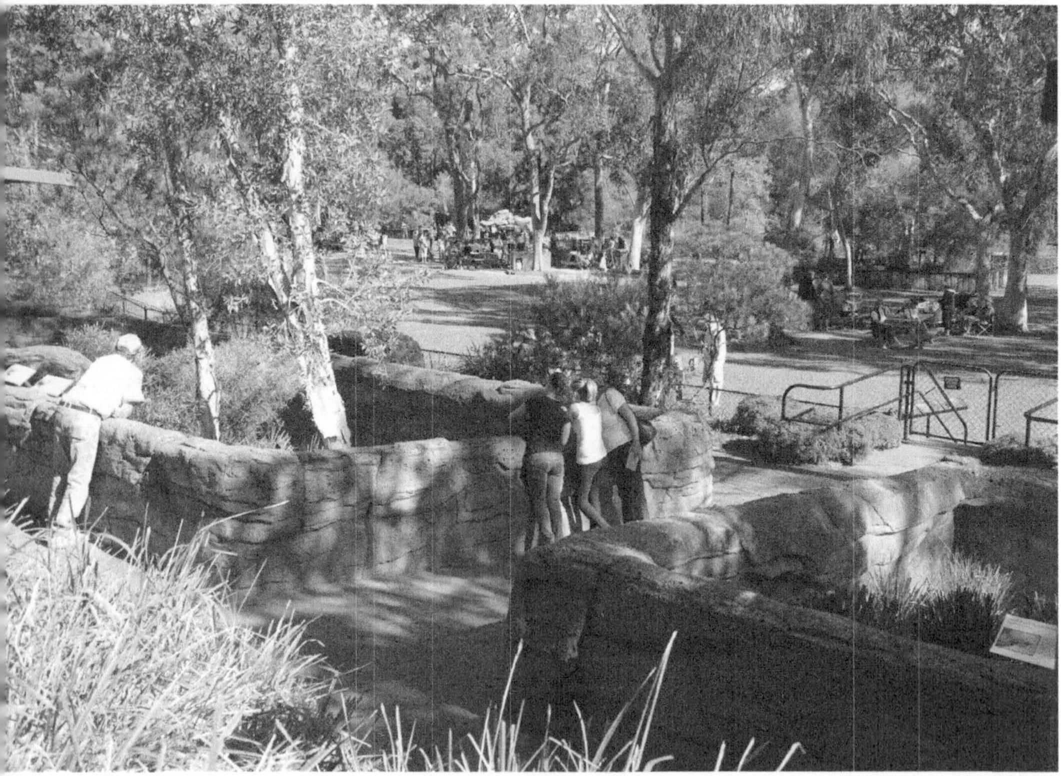

John Weigel

The park makes the most of its topography and bushland setting. In the foreground open reptile pits above the picnic area with its free-ranging kangaroos, 2010.

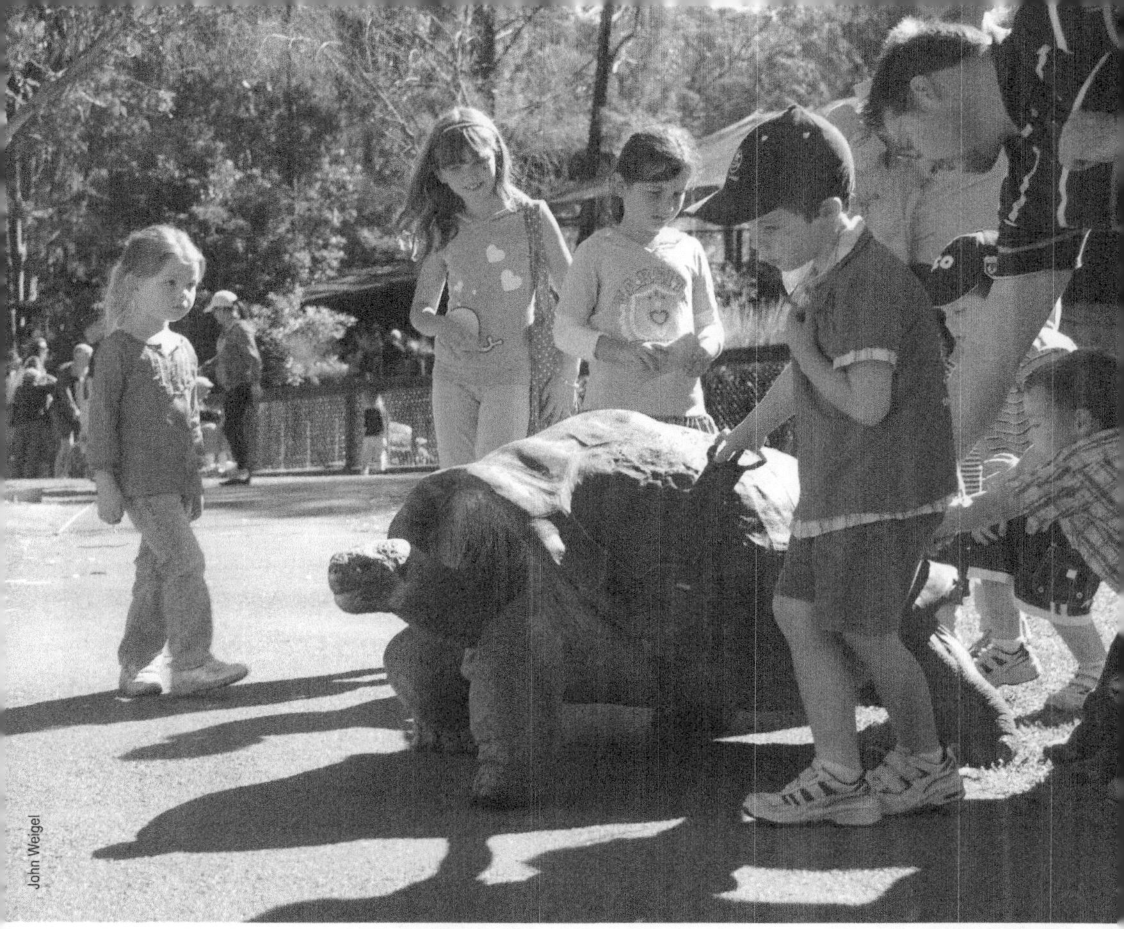

John Weigel

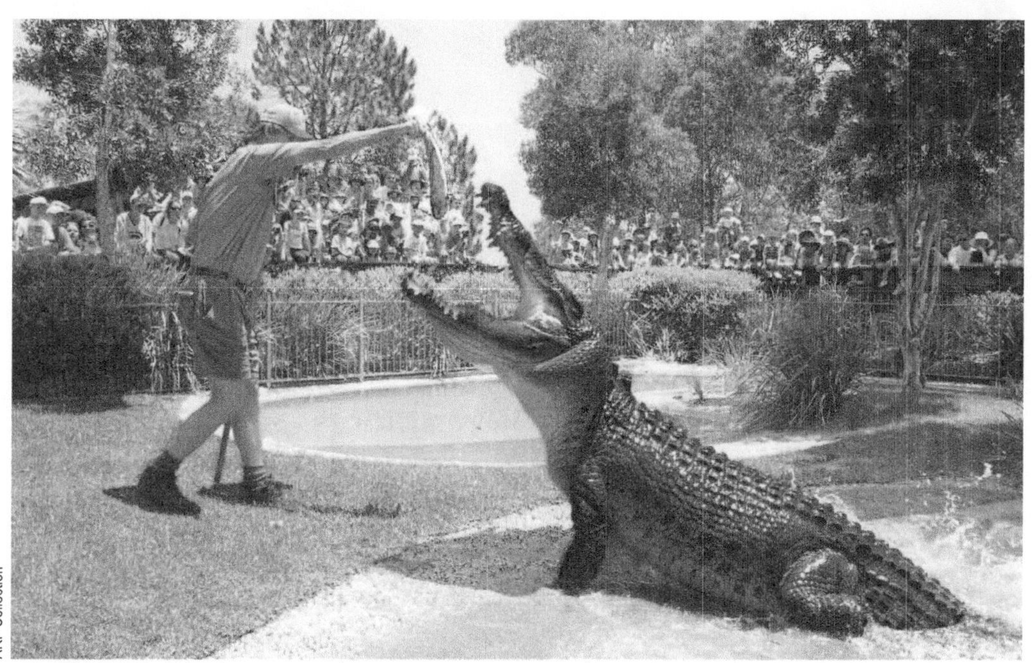

Children enjoy a rare opportunity to touch a Galapagos tortoise, 2010.

ARP Collection

John Weigel feeding Eric the crocodile.

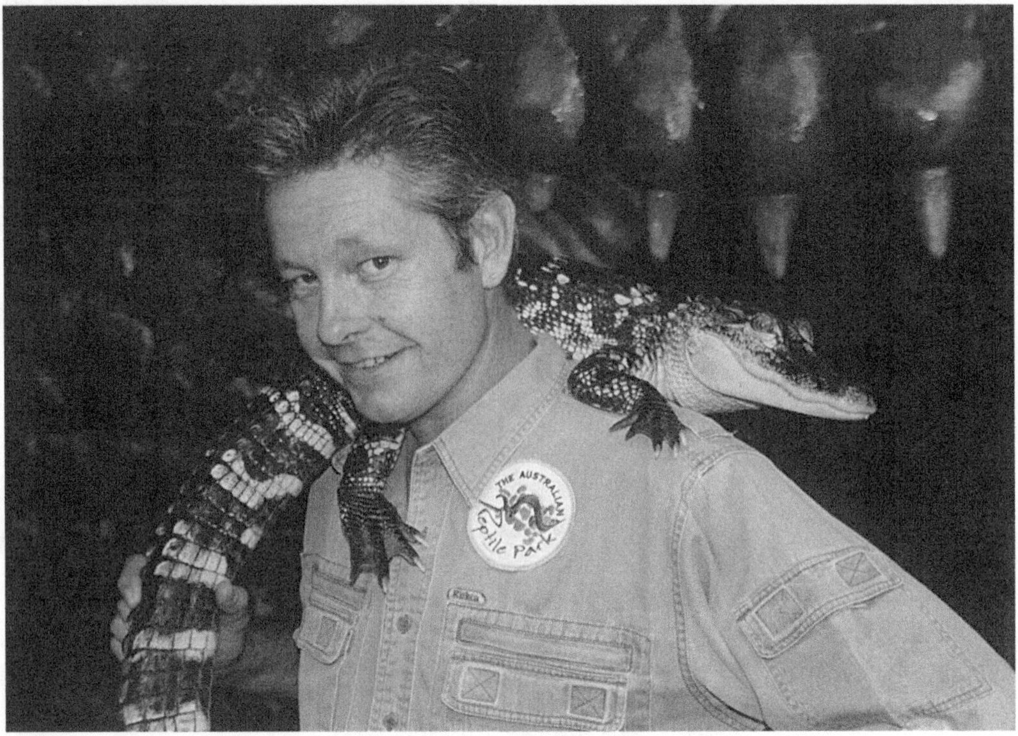


ARP Collection

Ranger Mick (Michael Tate) with alligator friend, Somersby.

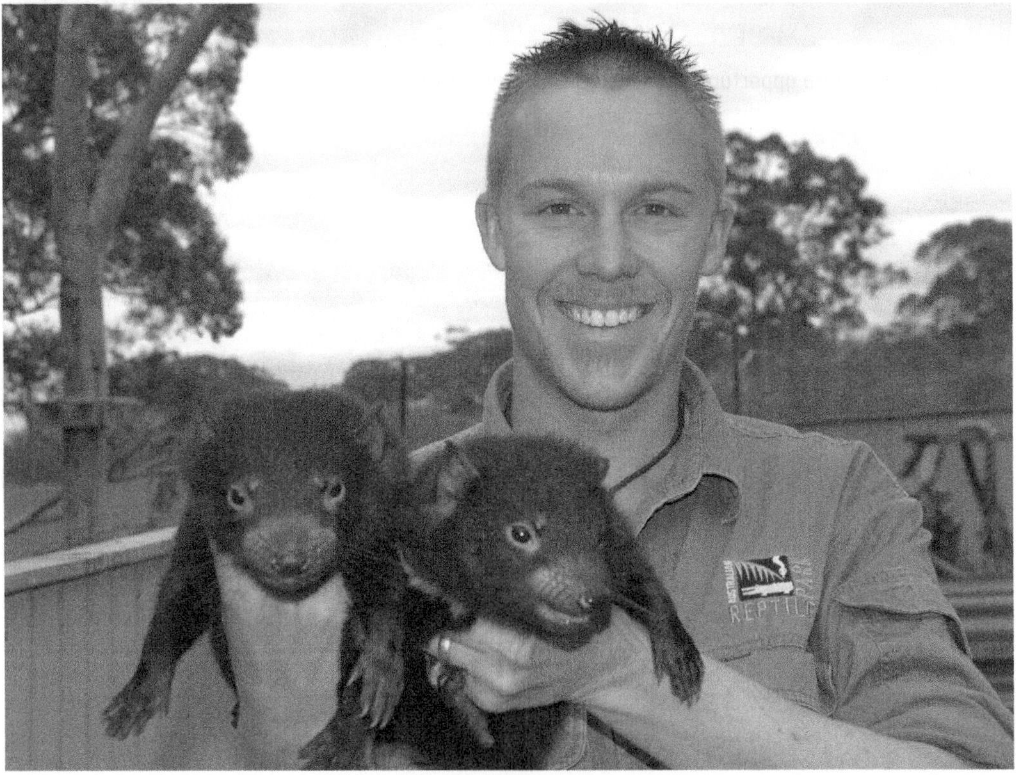

ARP Collection

Senior curator Tim Faulkner holds two baby Tasmanian devils, Somersby.

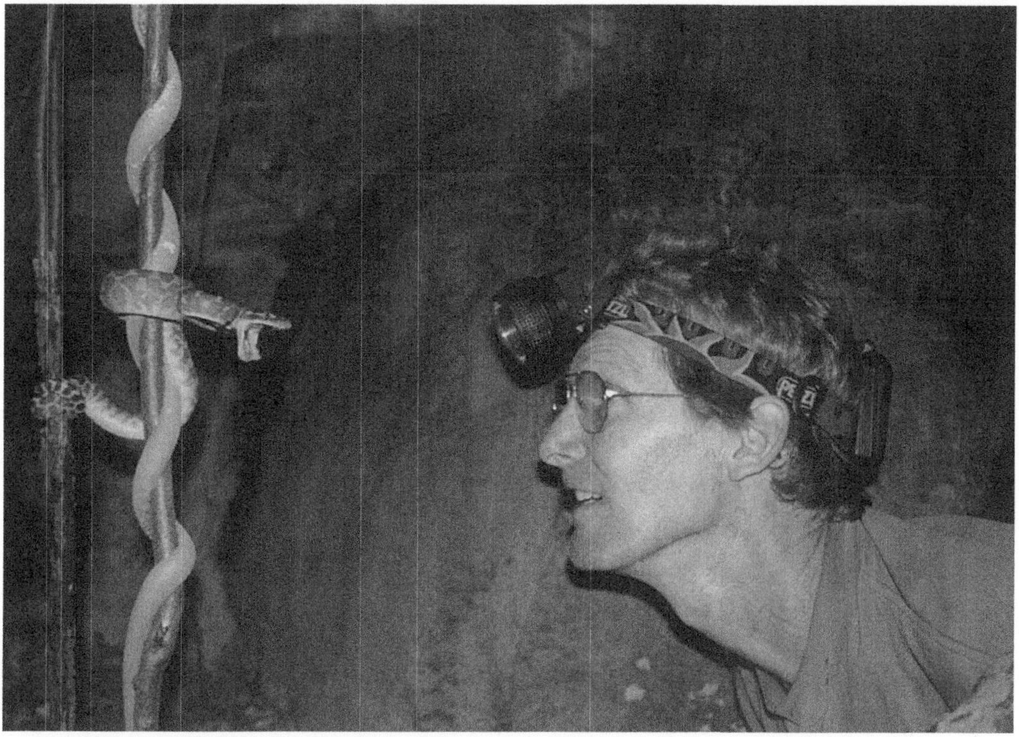

ARP Collection

John Weigel admires a less-than-friendly rough-scaled python in the Kimberley Ranges.

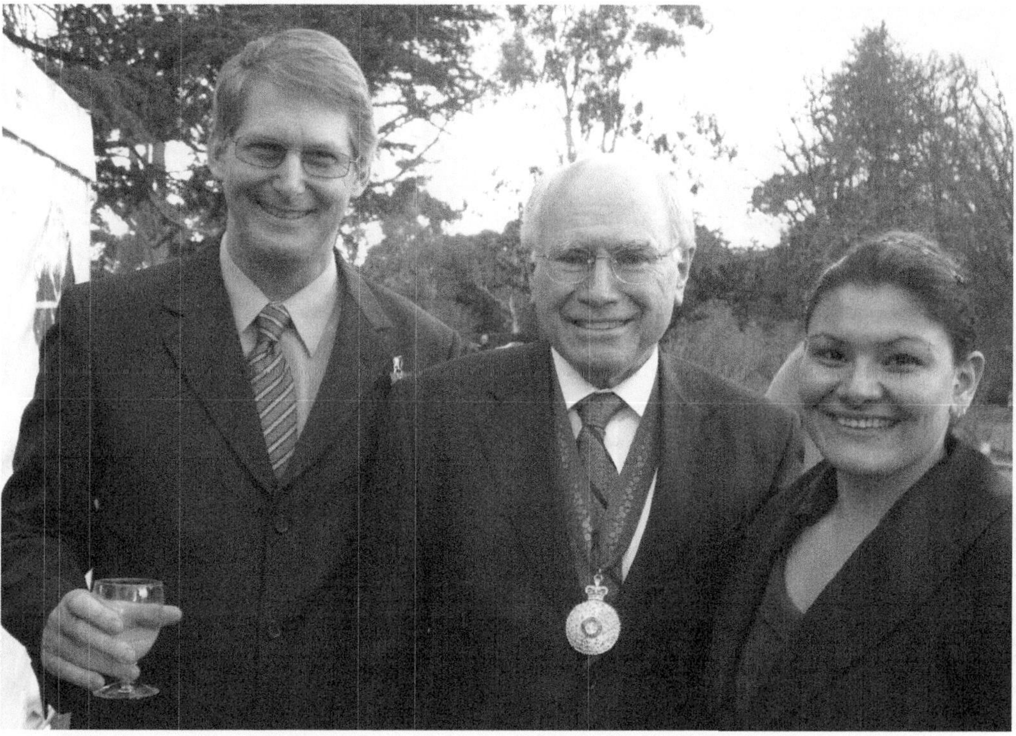

ARP Collection

John Weigel, former prime minister John Howard and John Weigel's daughter, Blanca, at the ceremony at which Weigel was conferred Member of the Order of Australia, 2009.

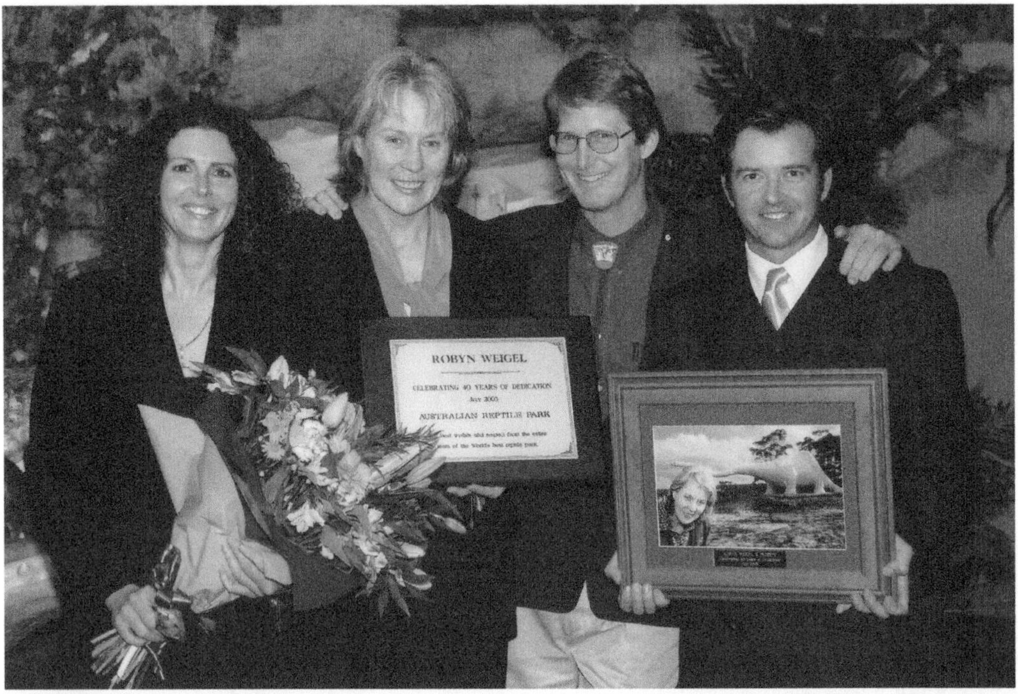


ARP Collection

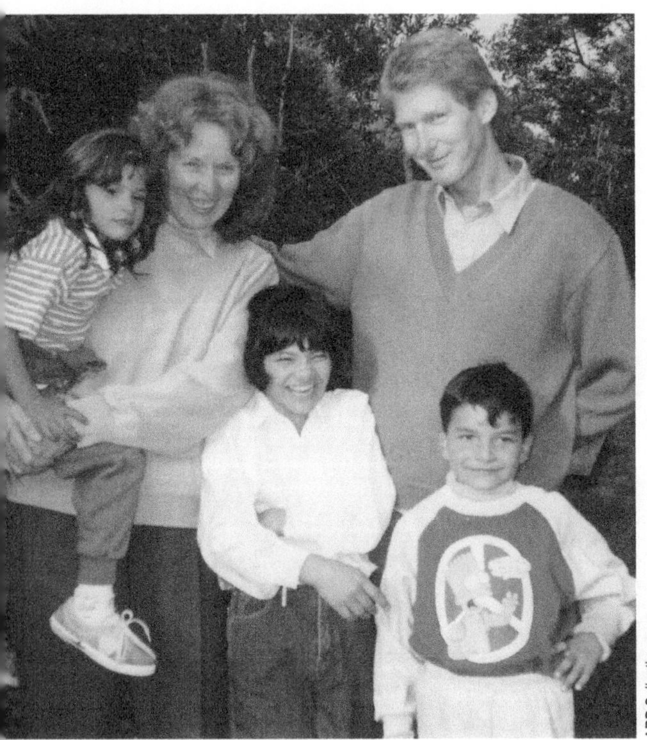

Robyn Weigel receives an award from Australian Reptile Park staff for 40 years of dedicated service to the park. From left: Mary Rayner, Robyn Weigel, John Weigel, Al Mucci.

ARP Collection

Weigel family: Robyn holding Johanna, Blanca, John and Arnie, late 1990s.

ARP Collection

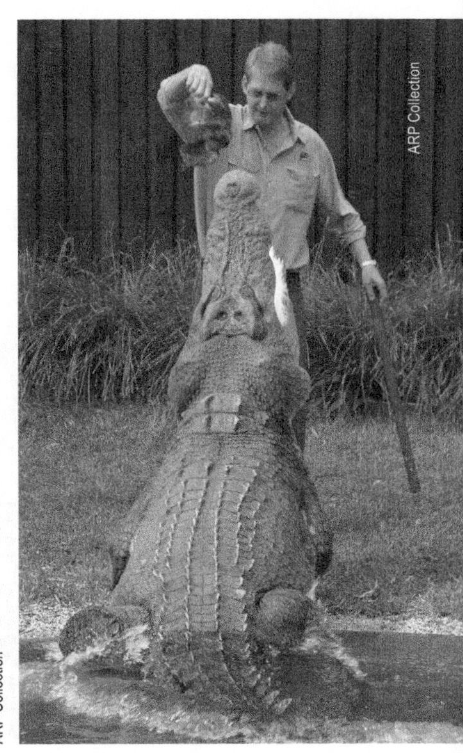

John Weigel offers Elvis the crocodile a portion of chicken, Somersby, 2009.

8

# The Gosford dinosaur

The lights went down in the cinema, and on the screen loomed a dinosaur, dwarfing a grown man at its huge feet. Rapid-fire British Pathe newsreel commentary rattled off facts: to make the dinosaur took 50 tons of concrete at a cost of £3000 (the equivalent of over two years earnings for the average man in New South Wales at the time). The real *Diplodocus* lived 70 million years ago, was 87 feet long, like this one, and weighed 40 tons. 'The Gosford dinosaur, concrete evidence of prehistoric times!' concluded the narrator. For some of the audience, the seed had been planted. They would drive to Gosford and see this dinosaur for themselves.

Car ownership in Australia had rapidly increased in the 1950s, opening the way for a new style of holiday making. The suburban family could pack their belongings, hook up the caravan and escape the city. The driving holiday was cheap and simple, often involving a drive only to the nearest coastal resort. The Central Coast was well suited, a comfortable drive from Sydney or Newcastle, with golden beaches, bushland, walking trails and sheltered lakes for fishing. But many holiday makers wanted some additional form of amusement or spectacle.

Many people remember walking the 2 kilometres from Gosford railway station to the reptile park, and Worrell mentioned the railway and the hourly bus service from the station to the park in his promotional material. Yet it was clear that he expected the vast majority of his visitors to arrive by car. The exterior of the park was designed to be viewed from the highway and there was enough parking for about 500 cars.

Almost as soon as Worrell had found what he thought was the ideal location for the park at Wyoming – with high exposure to traffic on the Pacific Highway – he learned that the Department of Main Roads intended to upgrade the Pacific Highway and build a separate high-speed road between Hornsby and Ourimbah so that motorists travelling between Sydney and Newcastle could bypass Gosford. For Worrell and other tourism operators on the Central Coast who relied on passing trade, this posed a significant threat to their livelihood.

As part of the roadworks, the Pacific Highway outside the reptile park was to be widened. Through traffic was predicted to decline by 50 per cent and the cars that did go past would be travelling at higher speeds. Worrell realised that countering the impact was best done by providing a high-quality tourist attraction and jointly promoting the region with other tourism operators. But he had also learned through his work with reptiles that to really pull a crowd, he needed something spectacular. 'With so many proposed detours of the Gosford District, business people will be forced to do something radical to defend their section of the Pacific Highway,'[1] he told a local newspaper.

His idea was radical indeed. He would build a life-sized *Diplodocus*, a dinosaur of the Jurassic period and one of the longest dinosaurs ever to live. The building of large-scale roadside attractions, now known as 'big things', was unknown in Australia in 1963. Coffs Harbour's Big Banana, usually noted as the oldest 'big thing' in Australia, did not open until December 1964. It isn't known why Worrell chose the *Diplodocus*. He wanted something dramatic; and it may have been a play on the link between Wyoming, New South Wales, and the US state Wyoming, where bones of the *Diplodocus* had been found since late in the 19th century.

Remaining true to the scale and dimensions of the original *Diplodocus*, the model was 6 metres wide by 26.5 metres long and 5 metres high. It was designed by sculptor Ken Mayfield, who had worked at the Australian Museum on a range of smaller dinosaur models, and engineered by AC Sullivan, an associate of Worrell's who was a local steel manufacturer. The model consisted of a steel skeleton which was covered on site with a skin of reinforced concrete. The technique of spraying the cement mix under high pressure over the steel frame to form the skin of the ancient reptile was new and Worrell organised a field day so that local builders could learn more about the innovative method. Once the concrete had dried, Mayfield painstakingly chiselled lifelike wrinkles and other fine details into the dinosaur's body. A separately modelled head with acrylic eyes provided the finishing touch. The posture of the animal – its head nearly level with its shoulders and its long tail snaking along the ground – was in keeping with scientific understanding at the time, though the way it held its tail has been debated by palaeontologists since then. According to an article in the *Central Coast Express*, it was 'Australia's largest figure of any life-form'.[2] Overall, the dinosaur took two months – 1200 work hours – to create. It was a masterpiece of engineering, design and sculpture, and its form was graceful and realistic.

The dinosaur's prominent position at the side of the Pacific Highway drew the attention of passing motorists and encouraged them to stop. Worrell was pleased if they then paid the admission fee to enter the park but he regarded the dinosaur as a distinct and separate attraction, ensuring that it had its own listing in Central Coast tourist guides as the 'Gosford Dinosaur'.

The Gosford dinosaur did indeed become the icon for tourists and locals that Worrell had hoped for. They gathered to have their photographs taken with it and bought postcards featuring it. Worrell's friend John Dwyer sculptured a small cast of the dinosaur so little golden dinosaur brooches, complete with opal inlay, could be sold through the park's gift shop.

The dinosaur was a great success in attracting holiday motorists to the park. By the late 1960s, the car park was regularly filled on spring and summer weekends and public holidays, and Worrell had to buy an additional lot for overflow parking. As in so many of the bold moves he took in his life, by creating one of the first 'big things' Worrell had introduced something that would become a fixture in Australian life. But he knew he had to keep the excitement up, so visitors to his park would soon be awestruck by a far more deadly giant reptile. This one was very much alive and instead of being a gentle herbivore, it had a well-developed taste for flesh.

9

# **Adventures with crocodiles**

Quite often when we had traced a river to its source in the morning, the tide had dropped by dark, leaving us stranded as the river became unnavigable for the boat and head-camp couldn't be reached. We slept on bare salt pans in our clothes and were attacked by hordes of mosquitoes and sand flies. One night we had captured a ten foot crocodile which we tied in the boat with my shirt covering its eyes. That night I shivered on the salt pan with the temperature down to 46 degrees [Fahrenheit], and supported the local population of mosquitoes and sand flies.

Eric Worrell[1]

To capture a 10-foot, or 3-metre, croc would be a feat for most people, but the quarry Worrell had his sights set on during this expedition was more than double that size. The animal was proving elusive, however. Several times over the course of the week, Worrell and the team of men he'd assembled had spotted the enormous black crocodile. One day, from a plane above, they had watched as he crawled towards the riverbank, a white goat in his huge jaws. He swallowed the goat in a couple of gulps.

Growing more than 6 metres long and weighing in at over half a tonne, the male saltwater crocodile is the undisputed ruler of the estuaries and wetlands of northern Australia. It is feared by those who live near it. For its part, the saltie fears nothing – except a larger crocodile.

One of Worrell's contacts had told him about this fearsome black crocodile, estimated to be 7 metres (23 feet) in length and nick-named Cassius. It had been seen up in the remote Burketown area in the Gulf country of Queensland.

The death of his first 4.1-metre male saltie just a few months after the park's opening had been a major disappointment to Worrell. Since then he had added other interesting specimens to his collection, among them an American alligator he received as a gift from Sir Edward Hallstrom earlier in 1962. The alligator, named Edward, had been at Taronga Zoo for 15 years, and later became something of a star of the Australian Reptile Park. His mate, Edwina, arrived a year or two later. In 1963, Worrell had also acquired two specimens of a different type of crocodile, the distinctive false gavial of Indonesia and Malaysia. He still hadn't got another big saltwater croc, though. This crocodile, Cassius, was the one he wanted for his park.

So here he was in the midst of an arduous nine-week, 10 000-kilometre trek that would take him and his party of three other crocodile hunters to the hostile inter-tidal mudflats and mangrove swamps of north-western Queensland. Worrell managed to convince the companies Toyota and Nissan, and the Japanese Consulate, to sponsor 'Operation Crocodile' as a way to promote goodwill between the Australian and Japanese people. Mobil Oil was another sponsor. The expedition was well equipped with two four-wheel-drive land cruisers, a fibreglass dinghy, two-way air-to-ground radios and specially made wooden crates to house the animals for transportation down to Gosford. They chartered a Cessna spotter plane for several weeks so they could patrol the rivers from the air.

The careful planning and resourcing of the expedition meant that the crocodile hunters conducted their activities at a high level of sophistication for the time. Worrell, seated in the light aircraft, would relay the location of larger crocodiles to the ground radio and the other men would then be dispatched by boat and four-wheel-drive. Even with such guidance, the trip involved hard and dirty work. Vehicles frequently got bogged up to their axles in thick, gluggy mud or the sun-hardened crust of the saltpans gave way beneath them. The nylon nets they used to catch the crocs were heavy and cumbersome, and it took the combined effort of three men to carry one down to the water's edge before it could be set across a river or billabong. Barramundi and sharks in the rivers sometimes got tangled in the nets, rendering them useless. A Cinesound cameraman filmed the expedition and about an hour of this remarkable film survives today, housed at the National Film and Sound Archives in Canberra. The grainy black-and-white footage shows the bogged vehicles, herds of buffalo thundering through thorny tropical scrub, nets full of barramundi hauled up from rivers festooned with water lilies, and of course the men's thrilling and dangerous work of capturing live crocodiles.

The expedition combined modern equipment with the knowledge and skills of an elderly tribal man who was paid to keep a watch for Cassius. The party also occasionally camped with professional crocodile hunters to gain an understanding of crocodile hunting from the perspective of the hunters. And local man Terry Arnold helped wrangle the crocodiles – despite having only one arm.

During the 1940s, when Worrell accompanied Jim Edwards and George Haritos on trips in the Top End, the great aquamarine rivers were the habitat of large numbers of big crocodiles. His attitude to crocodile shooting shifted abruptly from the late 1940s onwards when he observed that saltwater crocodile numbers had plummeted in a matter of a few years. In river systems where there had once been large numbers of crocodiles, the efforts of hunters, both commercial and amateur, had decimated these populations. He attributed it to spotlighting, which allowed the animals to be shot or harpooned more easily, and the increase in the price the leather industry was paying for crocodile skins. He came to see uncontrolled crocodile hunting as unnecessary, wasteful and unsustainable, and became a voice for the total protection of both the freshwater and saltwater species. Now, some 15 years or so later, he wanted to see how the crocodile populations were faring. What he saw wasn't good at all.

His aerial surveys of two river systems in the Gulf country, the Albert and Leichhardt, revealed only a handful of crocodiles in each river. Based on the size of the smallest of these crocodiles, Worrell concluded that no successful breeding had taken place for several years. Further explorations were made of rivers and billabongs that were known to have previously supported populations of the smaller freshwater, or Johnstone's, crocodiles, only to discover that they had been greatly reduced in number or in some cases completely annihilated. 'The air was foul with the stench of rotting crocodile carcases piled on the river banks,' Worrell wrote. 'Bleached crocodile bones bore sad testimony to years of slaughtering.'[2]

Though Cassius eluded him, Worrell and his party managed to collect six saltwater crocodiles, including one estimated to weigh half a ton, and 30 Johnstone's crocodiles, the largest of which was almost 3 metres in length. While the removal of 36 crocodiles from their native habitat might be seen as going against his conservationist stance, Worrell probably saw his actions as saving these few animals from the bullets of the hunters.

Worrell knew he had one more chance at a big crocodile on the way back to Gosford. Wirth's Circus had a small zoo, the Mt St John Zoo, at Townsville, where they kept their animals when the circus was not travelling. The zoo was being mildly terrorised by a large male crocodile that had managed to force its way in through a weakness in the netting of the crocodile enclosure and was amorously pursuing one of the zoo's females, Big Mamma, and behaving aggressively to Tarzan, Big Mamma's mate. The crocodile's avid pursuit of female companionship earned him the name Casanova. Mrs Wirth wanted this rogue crocodile removed and gave Worrell her blessing to attempt his capture.

Part of the billabong in which the crocodiles lived was covered by masses of floating water plants, which had to be removed so that the crocodile hunters could see what they were doing – and more importantly, what Casanova was doing. The party then set about rigging up the big nylon net in the water. They tied it to stakes cut from bamboo growing alongside the swamp to keep the net upright. Men standing on the bank also held lengths of bamboo and poked them into the water at various spots to locate Casanova and encourage him to swim towards the net.

'We finally managed to get him tangled up in the net, then the race was on to get him out of the water so he didn't drown,' recalled Arnold Gaunt, a local reptile enthusiast who helped in the capture. 'But we were in about eight foot of water and about eight foot from the bank and he was fully enmeshed in the net and yet he got his mouth open wide enough to bite the arse out of the boat on which Eric was standing. And that was with one bite!' Once Casanova was on land, Gaunt lay down beside the crocodile's front legs and pushed a rope underneath the croc to Worrell's friend Athol Compton on the other side. 'Athol had to reach underneath him to get the rope – that's how wide he was, his front legs. Then we put it around his front legs, around his nose, around his neck, around his front legs, his guts and back legs,' said Gaunt. The men winched Casanova up and pulled him into a wooden travel crate on a trailer. 'Then we packed all this duckweed around him and over him.' The capture had taken nine men eight hours, although the entire process took four days. Mrs Wirth was apparently so pleased to be rid of the big beast that she put on a champagne party for the victorious, if mud-encrusted and exhausted, crocodile wranglers.

Worrell and his group headed back to Gosford, stopping every 100 kilometres or so to pump water over the crocodile from roadside streams.

After Casanova arrived in Gosford in late August, there was much fanfare. The drive-in cinema at nearby Erina had already shown the Cinesound film of the party's adventures in the Gulf, which had screened around Australia and even in the UK. Now the drive-in showed a film of the capture of Casanova. A picture of him in his enclosure surrounded by park visitors ran on the front page of the *Gosford Star* under the welcoming headline 'Hello Casanova'. Worrell told the press that Casanova, which was around 5 metres in length, was the most dangerous crocodile he had encountered in 20 years of crocodile hunting. A fine large crocodile with a massive bulk, he made a spectacular display animal. His enclosure was a modest concrete pool, heated by an innovative solar-powered hot water system and surrounded by a chain mesh fence almost 2 metres high, located close to the main office building. He became the park's star attraction, drawing the very large crowds for which Worrell had been hoping. Casanova was Australia's first celebrity croc.

For wildlife parks and zoos in the southern half of Australia, exhibiting crocodiles is difficult. While they can live comfortably in a climate like Gosford's during the warmer months, they die during winter unless they have access to warm water. The park's picnic area had a heated swimming pool, so when the water in the large pond containing the other crocodiles and alligators began cooling down in autumn, staff waded through the pond feeling for submerged and hopefully lethargic crocodiles, which they then manhandled on to shore and transferred to the swimming pool. (Understandably, tourists declined to swim in the pool during the winter months!) Housing almost three dozen tropical crocodiles would have been beyond the park's capacity at the time, so while Casanova stayed, some of the other crocodiles Worrell captured in the Gulf were used in exchanges with other zoos and wildlife parks.

A few weeks after Casanova's arrival, Carol Worrell was reported as having heard a disturbance one night – a noise that sounded like fireworks followed by birds calling out and Casanova's bellows reverberating across the park. When she checked his pool the next morning, she found blood in the water. According to Eric, who had been away from the park on that night, someone had shot the crocodile between the eyes at close range using a .22 rifle. He was concerned that Casanova might die if the wound became infected. '"Casanova" shot – may die soon'[3] ran the no-nonsense headline on the *Central Coast Express*. Newspapers in Sydney picked up the story and large crowds came to pay their last respects to this famous crocodile. Queues of people stretched from the entry gate right through the car park. Casanova died on 22 September 1964. A photograph of him in his pool with Worrell appeared in the park's guidebooks up until the early 1980s, the caption stating that Casanova had been mortally wounded by vandals.

But did Casanova really die from a bullet? No one was ever charged with his death. No autopsy was conducted. Park employees Dekkers and Krauss both recalled seeing blood coming from Casanova's mouth *before* the disturbance reported by Carol Worrell.

During his eight-hour capture, Casanova could have become highly stressed. While the duckweed that covered the billabong where he was captured had been packed around him to try to insulate and keep him moist en route to Gosford, it would have been a long and uncomfortable journey for the animal to have endured. His own body mass would have weighed heavily on vital organs such as his lungs, which would normally have been supported by water. It was August when he was transported to Gosford – not a very warm month to be travelling – and he would have spent at least one full 24-hour period in sub-optimal temperatures.

Worrell had invested a lot in creating a celebrity out of Casanova and the crocodile had substantially increased gate takings and attracted much media interest. If Worrell had announced only a few weeks after his arrival that the crocodile had died possibly as a result of the stresses of capture or captivity, he might have diminished his own credibility and that of his reptile park. 'Many people have expressed disgust at the incident, and have been angered at the apparent senseless shooting,' said the *Central Coast Express*.[4] We cannot say for certain what Casanova died of, but there are at least some doubts about the story that he was the victim of a gunshot. As to what happened to Casanova's massive carcass, nobody could tell us. It seemed that the big crocodile just disappeared without trace. Perhaps Worrell sold Casanova's body to someone looking to own a large mounted crocodile. If that's the case, then it's very likely that the mounted skin of Casanova lies – no doubt with mouth forever ferociously open wide in anger – in someone's home.

10

# The finest reptile collection in Australia

As a boy in his early teens walked along a Melbourne beach one day in the mid 1960s, something strange caught his eye. Among the sand and shells and pebbles lay a snake. The boy had an interest in snakes, and this one didn't look to him like a sea snake. He watched the motionless creature for a moment to be sure that it was in fact dead and then leaned in for a closer look. While it looked like some sort of python, it was not like any Australian python he had seen before. With its orange-red markings, large head and eyes, the dead snake, a little less than a metre long, was a mystery. Curious, the boy took it to the National Museum of Victoria to find out more about it. The herpetologist who received it, John Coventry, was puzzled to discover that the snake was a Cook's tree boa, a snake native to the West Indies. Just how it had travelled so far from its homeland was a mystery.

After leaving school at the end of fourth form, local girl Robyn Innes enrolled in a secretarial and bookkeeping course at Gosford Technical College and then door-knocked Mann Street, Gosford, looking for work. On one of these job-hunting forays, 16-year-old Robyn bumped into Kim Worrell on the street. Kim mentioned that her father was looking to hire a secretary. After a short interview, Worrell, who knew Robyn's parents socially from playing carpet bowls with them, hired her as an office assistant. Tall and athletic, Robyn was a modest girl but capable of taking on a challenge.

'I don't mind taking crippled birds or spiders in bottles but when it comes to taking snakes offered by visitors to the park I quickly call for Mr Worrell,' she told the *Gosford Star* soon after starting the job.[1] The paper featured her in their 'Teenager of the Week' column, and many locals no doubt agreed when the reporter described her job as 'one of the most unusual in the district'. As an adult looking back on those days, she recalled: 'I enjoyed it from the first day I was there. I just loved every second of it. It was varied, it was vibrant.' Soon her role expanded. 'I was helping [Eric] write his books. I was doing letter writing … because he was doing other stuff. But I didn't think I was going to be there forever.' She was not to know, then, the extent to which her life was to become intertwined with Eric Worrell's and the future of his reptile park.

Buoyed by healthy gate takings and his growing national reputation as a naturalist, Worrell continued with his program of expansion. At the time, his friend Athol Compton and Athol's wife Marjorie were doing it pretty tough. A bushfire had destroyed much of their Dubbo property and they were struggling to get financing from a bank to rebuild. Worrell offered his friend, who had a strong interest in wildlife, work until he could sort out what he wanted to do. Compton was appointed park supervisor and was reunited with his namesake, Athol, the park's resident raven. On a trip to central Australia in 1962, Compton had taken the young raven from its nest, high up in the blades of a windmill. Ravens and crows were considered vermin, as were many species of Australian wildlife in the 1950s and 1960s, and Compton removed the bird and presented it to Worrell, thinking he might want it for the park. 'Can't stand the sight of them,' Athol, a sheep farmer, explained to Worrell, 'but they tell me they make good pets if you bring them up from young.'[2] Christened 'Athol', the young raven travelled some 5000 kilometres with Worrell and Compton and when given freedom to fly away, chose to return to the naturalists' party. As Worrell recounted, 'Every time we stopped for smoko or made camp "Athol" hopped out of his box and roamed around in the bush and was back by dark, when he perched on the front bumper bar of the land cruiser.'[3] Athol set up home at the reptile park, where he became a popular and eccentric character, living in an old apple tree near the house that Worrell built for his parents in the picnic area. He was an excellent mimic – his 'G'day mate' uttered in a long drawn-out country drawl – as well as a canny thief, stealing cutlery, shoelaces, food, cigarettes and even the occasional wallet. A photograph of Athol was often featured in the park's guidebook and he was a well-loved identity until his death in 1984 at the grand old age of twenty-five. Few ravens ever live to that age.

One of Compton's earliest unpleasant tasks was to terminate, on behalf of Worrell, the employment of Jon Dekkers, the first paid staff member at the reptile park, who was most likely sacked to make way for the subsequent employment of Peter Krauss.

In the following year, Worrell employed another of his friends, Jack Green, with whom he had remained in touch since the Ocean Beach Aquarium days. His initial full-time role was to manage the other three staff members, but Worrell's intention was for Green to be responsible for an aquarium that he was planning to build. In typical Worrell fashion, the aquarium was to be a major undertaking and they would exhibit sharks, which Green would handfeed as an added attraction. However, the aquarium remained on Worrell's wish list, never becoming a reality. Green felt uncomfortable being placed in a position above Krauss and Compton – after all, he was the new guy and had limited animal experience – and he was aware of tensions beginning to build. After a discussion with Worrell, his title was changed to aquatic curator and he was made responsible primarily for the growing number of aquatic exhibits, including, at a later stage, the crocodilians.


For Worrell, the 1960s were an extraordinary time of diversifying the park's animal collection. In October 1962, Worrell imported from Papua three tree kangaroos and three 'black wallabies' as well as 'three green tree snakes' – probably green pythons – 'and five pythons'.[4] Because the park did not have official quarantine status, the animals from Papua were required to spend 30 days in quarantine at Taronga Zoo, at Worrell's expense. In early 1963, Worrell also received two sea snakes from Taronga that had been collected in New Guinea. In 1963, he also acquired a large Indian python – 5.5 metres long – and a pair of that most regal of venomous snakes, the king cobra, slightly more modest at 4 metres. No other zoo in Australia exhibited the king cobra, or the lizard-like tuatara, which was presented to the park as a special gift from the New Zealand government in 1965. Cassowaries, brolga and black-necked storks were added to the collection thanks to wildlife parks in Queensland.

Frustrated by his inability to import reptiles directly, Worrell applied for a B Class zoo licence, which would allow the park to act as a quarantine station. Up until that time, only publicly owned zoos had been granted that status. His friend, Jock Marshall, professor of zoology at Monash University, wrote a letter of support without hesitation:

> This gentleman is doing quite remarkable work … both in the
> collection of venoms for the production of antivenenes and in
> the general education of the public in the need for the study and
> preservation of that precious, but unhappily dwindling asset,

Australian fauna … Worrell's primary interests – unlike most zoos
– are education and conservation. After seeing numerous such parks
during many years in Europe and the United States I am in no doubt
that Worrell's is superior to any that I have seen elsewhere …[5]

The application was successful, confirming the park's status as the
leading private collection of animals in the country. Worrell's first
major importation of reptiles was carried out by Peter Krauss. Peter
and Raija had married in 1963, and Worrell offered Peter a permanent full-time job the following year. The newlyweds travelled back
to Germany to decide whether they wanted to remain in Europe or
return and build their life in Australia. Optimistically assuming that
the couple would be returning, Worrell arranged with the Swedish
Herpetological Society to exchange a shipment of Australian reptiles
for a number of theirs, which the Krausses would be able to chap-
erone back to Australia. It turned out that Worrell was right: the
Krausses did return.

While in Europe, Krauss obtained additional animals from his
contacts at the Stuttgart Zoo and from an animal dealer in Swit-
zerland. The Krausses then booked passage out of Genoa, Italy, on
the Dutch freighter *Omen Kerk*. Housed in their cabin were approximately thirty species of reptile, including a Galapagos tortoise,
iguanas, boas and a number of highly venomous snakes such as the
black mamba, puff adder, rattlesnake, gaboon viper and cobra. Under
Krauss's care, the animals made it safely to Australia – excluding one
Cook's tree boa, which died as their ship steamed past the southern
coast of Australia. Krauss gave it a sea burial, assuming that it would
end up as fish food or quickly decompose. But the fish couldn't have
been biting that day and the currents must have been running,
because the boa washed up on a Melbourne beach to be found by a
curious young boy.

Krauss remembers the excitement bubbling out of Worrell when
he met their ship in Sydney and loaded up his Holden station wagon

with the exotic cargo for the one-hours-or-so drive up the highway to the reptile park. Peter managed these and other animals, and Raija worked at the park at various times over the next decade and a half, mostly caring for mammals and birds and developing considerable expertise in rearing orphaned marsupials. She shared the family home with many animals, including sick taipans that Peter brought home to look after in the warmth of their child's nursery.

Thanks to his friendship with Sir Edward Hallstrom, who would later become honorary director of Taronga Zoo, Worrell made further animal exchanges with the zoo. Late in 1964 he received six green pythons from New Guinea, two king snakes, two gopher tortoises and two gopher snakes, in return for a number of the park's venomous snakes: eight Papuan black snakes, one Papuan taipan and 11 common brown snakes. These snakes, which were still milked for venom by Worrell and sometimes Krauss, bolstered the zoo's exhibits while relieving a shortage of space at the reptile park. George Cann had retired from Taronga and British expatriate Bill Timmis, whom Worrell regarded as one of the best reptile men he had known, now managed the reptiles. Worrell received most of the zoo's dangerously venomous snakes in 1967, including timber rattlesnakes and water moccasins from America and several Australian species (presumably including the venomous ones he had given the zoo the year before). The zoo's then head keeper of reptiles, Graeme Gow, had received a bite from a rattlesnake and Hallstrom had decided to remove these snakes from Taronga's inventory.

Worrell received a shipment of North and South American snakes from Karl Switak of Steinhardt Aquarium in San Francisco in 1968 in exchange for a number of Australian species. Among the shipment of 31 specimens were an anaconda, two southern copperheads, a sidewinder, several species of rattlesnake including two Mexican diamondbacks, one fer-de-lance, a bushmaster, corn snakes, king snakes and boas. Another shipment of American snakes was received from the San Antonio Zoo in Texas. The

Ausralian Reptile Park now held the finest reptile collection in Australia.

Unfortunately, some visitors were not content with just viewing the reptiles, and theft was a problem. A variety of snakes, lizards and turtles were stolen from the park from time to time. Some thieves brazenly removed venomous snakes, lizards and other reptiles from the outdoor pits, hid them in backpacks and walked out of the main entrance, while there are also stories of others flinging reptiles that they had taken from the pits over the boundary wall to waiting accomplices.

The park's mammal collection was also growing. The first platypuses were on display by 1965, and by 1968 a new platypus exhibit had opened, based on designs developed by Green and also on the one at Healesville Sanctuary in Victoria. Visitors stood in a darkened viewing area to observe several platypuses, which had been trapped in local creeks with the approval of the NSW Chief Guardian of Fauna, in a long glass-fronted exhibit complete with waterfalls and a replica of a stream bank. A back-of-exhibit system of runs, tunnels and pools provided the range of living spaces the animals needed.

Towards the end of the 1960s, Worrell's expatriate contacts in Papua New Guinea offered him more animals, including tree kangaroos, cuscus (spotted and ground), long-beaked anteaters (echidnas), wallabies such as the nail tail (*Dorcopsis* species) and, of course, a variety of reptiles. He was also offered a variety of bird species, including some spectacular birds of paradise, but Australian customs regulations banned the importation of any birds in those days to prevent the introduction of threats to the poultry industry, such as Newcastle Disease.

Peter Krauss flew to Papua to arrange for the transport of the animals, a task that posed a considerable challenge. Although he was a trained zookeeper, he had no experience in organising the collection and maintenance of wild animals in a place such as Papua. While Krauss was in Bereina, several hours' drive north-west of Port

Moresby, he happened to hear a report on the ABC that a 3-metre Papuan freshwater crocodile (*Crocodylus novaeguineae*) had been caught at Port Moresby and was waiting for a representative of the Australian Reptile Park to organise collection and transportation back to Australia. This was the first Krauss knew of it, but nevertheless he managed to organise for a suitable crate in which to carry the animal back to Australia on the MV *Bulolo* with the rest of the specimens. Krauss remembers this animal with some fondness as it had a very quiet disposition and quickly accepted food by hand.

The last major building constructed on the Wyoming site was a groundbreaking exhibit that enabled visitors to watch rarely seen nocturnal animals going about their activities during the day in a dimly lit noctarium. Once the park had closed for the night, bright lights would switch on and as far as the animals were concerned, it was now daytime and they went to sleep. While such exhibits are now popular, at the time no other zoo in the southern hemisphere had such a sophisticated exhibit for displaying nocturnal animals. The Australian Reptile Park, through the expertise of Peter Krauss, was truly innovative in this regard.

The exhibit was funded largely by an American film producer who had intended to make a motion picture with the working title of *Boomer*, telling the story of a migrant family attempting to settle into Australia during a drought and being menaced by kangaroos. Worrell had agreed to supply the animals that were to be used in the film, which would be shot predominantly on the Central Coast. He had already supplied animals for television programs such as *Skippy* and movies including *Jedda*. Paul Kenworthy, the award-winning cinematographer for Disney's *The Living Desert* (1953) and *The Vanishing Prairie* (1954), was, according to newspaper and magazine reports, to direct *Boomer*. But unlike the noctarium, *Boomer* never got off the ground.

The noctarium enabled the park to exhibit delicate tropical species of cuscus and possum that Krauss had brought back from Papua and others he had trapped on the Atherton Tableland: the striped possum, Herbert River ringtail and green ringtail. It also housed long-beaked anteaters from New Guinea, brush-tailed possums, potoroos, small possums, quolls (including the eastern quoll, which is now extinct on the mainland), kowari, bandicoots and a number of grey-headed flying foxes, which were allowed to fly free inside the noctarium. Together they all created a very distinctive musty aroma.

At its peak, it was such an impressive and novel exhibit that the director of Taronga Zoo, Ronald Strahan, who was a friend of Worrell's, quizzed Krauss about its construction and maintenance prior to the zoo building its own nocturnal house. Worrell's noctarium was named in honour of wildlife photographer and bird observer Norman Chaffer, who at one stage Worrell was considering going into partnership with to 'farm' kangaroos for meat and skins. Just how they intended to manage these animals within a farming situation is not known.

In 1967, Lynette Abra, Jack Green's sister-in-law, joined the staff, initially as an administrative assistant. Clearly this was going to be no ordinary admin job, though. 'Eric had wanted me to type a letter for him,' she recalled. 'And I lifted some paper out from a drawer and underneath curled up was a black snake.' She dropped the paper, slammed the drawer and said, 'There's a black snake in there.' Worrell replied calmly, 'Oh, that's where it went.' A member of the public had brought in the snake in a bag but no one had checked for holes in the bag. After taking the snake out of the drawer, he told Lyn, 'You'd better go and put it in the black snake pit,' and handed it to her. 'I went out the door, and I'm going over to the black snake pit, and I'm trying to keep my eyes on it, and all these little kids came over and I'm saying, "Get away, get away!"'

People would hand in all kinds of animals that they had found in their backyards or on camping trips. There were reptiles, but also a host of injured and orphaned native birds and mammals. It soon became Lyn Abra's job, along with Robyn Innes and Raija Krauss, to care for the many orphaned marsupials that were brought in. These included a pygmy possum called Bartholomew, which was brought in as a tiny baby and lived in the office for more than four years, and a young wallaby that was taken from its dead mother's pouch and handed in to the park in 1967. As the wallaby grew, Worrell was puzzled by its appearance, so he showed it to Professor Geoffrey Sharman, a kangaroo expert at Macquarie University, in Sydney, who often visited the park in the 1960s to take blood samples from a number of the park's marsupials for his genetic research. Sharman was able to identify it as a young parma wallaby, a species that had been considered extinct in Australia (although a colony exported to New Zealand was doing well). Some searching of the area where the wallaby had been found – near Ourimbah, just a few kilometres up the highway from the park – revealed a healthy population of the animals. Parma wallabies have been exhibited at the park ever since.

Later that same year, Worrell took on his first reptile keeper. Up until this time, the reptiles had largely been the domain of Worrell and Krauss. Worrell was acquainted with Paul Horner, who had worked as a reptile keeper at Taronga Zoo, under the supervision of Bill Timmis and Graeme Gow. When Horner left Taronga, Worrell sent him a telegram offering him work at the reptile park. Horner jumped at the chance, although he was more a lizard man than a snake man, and was appointed as the keeper responsible for the reptile collection under Peter Krauss's guidance. Horner stayed for about 18 months and left in 1969, eventually making his way to Darwin, where he joined the herpetology section of the Northern Territory Museum.

The 1960s were remembered fondly by former staff and associates of the park. The animal collection expanded under the curatorial guidance of Peter Krauss and his considerable skills in animal husbandry and exhibition. Not only did the park boast Australia's largest reptile collection, including a substantial representation of south-east Asian and American reptiles, but it had also built up Australia's largest collection of macropods – kangaroos, wallabies and their relatives – and had constructed the first noctarium in the southern hemisphere. A stunning collection of little-known mammals from New Guinea and the Atherton Tablelands was also a significant element. The park had become a major tourist attraction and visitors flocked to see the giant Gosford dinosaur, picnic among friendly kangaroos and waterfowl, and view the dangerous reptiles and unusual mammals and birds on exhibit.

Worrell's friends such as zoologist Jock Marshall and popular naturalist Vincent Serventy were vocal in their admiration of the park and its work, and helped to consolidate the reputation of both the institution and its owner. Vincent Serventy, who had first met Worrell in the late 1950s, was a lifelong friend and supporter and he often wrote about Worrell's adventures or happenings at the reptile park in his weekly newspaper columns.

Worrell had assembled a dedicated and loyal team, who enjoyed their work and were willing to go above and beyond the call of duty when needed. These people helped Worrell realise his dream. Key staff worked well with each other and created a community built around their interest in wildlife and their commitment to the park and to Worrell. A general sense of optimism coloured much of this decade for the park, just as it did Australian society more generally.

In southern states it was difficult to maintain appropriate temperatures for taipans, snakes of the tropics. After the initial challenges of keeping taipans and the loss of eight of them to pneumonia at the Ocean Beach Aquarium, in 1965 Worrell commissioned John McLoughlin, his north Queensland agent, to extract taipan venom for him and transport it down to the reptile park. McLoughlin, who had been sending Worrell taipans that men such as Vince Edwards and

Frank Little had caught, began maintaining the snakes at his place in Kuranda on the Atherton Tablelands and sending the venom to Worrell, who sold it on to the CSL. At first the venom was transported frozen, but later McLoughlin began freeze-drying it himself. Then, in 1968, Worrell made a significant proposal: that McLoughlin take over the responsibility for supplying most of the park's snake venom, not just taipan venom. He accepted. The first stage of this ambitious plan involved McLoughlin purchasing a block of land at Trinity Beach, a small coastal village about 5 kilometres north of Cairns.

Each of McLoughlin's snakes was kept in its own individual case, which enabled him to keep a close watch on their feeding patterns and general health, something that was not easily done in the pits at the reptile park. Once individually housed, a snake was also less susceptible to parasite or disease transmission. McLoughlin supplied venom from taipans, tiger snakes, Papuan blacks, king browns, death adders and browns. From this time until the late 1980s, the venom extraction carried out at the reptile park would be done largely as a performance for tourists, although the venom collected from the park's tiger and brown snakes was processed and added to the venom supplied by McLoughlin for consignment to the CSL.

Worrell's grand vision was to establish the Trinity Beach site as the reptile park's Cairns outpost, which would eventually open to the public as a tourist attraction. However, John had no interest in managing a facility that was to be open to the public and neither he nor his wife Grace enjoyed living at Trinity Beach, preferring the quieter and cooler atmosphere of Kuranda. After a couple of years they abandoned the idea altogether and shifted back to Kuranda. McLoughlin, without ever having been bitten by a venomous snake, continued to reliably supply venoms to Worrell up until 1982, when the association was brought to a mutual, and according to McLoughlin, friendly, end. McLoughlin thereafter supplied venom directly to the CSL until the late 1980s, when he resumed supplying venom to the park when it was under different management.

From his Ocean Beach Aquarium days, Worrell had continued to collect Sydney funnel-web spiders from locals and send them on to the CSL in Melbourne, which was sporadically conducting research into a funnel-web antivenom. Their progress toward understanding the venom was slow and it had become obvious that to further advance their research they needed greater volumes of venom than the voluntary collection and dispatch system was delivering. Many spiders died during their journey south, while in the laboratory, females lived on average only four months and males an even shorter period of time. So the CSL decided to establish a large funnel-web colony in Melbourne rather than relying on the piecemeal supply of spiders from around Sydney. In 1960, they called for tenders for the collection, breeding and supply of 2000 each of mature male and female funnel-web spiders over six months. Worrell considered tendering but it was not feasible for him to acquire such a large number of spiders as he was not breeding them in captivity. He had a more long-term plan: to set up a venom-milking program for the funnel-webs, just like the one he already had in place for snakes. It would take some time to come to fruition and in the meantime, no one else was able to fill the order for 4000 funnel-webs, so the CSL continued to rely on public appeals through the NSW media to secure spiders for its research.

After years of frustration, CSL scientist Saul Wiener gave up on his research in the early 1960s, but his colleagues, biochemists Merle Gilbo and Dr Norman Coles, continued work on isolating the toxic components of the spider's venom. The work was reinvigorated by the CSL's newly employed head of immunology research, Dr Struan Sutherland, in 1967. The CSL's deputy director (medical), John C. Trinca, visited the park in July of that year to inspect snake venom production but also to investigate the possibility of Worrell supplying funnel-web venom.

Sutherland himself visited the park late in 1967 and afterwards wrote to Worrell, 'I feel confident that with time, luck and plenty of hard work we shall together eventually find a solution to the funnel-web.'[6] The two men had struck up a strong friendship almost immediately. There had also been discussion of the park supplying the CSL with blue-ringed octopuses for use in research on their venom, and early in 1968, Sutherland wrote to Worrell about the feasibility of it: 'I doubt if you need any guidance at all. Over the Xmas period I managed to read a fair deal of your published works. Very pleasant reading I might add.'[7] As a laboratory scientist, Sutherland particularly admired Worrell's practical expertise. He wrote, 'Without your expert help current studies on the diagnosis of snake bite by *in vitro* methods would be academically satisfactory but perhaps lacking in practicality.'[8]

Sutherland arranged to have the CSL approach Worrell for a formal quote for supplying funnel-web venom. Coming up with a price needed some careful consideration. Snake venom production costs had tripled since he had set prices with the CSL in 1951 and negotiations over new pricing in 1967 had been difficult. Sutherland knew this, so he advised his friend not to set a price per milligram of funnel-web venom until he had a better idea of the actual production costs. At Sutherland's suggestion, Worrell asked the CSL for $4000 to cover the setting up of a funnel-web lab with a dedicated staff member and the purchase of 1000 spiders. After an establishment period, the park would supply 100 milligrams of venom each month for 12 months without additional payment. This arrangement would give Worrell a good idea of how much the venom cost him to produce before he put a price on it, avoiding some of the problems he'd had in setting costs for snake venom.

Worrell agreed in February 1968 to move from the establishment period to active supply of funnel-web venom no later than the end of April 1968. He was initially very optimistic, believing three months would be sufficient to refit an existing building with air conditioning and purpose-designed shelving for the beakers to house the funnel-webs, acquire 1000 spiders by offering the public 20 cents per spider and get venom production under way. Everything took longer than expected and when it came time to collect the spiders, winter had arrived and few were being found by the public. A dedicated staff member for the funnel-web work – spider keeper Kay Brien – was not hired until October 1968. Kay had been working as an office assistant until she was offered the role. Although she had no training, experience or even interest in funnel-webs, she, like the other staff at the park, was willing to have a go. By December 1968, the park had just 200 funnel-webs and venom yields were still low.

Correspondence between the park and the CSL was intense, with the CSL by turn offering practical advice on keeping and milking the spiders, and threatening to end the arrangement and recoup the money they had invested. Efforts to increase the venom yield drove experimentation into the substrate on which to hold the spiders at the park. The initial use of damp cotton wool was discontinued when potting mix was found to be more suitable, retaining moisture well and simulating the natural habitat of the spiders.

The painfully slow and careful milking of funnel-web spiders for venom took place once a week, on Wednesdays. Because these spiders are aggressive when disturbed, all that was required to make them rear up was a gentle breath into the jar or a tickling with forceps. (This approach was found to be better than another technique the CSL had experimented with: using electrodes to stimulate the chelicerae – the mouthparts, including the fangs – of anaesthetised spiders.) When drops of venom appeared on their fangs, it was suctioned up in a pipette. It was then flushed from the pipette with a weak acetic acid solution into a glass vial to be frozen, packed in dry ice and sent to the CSL. The pipettes the park was using to collect the venom were deemed inefficient so Sutherland provided custom-made milking pipettes in the hopes of improving yields. He also offered the park advice on milking techniques (suggesting that while any interference would produce venom, a full five minutes of provocation would give the greatest yield) and diet (a small piece of raw rabbit liver once a fortnight).

In December 1969, Worrell sent three vials of female funnel-web venom to the CSL. He counted this as the end of the establishment phase and the beginning of venom production, writing, 'Our actual contract has started today.'[9] An immense amount of labour had been required to produce this venom. One vial – just 9.45 milligrams – contained the output of 400 painstaking milkings.

While the milking process was slow, the urgency of the work was reinforced by the death in 1970 of 17-year-old Sandra Burgin after a funnel-web bite she suffered while camping at Nowra, on the NSW South Coast. So refinements continued to be made to the park's funnel-web program. They experimented with using blowfly pupae as food and eventually settled on a once-a-week feeding regime that included mealworms, fruit flies, small cockroaches and grasshoppers, liver and blowflies, and a fine spray of water to maintain humid conditions similar to those in an underground tunnel. Funnel-webs release venom whenever they are provoked, so handling of the spiders was minimised to once a week to avoid wasting venom at non-milking times. It became evident that the milkers' skills played a significant role in production, too, and following Kay Brien's departure from the park, Robyn Innes and Lyn Abra developed exceptional skills in the care and milking of these deadly spiders. After a prolonged establishment period, the park finally began to provide the CSL with the venom it needed to intensify its research.

For the most part, the 1960s were remembered fondly by Worrell's dedicated and loyal team. Yet as they neared the end of this decade of building and expansion, trouble was brewing. The delays and setbacks in the funnel-web venom program were perhaps the first signs that all was not as it seemed. The public knew Eric Worrell as a success story, but beneath his confident exterior, Australia's Snake Man was struggling with some of his own personal demons.

11

# Managing the deadly
# and dangerous

The scrub or amethystine python is found in the tropical forests and savannas of north Queensland and is Australia's longest snake, growing up to 5 metres in length. Pythons are non-venomous but the scrub python is known to be rather snappy and, with a mouthful of backward-curving sharp teeth, can inflict quite a nasty bite. It is also the only Australian python known to be linked to the death of a human, which occurred in 2005, in South Australia. Circumstantial evidence suggested that the pet python was implicated in its owner's death by asphyxiation.

Handling one while standing chest deep in water was a somewhat daunting prospect for a 15-year-old boy, even when that boy was Steve McEwan. A keeper-in-training at the reptile park, he already had experience catching dangerous snakes, such as eastern browns, in the bush around his home in the western suburbs of Sydney.

The idea on this April Sunday morning was that bringing the snake into the park's swimming pool would help to soften its old skin, easing the shedding process over the next few days. McEwan kept a firm hold on the python's neck but no doubt the snake wasn't all that appreciative and struggled against the boy's grip.

McEwan lost control of the snake and became distressed as it coiled around his upper chest and neck. Worrell jumped into the pool and struggled with the python until he had rescued the boy – who was by then close to slipping beneath the water – from its grip.

At least, that is how events unfolded according to a series of dramatic photographs appearing in *The Sun* newspaper. Fairfax photographer Russell McPhedran was at the park that day and managed to capture a sequence of spectacular images of the boy, the python and Worrell struggling in the pool. The look of distress on McEwan's face was genuine but there is also the suggestion by a staff member that Worrell spiced up the skirmish to some extent. The incident may have got somewhat out of hand, turning into a more hazardous situation than Worrell had intended. Still, he knew that drama drew publicity – and it paid off in this case. Most of the newspaper's front page the next day was dominated by the most breathtaking of the photographs, which earned McPhedran a prestigious Walkley Award for best news picture of 1976. Another six photos appeared inside. The captivating shots were also wired around the world and appeared in newspapers in the UK, Europe and the USA.

For young Steve McEwan, things went from bad to worse. A few days later, he suffered a bite from a deadly tiger snake. He survived the snakebite, but his relationship with Worrell began to deteriorate. Reluctantly, he decided to resign some months later. From working in a job that was, in his words, initially 'as good as being at Disneyland', he became a car detailer for a used car yard in western Sydney.

The new decade had begun well for Worrell. On New Year's Day, he became a member of the Order of the British Empire for his services to herpetology. The MBE was well deserved, as Worrell's contribution to popularising herpetology was considerable: his writings were very influential, especially *Reptiles of Australia*, the first book of its kind in Australia, which served as the 'bible' for amateur and professional herpetologists for more than a decade. Originally published in 1963, it was reprinted in 1964 and 1966, and a second edition was published in 1970, indicating the high and sustained level of demand for the book.

As it continues to do to this day, the reptile park played a significant role in promoting reptiles to a broad audience. To reptile fanciers, the park meant even more; to them it was something of a spiritual homeland, a kind of Mecca. For a child who already had a fascination for reptiles, a visit to the park was a nearly overwhelming experience. Some people we interviewed told us of the butterflies in their bellies as their parents pulled up in the car park, under the shadow of the dinosaur. Once they got inside, it was almost as if the photographs from Worrell's books had come to life. These young reptile enthusiasts could barely contain their excitement when they came face to face with a king cobra, a gaboon viper or a pit of black tiger snakes. Perhaps the park created excitement, especially for amateur herpetologists, because it was one man's reptile-keeping dream made real. The park gave hope to those whose ambition was to work with reptiles.

Professor of Ecology at the University of Sydney, Rick Shine, now a foremost expert in the ecology of reptiles and amphibians, recalled that the reptile park was part of his consciousness from a young age: 'I used to catch the train up to Gosford and get off at the station and walk up to the park and spend the day there just sort of hanging over the concrete pits and marvelling at everything.' For Shine and many small boys like him, it was a pilgrimage. 'I'd go probably a couple of times a year and it was all pretty exciting stuff to see these very exotic animals, because at that stage they had things like king cobras.' Shine was also heavily influenced by Worrell's books. 'We didn't really have much in the way of books that exposed young Australians to Australian reptiles, so in fact Eric's book *The Reptiles of Australia*, in 1963, was actually quite significant … I can remember looking at this and realizing [that there is] a group that's now called *Simoselaps*. I didn't even know they existed in Australia, these coral snake type things with spectacular radiations – the tropical stuff – island snakes, creatures of mystery …'

Worrell was featured in the *Sydney Morning Herald* on 1 January 1970, along with other recipients of the Queen's Honours, including Slim Dusty. The next week, a brown snake fatally bit a 20-year-old NSW man and Worrell was again featured in the *Herald*, this time discussing the dangers venomous snakes posed in high summer. Such was his stature, it was Worrell to whom the media turned when they wanted comment or expert opinion on venomous snakes and spiders.

His park was one of the state's iconic tourist attractions. Worrell was a successful author, having ten books released by leading Australian publisher Angus and Robertson, and he regularly appeared on popular television programs and in films. He was an honorary ranger with the NSW National Parks and Wildlife Service, a life member of the Royal Zoological Society of NSW and he was often used as an expert witness by the RSPCA in cases of cruelty to wildlife. He was also a founding member of the Australasian Association of Zoo Directors of Australia and New Zealand. Locally, he was a Rotarian and, as a prominent citizen, lent his patronage to the Gosford Sea Scouts, Gosford Diggers Swimming Club and Gosford RSL fishing club. He was a generous host and charming and witty when he was among his friends. Yet while on the surface all was positive, aspects of his private life were troubled and over time those troubles spilled over into his work at the reptile park.

When Worrell, a monarchist, proudly received his MBE from Queen Elizabeth herself at a ceremony at Sydney Town Hall, his son,

Mark, was there – but notably absent was Carol. Their marriage, which had borne three children – Kim, Mark and Brook – ended in bitter divorce that same year. Worrell had been using his office at the park as a bedsit throughout the mid- to late 1960s, having all but moved out of the family home that he had built in 1963 overlooking Wingello Creek. After the divorce, Carol moved to the North Coast of New South Wales, where she continued to work in the tourist industry, running a motel.

Although Worrell does not appear to have been a drinker in his earlier years up in the Northern Territory, Russell Drysdale's diaries of his bush expeditions with him are peppered with references to alcohol, from frequent long stops at country pubs, to the merits of local mixed drinks, to the virtues of rum for covering up unpalatable water supplies. From the early 1960s, Worrell became a part of the small social set of local businessmen who drank regularly at the Gosford Aquatic Club. When the Grange Hotel opened just up the road from the reptile park in October 1972, Worrell became a regular there, earning his own corner of the bar, named Funnel-Web Corner. He donated several bottles of preserved snakes, which were placed on the shelves along with the bottles of bourbon and whisky. By the late 1960s, Worrell had a drinking problem.

Staff began to notice. His drinking began to affect his ability to make prudent decisions, on occasion putting him, and sometimes others, at risk. Over the course of several years, into the late 1970s, his behaviour was sometimes irrational and he gradually became more removed from the daily operations of the reptile park. Bit by bit, his problem with alcohol began to change the man and these changes began to show themselves to those who knew him closely. Some staff and colleagues were affected by his unstable behaviour or ill-conceived and inappropriate decisions. It increasingly fell to senior personnel to convey those decisions to staff. By now, the parks' full-time staff had grown to around ten, with Peter Krauss as general curator and Jack Green as curator of aquatic exhibits.

Worrell was a hard taskmaster and expected a lot from his staff. He maintained good working relationships with some staff but others found him to be a difficult man to work for. In particular, it seemed that nobody could do the job of maintaining the reptiles to his satisfaction. Yet Worrell seems to have had little awareness that some of his employees perceived his management style negatively. Indeed, he considered himself a generous employer, and in the first decade or so of the park's life, he undoubtedly was:

> We could not afford high wages. Our family only drew living expenses and were not insured. Every year we held our staff Christmas party where guests of staff were also invited. There was an annual bonus for all staff. We gave four weeks annual leave where everyone else gave one to three weeks. There was generous superannuation. I put selected employees through three to four year courses on Animal Care at Sydney Technical College where I was a part-time teacher.[1]

He wrote that his staff, whom he regarded as family, were happy and that he had achieved 'a compatible harmony of management and workers'. It is the case that in the 1960s and early 1970s, staff recalled many good times, including staff barbecues and parties hosted by a generous Worrell; and modest Christmas bonuses were also paid to senior staff. But those glory days began to wane.

Relations between Worrell and general curator Peter Krauss sometimes became strained. Worrell made a point of reminding Krauss that he could take away any extra benefits or conditions attached to his employment if he felt the need to do so. Worrell would write tasks that Krauss's staff needed to complete on a blackboard. Krauss recalls one day when most of the jobs had been completed but Worrell demanded to know why one of them hadn't. Krauss assured him that his staff were doing the job right then and it would be finished by nightfall. 'He could be very argumentative,' recalls Krauss. 'He made an issue out of something that wasn't an issue … he started to argue and screamed out that I was undermining his authority.' Worrell's deterioration at this time was noticeable, according to Krauss: 'Even when he wasn't intoxicated, he was displaying the effect of it … It was sad and left me upset.'

In 1972, Krauss resigned. His marriage to Raija had ended, too, and he decided to leave the Central Coast. He had purchased a few hectares on the Atherton Tableland and began his next career as a cane toad farmer, supplying universities and other biological research and teaching institutions. His departure was the most significant staff loss up to that time. His patient, rational and systematic approach, coupled with an absolute thoroughness in ensuring animals were kept clean and well maintained, made him first-rate. Without Krauss, it is unlikely that the park would have flourished and expanded as it did in the 1960s and into the 1970s. Krauss had helped Worrell realise his boyhood dream of creating a nationally recognised reptile park and he left with an excellent reference from Worrell.

Later that same year, the park experienced its first serious industrial relations problem. Worrell knew that there was a long list of people hopeful of working at his park. Many young men and women would have done anything to get an opportunity to work among the diverse and rich collection of Australian and exotic animals that the park held. If a staff member was not prepared to work for the low pay and basic conditions he provided, they could be replaced. He was strongly opposed to the unionisation of labour. When a young keeper joined a union, he dismissed her. Appalled by the way Worrell so readily sacked this keeper, and not a little concerned for their own job security, about a quarter of the keeping staff decided to join the Australian Workers' Union. Worrell wrote to each keeper stating that if they did not resign from the union they would be stood down. The keepers took the letter to the Central Coast Trades and Labour Council, which then supported the workers to go on strike for the right to unionise. Before they could begin their strike,

Worrell sacked the keepers and the case was heard before the NSW Industrial Relations Commission in Sydney. He was directed to reinstate the workers and allow them the right to belong to unions if they so wished. Within a few weeks, the keepers had returned but found working at the park difficult and a number of them resigned. This incident rankled Worrell. More than a decade later, when writing an addition to the *Song of the Snake* that was never published, in a chapter he called 'The Snake's Strike', he told a different story than the people we interviewed told us. In his version of the events, the judge found in his favour and held that Worrell was entitled to sack the worker who had led the industrial action.

Worrell found staffing matters, and in particular, the drawing up of suitable terms and conditions, duties and appropriate pay schedules, a headache and he sought advice from time to time from his friend Ronald Strahan, who by this stage had become director of Sydney's Taronga Zoo, replacing Sir Edward Hallstrom. In January 1970, Strahan provided him with details of Taronga's staffing agreements as well as giving him feedback on Worrell's own drafts, suggesting he tighten up the wording of specific clauses and identifying areas where he felt Worrell should seek legal advice. In 1972, the basic wage for keepers at the reptile park was set for males at $41.10 per week, less than half the average weekly wage at the time, and for women at $32 per week. (Equal pay legislation would soon put a stop to that kind of overt gender imbalance in pay rates, common across the country at the time.) However, the take-home pay was increased through margins that were added depending on the level of responsibility of the position: supervisors and head animal attendants received an additional $28.90 per week, animal attendants $18.90, animal attendants in training, $16.50 and the groundsman, $14. Workers under 21 were paid a percentage of the weekly rate, with those under 18 receiving only 55 percent of the adult rate. By employing juniors at animal-attendant-in-training level, Worrell made considerable savings on salaries.

Worrell saw the park as the idealised haven he had dreamed of in his youth. In his eyes, the appeal of working with the park's collection of animals should be sufficient reward. Although he prided himself on creating a *modern* wildlife park, his approach to managing employees was not quite so modern. From the mid-1970s onwards, this attitude contributed to a high turnover of staff, many of whom were a significant loss to the park's operations. Such an attitude, however, was not uncommon in business at the time, especially in owner-operator enterprises; the substantial changes that occurred in Australian industrial relations, and for that matter, occupational health and safety, were still some way off.

For Worrell, there were some pleasurable departures from the challenges he faced at the park, when he went on sea snake collecting trips around the Whitsunday Islands. In 1971, Harold Baxter, the head of biological chemistry research at the CSL, was conducting experiments to determine whether tiger snake antivenom provided protection against sea snake venoms and he asked Worrell to provide venoms from a range of sea snakes. Worrell was more than happy to do this. He sailed on a charter vessel known as the *Coralita*, built in 1969 to take divers out to the Great Barrier Reef. These trips were as enjoyable as one would imagine cruising the outer reef would be: aquamarine water, sunny days, balmy temperatures and plenty to eat and drink in the company of friends. And for Worrell, the thrill of collecting and milking sea snakes only added to the enjoyment.

In June 1973, Worrell took a very special cruise on the *Coralita*, with Robyn Innes, her parents and a select group of friends: Russell Drysdale (now Sir Russell), Vincent Serventy, underwater photographer Ron Jenkins and two mates from Gosford, Gordon Martin and Barry Hanks. Worrell and Robyn had been seeing each other for some time and now they were to wed just off Great Keppel Island. While the couple had let Robyn's parents and the captain in on the secret, the wedding came as a surprise to the guests. They had spirited a priest on board at Yeppoon, on the central Queensland coast, and Martin was drafted to be the best man. The shipboard wedding was perfect for the adventurous couple. Worrell recalled, 'Robyn and I cut the cake with her diving knife. Festivities continued throughout the night as the *Coralita* headed into the Coral Sea. Denise [a crew member] played the piano she had brought aboard for the occasion.'[2]

Worrell organised for colour photographs of the wedding to appear in the following issue of the *Australian Women's Weekly*, keeping him and his new wife in the public eye.

Working with highly venomous snakes and spiders as well as other animals that have the capacity to inflict serious injuries, such as large alligators, crocodiles and cassowaries, is an extremely hazardous occupation. The fact that venomous snakes were regularly milked at the park meant that there was great potential for accidents. Worrell himself suffered about half a dozen bites that he considered dangerous and various other staff also took bites from venomous snakes, but no one ever suffered a fatality from any of the deadly or dangerous animals held there. One young employee, Tim Hawkes, a skilled keeper, was bitten a couple of times by dangerous snakes. He even suffered the indignity of being bitten on the nose by a red-bellied black snake. Worrell often recalled an instance where he was bailed up by a taipan with mouth canker that literally flung itself open-mouthed at him as he opened the door of its case. A tense couple of minutes followed as Worrell fended off the snake and manoeuvred it back into its case.

By the mid-1970s the park had an amazing collection of venomous snakes and the inventory of exotic species included king cobra, krait, water moccasin, black mamba, monocled cobra, spitting cobra, several species of rattlesnake, sidewinder, rhinoceros viper, gaboon viper, puff adder and a bushmaster. It takes a certain kind of person to be able to look after the needs of such animals. Many of those who worked at the park were motivated not by money or ambition but by their love of animals. During the almost four decades the park was located at Wyoming, a large number of people cared for the animals either as paid staff or as volunteers. Some staff arrived already having experience working in a zoo, such as Peter Krauss and later, Paul Horner and Lyall Naylor, while others were enthusiastic reptile fanciers, willing to trade off rather low wages in exchange for the opportunity to work with Worrell and his amazing animals. Some stayed for only a few months, while others, particularly core staff such as Jack Green and Lyn Abra, stayed for much longer, or, in Robyn's case, are still there.

Many park employees or volunteers have gone on to carve out distinguished careers, such as Bev Drake, recently retired from her senior curator position at Werribee Zoo in Victoria; Paul Horner, recently retired as herpetologist at the Northern Territory Museum; Michael O'Brien, manager of Cairns Tropical Zoo; Lynda Vereyt (now Croxford) and Grant Husband, senior keepers at the Territory Wildlife Park, south of Darwin; and Trent Russell, manager of the National Zoo in Canberra, to name but a few. Steve McEwan, who famously tussled with a scrub python in the park's pool as a teenager, is now back doing the work he loves, running a mobile educational reptile display on the NSW Mid-North Coast.

The park was a performance space in which the animals were the stars and the staff had the responsibility of ensuring that they were shown off to best advantage. To visitors, there was an ease and naturalness about the operation, but that was an illusion maintained through a tightly timetabled behind-the-scenes schedule of maintenance and animal care. The day-to-day working life of the keepers at the park was based around a strict routine – though compared to other workplaces it was far from a conventional one.

The keepers were always busy feeding and attending to their animals and cleaning their exhibits. There were bags to be filled with puffed wheat that visitors bought to hand-feed to the animals. Baby wombats, kangaroos and koalas had to be bottle-fed. The menagerie of mice, rats, mealworms and other animals that were bred for reptile food needed to be fed, watered and have their enclosures cleaned. The mammal and bird staff were also kept busy preparing food for their animals. Occasionally a local farmer would offer an entire calf or even horse that had died. The animal would be butchered at the park and its meat stored for future use. The grisly task of dismemberment would become the task of one of the keepers or of Alec Innes, Robyn's father. Both of Robyn's parents assisted at the park in many ways and provided ongoing support to Robyn and to Eric, as did Robyn's brother, John, who offered Eric financial advice on occasions. Fruits and vegetables were prepared for lizards, land tortoises, noctarium animals and some of the birds. Surplus day-old male chicks from nearby poultry farms had to be sorted and bagged up, then frozen for later feeding to many of the reptiles as well as to eagles, owls and carnivorous mammals such as quolls. Eucalypt leaves for the koalas, possums and other leaf-eating mammals were gathered in local state forests. The park paid an annual fee for a permit to collect these leaves from lands managed by the then Forestry Commission of New South Wales. For some staff, it seemed the work never really stopped, as they would rise at regular hours throughout the night to feed baby marsupials in their care.

To ensure the smooth operation of the park, tasks were allocated down to the quarter hour, which meant there was little leeway if something unexpected happened. Yet the very fact they were working with live animals meant that there were often unusual or unpredictable elements to the day. And at Worrell's reptile park, there was always the risk that the unpredictable could turn deadly.

The park's two king cobras provided a very impressive display and Worrell was justly proud of them. At around 3.5 metres, they were possibly the longest such snakes in captivity in any zoo outside south-east Asia in the 1960s and 1970s. The venom of king cobras is not as deadly as that of some other snakes, but the large amount of venom they deliver makes them amongst the world's most dangerous snakes. Keepers who have worked with king cobras speak about them as having a greater intelligence or at least awareness of their environment than other snakes, which also means they can be much more difficult to manage in a captive situation. Greg Miles was employed by Worrell as reptile keeper in the early 1970s. He sometimes had nightmares about them escaping.

Another characteristic of king cobras is that they eat only other snakes. Pythons that died at the park or were brought to the park dead were kept in the freezer and defrosted before being fed to them. 'The actual act of feeding king cobras was quite something,' Miles remembered. 'You had to get it exactly right, couldn't make a mistake.' He would coil the thawed python around his arm in readiness, then quickly open the door to the king cobra enclosure, throw in the python and slam the door shut. Sometimes part of the python jammed in the door and the cobras rushed over before Miles could safely close it again. 'Once you'd shut the door and you'd look in the flap and see what they were doing and there'd be this face looking at you. Red eyes burning into your brain.'

Worrell's friend John Dwyer recalled a time when he was standing with Worrell in front of the king cobra case admiring these snakes when Worrell noticed some marks on the inside of the glass on the front. Both cobras were lying along one side of the enclosure, well away from the front. Worrell decided to wipe off the marks. By the time he had entered the exhibit, walked down to the glass and started to wipe away the offending marks, both snakes had slithered quickly down to near where he stood and raised their magnificent bodies a metre or so from the ground, gazing at him as only a king cobra can gaze. By this stage a heavily perspiring Dwyer was thinking he would have to break the glass to rescue the snake man, but Worrell stayed absolutely still for a couple of minutes and the cobras lost interest and slithered away. A rather close call.

Jack Green also had a frightening experience when he was attempting to introduce a new king cobra into the exhibit. In the small service area behind the enclosure, he brought the snake's travel box up to the entry of the exhibit and opened both doors so it could crawl in. Something had gone wrong, though. The openings of the travel box and the exhibit had somehow misaligned. Green felt the large head of the king cobra slide up along his arm, over his shoulder and against his neck. 'Inch by inch it slid over my shoulder and down my back,' he wrote in *Alligator Man*. 'I thought it would never stop coming out of the box. It seemed endless and I had no idea where its head was positioned behind me.'[3] The snake slithered to the opposite corner of the service area and coiled itself about 2 metres from Green. It had raised part of its forebody off the ground and its neck was flattening out to form its hood, indicating its displeasure at being threatened. Eventually, Green was able to subdue the angry cobra using a noose attached to a 2-metre-long rod and relocate it into the main exhibit.

Green was no stranger to risk, being in charge of the crocodilians. Worrell had been fortunate to receive several dozen young alligators from Major MacGogy, a former US Air Force officer who had wanted to establish an alligator farm in Australia. He had managed to import the animals but was apparently refused permission to breed them commercially. MacGogy subsequently donated them to Worrell and they formed the basis of the large alligator colony that is so much a feature of the park today.

Alligator feeding became a regular performance from the late 1960s onwards and Green was the star of the show. On Saturday afternoons from the October long weekend through to about Easter, Green would hand-feed the alligators, which were housed in two adjacent ponds behind the main reptile complex. The smaller of these ponds contained the alligators donated by Sir Edward Hallstrom in the early 1960s, Edward and Edwina, which were very large but generally well-mannered.

Carrying a bucket of fish or chicken, Green would enter their enclosure across a narrow log bridge leading to a central island. The usually lethargic saurians would suddenly become highly animated, propelling their massive bulk up out of the murky water to grab the morsels of food as Green offered them. Visitors crowded around the edges of the enclosure as open-mouthed as the alligators – which were open-mouthed for different reasons – as the animals chased Green around the island. After a few too many slips and falls, he wore spikes attached to his shoes for greater grip. For Steve McEwan, who helped Green prepare for the show, the excitement was in 'the build up to it. It was coming. The crowds starting to build up. Jack putting on his shoes. Like going into the 100-metre world championship or the Olympic Games.'

After these alligators had been fed, it was the turn of the 30 or so smaller alligators and, at times, crocodiles, in the adjoining, larger pool. A fenced feeding jetty was constructed in the early 1980s, which gave much more protection to the feeder, but for years Green simply entered the enclosure with a bucket of quartered chickens or fish. In an instant, he would be faced with the open and snapping mouths of dozens of alligators and a smaller number of saltwater crocodiles ranging in size from 1.5 metres through to 3 metres long. Stepping quickly through and around this frantic pack of hungry reptilians, Green was as skilful and as dextrous as any football player or ballet dancer. Green's show was captured in Jeff and Mare Carter's 1972 film *Tigerman* and, not withstanding the late Steve 'Croc Hunter' Irwin's manoeuvres with crocodiles, Green's work was most impressive.

During the early 1970s, Robyn Innes and Lyn Abra volunteered to be trained in the milking of venomous snakes. Worrell saw it as an opportunity to help them develop new skills while creating a new attraction for the park. He had a progressive attitude to the employment of women and never saw gender as an impediment to working at his park. The women would mostly milk tiger snakes for demonstrations on Sundays and during school holidays. Women milking snakes evoked the long history of show performers such as Paula Perry dancing semi-erotically with snakes – but now the show had been transformed into a more worthy scientific endeavour, the women costumed in khaki rather than sequins. Val Lawson of Koombal Park in Cairns had milked taipans and other snakes such as death adders in the early 1960s, but Innes and Abra were now the only active female snake milkers in Australia. Two young women milking deadly snakes was an unusual and gripping attraction and Worrell ensured that the news got out. A number of newspaper and magazine articles followed, as well as the filming of their tiger-snake-catching trip with Worrell to the flooded Lachlan River, in southern New South Wales, for *Tigerman*.

Both women suffered a bite from a tiger snake during their dangerous milking work. After her experience of snakebite, Lyn decided not to milk snakes any more and restricted her milking activities to funnel-web spiders. Robyn was bitten one Sunday afternoon during a milking demonstration for visitors. Soon after being bitten, she developed a very painful headache. Panic broke out among visitors, with some of them fainting. This made getting Robyn from the milking area to the main office rather difficult. Worrell drove her to Gosford Hospital, explained to the doctor on duty that she had been bitten by a tiger snake and detailed the appropriate treatment. The doctor ignored this advice and left Robyn waiting while he looked into how she should be treated. He came back some time later carrying Worrell's own *Dangerous Snakes of Australia and New Guinea*, opened at the section on treating snakebites! After Worrell told him, 'I wrote the bloody thing,' the young doctor started taking his advice. Worrell also asked his own GP to come in to assist with Robyn's treatment. A week later, Robyn reacted to the antivenom that she had been given and became quite ill and completely covered in hives. Fortunately, that was the only serious bite she was to receive and Robyn continued milking until the early 1980s.

While he was never bitten by a deadly snake during his time at the park, Greg Miles had some close encounters with some spectacularly venomous snakes, such as the African black mamba, a snake whose bite, unless treated with appropriate antivenom, leads to close to 100 per cent mortality, and the bushmaster, a large snake from Central America with a rather nasty reputation for being permanently bad tempered and difficult to manage in captivity. He once came close to being struck in the face or neck by the black mamba. He was hosing down the inside of the glass front of one of the snake exhibits, which required him to lean across the closed bottom half of the door in the service room; his upper body was partway inside the enclosure. As the jet of water hit the glass front, he looked around and registered the puff adders inside the enclosure. 'And I thought, "Puff adders, they're in with the mamba aren't they?"' Miles recalls. The mamba's hide box was on the back wall of the enclosure. 'So I looked sideways and there he is with his tongue, and his head cocked, on top of his box, going, "Hullo",' says Miles. 'I froze, the hose going, and sort of looked at him and started moving slowly, backing out, knees trembling, until I was clear. He left me alone, so I loved that snake.'

Jack Green had a similar experience when he opened the door of the red rattlesnake case a few minutes after looking at the two snakes from the front of the exhibit and seeing that they were a safe distance from the door. One of the snakes was still where he had seen it, but the second had moved. It was the ominous sound of the snake's rattle that alerted him to danger:

> Slowly I forced my eyes to swivel to the left and there, coiled
> on a cactus high in the corner was the second rattlesnake; his
> whole body tensed and coiled for a strike. There was no way
> in the world that my reflexes could beat the snake's. It was
> positioned at face level and had enough length to reach me
> with ease.[4]

Green managed to ease his upper body out of the case at an excruciatingly slow pace, much to the pleasure of four young men who had gathered at the front of the case, enjoying the drama.

In another close encounter, Miles opened the door to the bushmaster's case and the snake, which normally did not move very much at all, lunged straight for his hand. Bushmasters are capable of injecting very large quantities of toxic venom and the victim may die if left untreated. The snake wrapped its fangs, which were huge at close to 4 centimetres long, around his fingers. Miles felt a slight tug and then the snake relaxed its grip and retreated. Two small drops of blood betrayed the fangs' scratches on his fingers. Thinking he had received a venomous bite, he made his way to the refrigerator in the office in which the park's store of antivenoms was kept. There were three ampoules of the appropriate antivenom sitting in the fridge – but all were out of date. Miles was simply lucky that the snake had failed to envenomate him.

12

# Consolidation
# and challenge

It is 2.00 pm on a balmy Sunday afternoon in late spring and a crowd of about a hundred visitors has gathered expectantly along the walls of the large snake pit. Casual conversations stop and the visitors watch with interest as a man in his early 50s with a distinctive white beard eases himself over the wall and casually drops a snake stick and small flat pad onto the grass below. Opening up a well-made wooden case, he brings out a small glass vial over which a thin membrane cut from a blue rubber glove has been attached with an elastic band. He places the vial on a log and then looks around the pit. He walks casually across the grass and in an instant grabs hold of a snake, about a metre long, by its tail. The snake objects fiercely by thrashing about in his grip. A collective gasp can be heard from the crowd as the flailing snake is carried over to the pad. With the skill that comes with repeated experience, the man carefully pins the snake's head against the soft surface with the snake stick and then takes hold of the neck. Not getting this part right could give the snake a chance of sinking its fangs into a hand or finger. Walking close to the wall of the pit and making sure that everyone can see the snake's open mouth, he brings the glass vial up to the snake, which then bites into the rubber sheet with such force that some of those standing closest hear a popping sound as its fangs penetrate the thin membrane. Drops of yellowy venom slide down and pool at the bottom of the vial. The risky performance that Eric Worrell made his own has begun.

Visitors were not disappointed when they came to see Worrell in person at the reptile park. Rather than seeing only energy-conscious reptiles curled up in display cases or quietly sunning themselves, they witnessed his regular demonstrations at weekends and during school holidays. His style was inspired in part by the performances of the snake men and women of the past but was tempered by his more scientific understanding of reptiles. Worrell had a real affinity for the show tradition, even adopting the pseudonym Sapengro, a Romany word for snake handler, for some of his writing. One of the articles he wrote as Sapengro was an account of a winter trip to Queensland with a show train. Published in *People* magazine in April 1957, it portrayed beautifully the many amazing characters who performed in the sideshows or worked behind the scenes. Worrell evocatively conveyed the day-to-day realities of this strange, nomadic life on the margins of society.

An element of the showmanship of these entertainers ran through Worrell's reptile displays. When he milked a snake in front of a crowd, he held it up in front of his face, where it could be clearly seen and was visually framed by his arms. He moved quickly, deftly handling his dangerous charges. Viewers were kept at a distance in stadium seating but pressed up against the edges of the pit, only metres away from the action. The appeal of these astonishing demonstrations was written on the faces of those watching – faces marked by concern, awe, fascination and total engagement.

Although Worrell later wrote of his ambivalence towards keeping snakes in pits, citing in particular the difficulties associated with feeding and disease control, seeing numbers of snakes together in a pit captivated visitors. Each of the main pits was mounded towards the middle, giving the onlooker clear lines of sight. A narrow water-filled moat surrounded the central island. Rocks, logs and some shrubs and ferns lent a more naturalistic, albeit stylised, feel to the enclosures. Visitors enjoyed looking up close at the single snakes exhibited in cases inside the main building, but the pits produced a very different experience for them. A visual spectacle was created out of the tangled bodies of upwards of 100 venomous snakes, entwined, snake upon snake, interacting with one another, sunning themselves or exploring their surroundings. A case containing one snake had limited visitor staying power, but the sight of the effortlessly moving bodies of dozens of snakes was compelling.

The pits of highly venomous snakes became performance spaces in which Worrell and other staff demonstrated their mastery over the animals. Worrell had a flair for showmanship but also a scientific and rational approach. He stressed how unlikely it was to be bitten by a snake but equipped his audience with the knowledge that might help save lives in the event of being bitten. He ensured that visitors understood that their holiday fun was the first step in a series of procedures by which the CSL produced life-saving antivenoms. This was snake handling with a moral purpose and visitors were not just entertained, they were offered an insight into the esoteric world of anticoagulants, antibodies and antidotes – the scientific research that had tamed the terrors once held by the Australian bush.

Worrell saw the snake-venom business as a defining activity of his park. It had provided the economic base that launched the Ocean Beach Aquarium and it gave the Australian Reptile Park a serious purpose beyond education and entertainment. The park's role in the production of antivenom by the CSL provided the justification for Worrell to make the claim that the Australian Reptile Park was the only zoo in Australia that saved lives. One of this book's authors, Kevin Markwell, was the beneficiary of some of the venom extracted by the park in 1980, after he was bitten on the finger by a red-bellied black snake that he was holding somewhat carelessly. He was a cocky 18-year-old and the snake had been about a metre long, so he didn't think it would pose too much of threat and that he could 'tough it out' – he changed his mind an hour or so later when he was dry retching and passing blood in his urine. Once at hospital, he was eventually given black snake antivenom. After an initial feeling that he was lying on sandpaper, he made a full recovery, much to the relief of his anxious parents.

While the antivenom program was clearly of great benefit to the community and integral to Worrell's vision of the park, the business side of venom production caused him headaches. The prices that Worrell charged the CSL remained unchanged for 16 years, from 1951 until 1967. By then, production costs had tripled and Worrell was producing 'all venoms at a financial loss', as he told the CSL when he foreshadowed price rises.[1] His losses were somewhat offset by the fact that he was by then also supplying researchers and other laboratories within and outside Australia with snake venoms, usually for higher prices than he charged the CSL.

No doubt Worrell had held off from renegotiating his fees with the CSL in the hopes of maintaining the good relationship that he had developed with them. The fragility of that relationship was made evident when the park accidentally quoted the CSL the general price of $400 per gram of taipan venom, rather than the lower CSL price of $240 per gram. The CSL replied that they were going to switch to an alternative source: Charles Tanner, with whom Worrell had hunted taipans outside of Cairns in 1956 and who had established his own snake venom supply business at his property in Cooktown. Worrell quickly responded by telegraph to rectify the error. He certainly would not have wanted to have lost his taipan venom sales to another supplier.

Worrell felt that the original price that had been set by the CSL for Papuan black snake venom was unrealistically low, because it had been calculated based on 'an abnormal yield from a specimen that cost nothing to obtain as it was collected in Government time'.[2] Papuan black snakes were relatively difficult to obtain, as they had to be caught by locals in Papua, purchased by an agent there and then transported down to Gosford. The snakes that reached Gosford were often not in very good condition and frequently required force-feeding, making them expensive to maintain. Indeed, in 1967, Worrell claimed that he had lost between $2000 and $4000 per year on the production of Papuan black snake venom. He requested a substantial price hike from $80 per gram to $500 per gram. Initially the CSL was not willing to consider paying that much, but after Worrell had discussions with the deputy director, John Trinca, they ultimately agreed. However, Worrell received a letter noting: 'Other suppliers have been requested to quote for these venoms and when these [quotes] have been received you will be advised what venoms in addition to Papuan Black Snake, we wish you to supply over the next three years.'[3] This seems to have been a not-so-subtle hint that if Worrell set his prices too high, the CSL would simply go elsewhere.

Worrell shared the bulk of venom supply to the CSL with Charles Tanner, who supplied venoms from king browns, browns and death adders as well as taipans. The pair came to an agreement that they wouldn't compete with each other.

At the beginning of 1974, Worrell wrote to Tanner asking for his opinion on raising the price of venoms. Tanner wasn't sure that another increase was a good idea. He was concerned about the possibility of creating dissatisfaction among researchers on fixed budgets, as well as the fact that their old contacts at the CSL were disappearing, which might make both of them more vulnerable to changes in policy. Tanner pointed out that he also thought that the venom business still paid 'a good wage for very little work' and asked Worrell why he didn't show a reasonable profit.[4]

However, Tanner must have had a change of heart, for in June 1975 the men cosigned a letter to the CSL in which they set out a new schedule of prices for tiger, brown, death adder, taipan, king brown and Papuan black snake venoms. The proposal was for an across-the-board increase of 20 per cent, except in the case of the brown snake, where a flat rate of $1000 per gram was maintained. Tanner had suggested that the prices they charged should be pegged to the Consumer Price Index. Inflation was rampant and they wanted to avoid the need to keep reviewing their pricing strategy. The CSL agreed to all of this in October 1975.

Even given these uneasy negotiations over pricing, the relationship between the park and the CSL was generally a very amicable one. The CSL often came to Worrell when they needed more unusual deliveries. On one occasion, they asked for paper wasps; Robyn Innes and Lyn Abra were given the hazardous job of collecting them. They worked out that the wasps were asleep at night and so, after locating nearby nests, they did their collecting after dark. In 1971, the CSL wanted dried fleas and again the park came to the rescue, obtaining fleas from dogs that had been euthanased at the RSPCA in Sydney. A request for mosquito eggs was no problem for Worrell. In the late 1960s he had even obligingly tethered a goat in a tick-infested area so that once it had at least 50 engorged ticks hanging off it, CSL technicians could take a blood sample from it in order to prepare an antiserum to the paralysis tick. The park also supplied the CSL with frozen blue-ringed octopus.

Snake venom remained the principal product. By the mid-1970s, the reptile park was supplying venom from 17 species or subspecies of Australian venomous snakes. On occasion, researchers requested Worrell to obtain venom from rarer snakes whose venom was of unknown potency and composition. Dr John Trinca of the CSL had

Worrell provide venom from the rough-scaled snake, which is found in a region north and south of the Queensland–New South Wales border and has a potentially fatal bite. Such requests increased once Dr Struan Sutherland was appointed head of immunology research at the CSL in 1967 and began his own research program into the venoms of Australian snakes. Together with colleagues Alan Coulter and Allen Broad, Sutherland undertook an ambitious program of research to investigate the composition of snake venoms; develop methods to detect snake venom in a person, living or dead; improve first aid; reduce the risks of reaction to the antivenoms; and to improve care of the envenomated patient. Worrell and the Australian Reptile Park, as well as Charles Tanner, provided valuable venom and expert knowledge of Australia's venomous snakes, so much so that Sutherland suggested that Tanner and Worrell be co-authors on a paper, with Allen Broad as lead author, relating to the biochemical 'fingerprinting' of Australian venomous snakes. Although Worrell and Tanner were often given credit in scholarly papers as the source of experimental venom, Worrell's own writing by this time was exclusively for a popular audience.

The park's funnel-web lab was in good order by the mid-1970s and the public were responding to calls to leave any funnel-webs they encountered at depots in Sydney, the Central Coast and later, Newcastle. Yet funnel-web venom and its effects were much harder to understand than those of the snakes in the venom extraction program. It would be over a decade from when the park established its funnel-web lab in 1970 until the CSL developed an effective antivenom. Not to see any real progress in the development of an antivenom was terribly dispiriting at times for staff, especially Lyn Abra and Robyn Worrell, who invested so much of themselves over a long period in collecting, maintaining and milking spiders. This was not only tedious and time consuming but dangerous. Unlike the milking of snakes, which was undertaken in the knowledge that antivenom would quickly be used in the event of a bite, working with funnel-webs exposed the Worrells and their staff to the very real threat of death. Bites did occur, like the time in February 1969 when a large female bit Worrell on the thumb. The fangs had to be prised out and blood flowed freely, but the only symptoms he experienced were a headache and a deep ache in his hand. On another occasion a funnel-web scuttled up Abra's scarf, which was dangling into the spider's beaker. Luckily, she managed to brush it off before it reached her neck.

This work was also done in the knowledge that it was not bringing any income to the park. After the tensions between the park and the CSL over the funding of the set-up of the funnel-web lab, Worrell decided not to charge for the venom. The CSL had treated the park well in other matters, specifically in buying large quantities of snake venom, and Worrell's long friendship with Struan Sutherland may also have played a part in his decision. Worrell in effect supported the CSL's funnel-web project by investing some $500 a week in main-tenance of the air-conditioned funnel-web laboratory, the production of insects to feed to the spiders, wages and freight. Worrell weighed these costs against the urgent need for a life-saving antivenom. In January 1979, Christine Sturges, a healthy 32-year-old woman, was bitten by a funnel-web in her home in Helensburgh, south of Sydney. Despite the application of a tourniquet and rapid hospitalisation, she died. One year later, in January 1980, came the death of two-year-old James Cully from a bite suffered while holidaying on the Central Coast. Preventing further tragedies was what motivated the reptile park's work with funnel-webs. Only when the CSL developed an antivenom would the park begin to charge for the venom.

Sutherland had been able to step up his research into the funnel-web venom and development of an antivenom once the park had begun supplying adequate quantities of quality venom in 1972. In 1973, Sutherland discovered that the main toxin, which he named atraxotoxin, was readily adsorbed by paper, cellulose and glass. This meant that the venom they had been working with in the lab, which had been in contact with glass, was much less potent than the toxin spiders injected into their victims. Lining the laboratory's glass-ware with silicone solved the problem and was a major step towards solving the mysteries of funnel-web venom.

By mid-July 1980, an experimental antivenom had been developed by Sutherland and his team. It was shown to be successful at treating newborn mice injected with funnel-web venom and, later, monkeys. The testing of this antivenom required Robyn Worrell and Lyn Abra to step up the output of venom. In December of that year, antivenom was sent out to selected Sydney hospitals as a clinical trial. An engineer bitten on the foot was treated with the antivenom at the Royal North Shore Hospital in Sydney in February 1981. The following May, a toddler from Terrigal, on the Central Coast, was rushed to Gosford Hospital in a critical condition and received the antivenom. In both cases, it rapidly reversed the toxic effects of the venom and the victim survived. Of hearing by telephone that the antivenom had worked on the Sydney man, the CSL's Struan Suther-land said: 'I was swept by deep elation comparable to that felt after passing final year medicine. This was then followed by thoughts of gratitude to the handful of people who had unfailingly helped in their different ways to achieve this happy result.[5] The important role of the Australian Reptile Park in this achievement was recog-nised when the National Australia Bank presented its Humanitarian Award for voluntary work jointly to Eric and Robyn Worrell in 1980.

Between 1981 and 2006, 75 cases of funnel-web spider bite were treated with the antivenom and there were no deaths. The amount of time victims spent in hospital dropped from an average of two weeks to just one to three days.[6]

The park continued to supply funnel-web venom for the CSL's antivenom production and for other researchers. They required large numbers of spiders for milking but breeding had never proved successful, so the park continued to rely on the public to locate funnel-webs and bring them to drop-off points throughout the range of the spider. Fortunately, nobody seems to have been bitten while collecting spiders for the park – in contrast to the many people who are bitten by snakes when trying to remove or kill them. The population of the funnel-web colony at the park varied depending on the season and the weather, with just 50 in January 1982, 150 in June 1985 and 100 in 1987.

In 1982, with the CSL's orders for snake venom at their lowest since 1950 and at the urging of his accountant, Worrell approached the CSL to negotiate a price for the funnel-web venom. He proposed a very modest $1 per milligram of dried venom, but payment was not implemented until 1986, when the financial future of the park was looking perilous. The park did introduce a charge for other users of funnel-web venom, such as Professor Tom Torda of Prince Henry Hospital. In 1973, a team at Macquarie University led by Dr Merlin Howden began research on a new type of funnel-web antivenom and, later, a funnel-web venom vaccine. The researchers initially called for people to bring spiders to them but the park took over the supply of venom for that project from 1983. The university was charged $15 per milligram for the dried venom, making a small but welcome contribution to the finances of the park. Funnel-web venom was also sent to a researcher in Japan in 1983.

The collection of funnel-web spiders and the painstaking extraction of minute amounts of venom continues today at the park so that the resulting antivenom, which has proven to be so effective against the bites of this spider, will always be available to those who need it. Since the antivenom was first made available, no one has died from the bite of the Sydney funnel-web spider.

When the NSW government needed advice on developing legislation to protect native reptiles in the early 1970s, Worrell was one of the experts to whom they turned for advice. Worrell had forged good working relationships with the National Parks and Wildlife Service since its establishment in 1967 and he had conducted a few training sessions for rangers seeking to learn snake identification and handling techniques. At that time, in New South Wales, native reptiles could be killed, collected from the wild or kept without the need for any kind of permit or licence. Worrell had advocated for the legal protection of native reptiles, particularly lizards, harmless snakes, freshwater crocodiles and turtles and tortoises for some years, most notably in his chapter 'The Unpopular Ones' in Jock Marshall's *The Great Extermination*, published in 1966.

The NSW government set up a committee to explore what kind of regulations should be placed on amateur reptile keeping within the planned wildlife protection legislation. Hal Cogger, deputy director of the Australian Museum, was on that committee. Worrell drew up for Cogger a list of the reptiles he felt deserved rare species status in New South Wales, based on their current rarity or limited or decreasing habitats. These were species that Worrell believed no one should be allowed to keep without a special licence. He assembled another list of species that he felt unlicensed amateur reptile keepers should be allowed to keep in numbers up to five. This list included the larger and more common skinks, dragons and the swamp snake and carpet python. He also proposed setting minimum cage sizes for each species or group of species. Worrell wanted to prevent the housing of reptiles in enclosures that were too small for their health, a practice he referred to as 'the cocky's cage tradition'. Cogger felt that imposing cage size restrictions on keepers was probably 'presumptuous and unnecessary'.[7] Instead, he suggested restricting the number of individuals of a single species that could be kept to two, rather than the five that Worrell had put forward.

The ensuing *NSW National Parks and Wildlife Act (1974)* made all of the state's reptiles protected fauna. Amateur reptile keepers were allowed to have up to two animals of each species from a list of nine, which included eastern blue-tongue lizard, carpet python and eastern long-necked turtle. These animals had to be obtained from a licensed keeper and could not be taken from the wild. The regulations were much stricter than either Worrell or Cogger had recommended. It became a crime to hold any other reptiles without a licence. The legislation, and the astonishing vigour with which the NSW NPWS was willing to prosecute, had a negative impact on the hobby of reptile keeping for well over a decade. Worrell had contributed to the background formulation of this legislation, but he, too, would find himself targeted by it. The new obligations it imposed on the park as a collector and custodian of reptiles would prove to be difficult for Worrell to accept.

The 1970s were a difficult time economically throughout Australia. The OPEC oil crisis fuelled inflation within a stagnant economy, leading to increased costs and the erosion of spending power. With it, the leisure dollar diminished. Income was not flowing into the park as it had in the boom years of the 1960s and Worrell was economising by relying more on volunteers, employing more lower-paid junior staff, and allowing positions to go unfilled. As a result, it would be fair to say that by the late 1970s, the park was starting to look tired.

Worrell was still building the collection even at this time, when things were becoming quite tight financially. In 1976, he organized for another pair of king cobras to be imported to the park from Malaysia as well as a number of other species, including kraits. Other acquisitions around this period included: an Asian shipment including gliding geckos, tokay geckos and bent-toed geckos; an African shipment of a pair of giant horned toads; and from Taronga

Zoo, some small reticulated pythons, corn and yellow rat snakes, and a mangrove monitor. Another importation arrived from the Transvaal Snake Park, in Johannesburg, South Africa, which included twig snakes, shield-nosed snakes, black-necked spitting cobras, booms-langs, common puff adders and horned vipers. A shipment from Thailand included butterfly lizards, mangrove snakes and monocled cobras, while an importation from the Hawaii Zoo included several species of rattlesnake. Some of these animals were obtained through exchanges of animals from the reptile park, while others had to be bought.

Despite the looming financial difficulties, the park's collection was at its most diverse and interesting in the 1970s. Its reptile and amphibian collection was especially impressive, as this inventory for January 1979 suggests: [8]

**SINGLE ANIMALS**

| | |
|---|---|
| 1 frilled lizard | 1 tuatara |
| 1 Bengal monitor | 1 black head python |
| 1 tegu | 1 carpet python |
| 1 mangrove monitor | 1 boa constrictor |
| 1 western rattlesnake | 1 Indian python |
| 1 western diamondback rattlesnake | 1 amethystine python |
| 1 black mamba | 1 Haitian boa |
| 1 sidewinder | 1 tokay gecko |
| 1 speckled rattlesnake | 1 red blacksnake |
| 1 rhinoceros viper | 1 juvenile brown tree snake |
| 1 Gaboon viper | 1 Gould's monitor |
| | 1 Gippsland water dragon |

**PAIRS**

| | |
|---|---|
| 2 taipans | 2 King Island tiger snakes |
| 2 king browns | 2 Flinders tiger snakes |
| 2 blue-bellied black snakes | 2 kraits |

<table>
<tr><td>2 wagglers pit vipers</td><td>2 anacondas</td></tr>
<tr><td>2 black mambas</td><td>2 prehensile-tailed skinks</td></tr>
<tr><td>2 monocled cobras</td><td>2 pink-tongue skinks</td></tr>
<tr><td>2 speckled rattlesnakes</td><td>2 death adders</td></tr>
<tr><td>2 western diamondback rattlesnakes</td><td>2 Burtons legless lizards</td></tr>
<tr><td></td><td>2 swamp snakes</td></tr>
<tr><td>2 red rattlesnakes</td><td>2 bullfrogs</td></tr>
<tr><td>2 Gila monsters</td><td>2 green iguanas</td></tr>
</table>

**MULTIPLE**

<table>
<tr><td>3 king cobras</td><td>4 southern blotched blue-tongues</td></tr>
<tr><td>3 water moccasins</td><td></td></tr>
<tr><td>3 Chappell Island tiger snakes</td><td>4 Children's pythons</td></tr>
<tr><td></td><td>5 sand monitors</td></tr>
<tr><td>3 western brown snakes</td><td>5 diamond pythons</td></tr>
<tr><td>3 puff adders</td><td>5 Gillen's pygmy monitors</td></tr>
<tr><td>3 reticulated pythons</td><td>5–6 shinglebacks</td></tr>
<tr><td>3 boa constrictors</td><td>9 lace monitor</td></tr>
<tr><td>3 mangrove snakes</td><td>5 brown snakes</td></tr>
<tr><td>3 carpet pythons</td><td>25 tiger snakes</td></tr>
<tr><td>4 African pythons</td><td>30 black snakes</td></tr>
</table>

Of all the snakes used in the venom production work, only tiger snakes were kept in numbers large enough to maintain venom supply. The venom of the others – taipan, eastern brown, king brown and death adder – continued to be supplied by John McLoughlin in Cairns. Those venoms brought in $20 000 to the park's coffers in 1979–80.

By the end of the 1970s, the Australian marsupial collection that had been built up by Krauss had begun to diminish. Nevertheless, the park still had more species of kangaroos, wallabies and certain marsupials than any other zoo in the country, and a very wide array of nocturnal mammals, as well as rarely exhibited species such as the platypus.

The park was also having some success breeding certain of their animals, including species never before bred in captivity. They had been trying for some time to breed cassowaries but while eggs had been laid, they were usually infertile. In 1972, the park overcame these difficulties and became the first zoological institution in Australia to breed this species. Five eggs hatched but the mother trampled four hatchlings. The survivor, Charlie Brown as he was named, was removed and reared by keepers Bev Drake and Raija Krauss. The two women subsequently wrote an article on the cassowary-breeding program, which was published in the *International Zoo Yearbook* with Worrell as lead author. The park also bred the American alligator and the beautiful spotted cuscus, again both firsts in Australia. Peter Krauss had bred common wombat in an off-exhibit area in the early 1970s, just prior to his departure.

Worrell made many of his animals available for scientific studies conducted by university-based researchers. Greg Mengden, an American herpetological researcher based initially at the Australian National University and then later at the Australian Museum, was given access to the extensive collection of venomous snakes in order to take blood samples for DNA analysis. Ironically, though, it was often mammals that the researchers were interested in rather than the park's extensive reptile collection. Research was carried out on some of the Papuan mammals, such as the cuscus, in the early 1970s. Geoffrey Sharman, professor of biology at Macquarie University in Sydney, conducted a long program of research on a variety of kangaroos, wallabies and other mammals kept at the park. The park was also involved in a research project to explore the possibility of using common kangaroos and wallabies as surrogate mothers for more endangered species of macropod, a technique known today as cross-fostering and which is being used to accelerate the captive breeding of the highly endangered brush-tailed rock wallaby in South Australia. Clearly, the scientific research component of the park had been very important to Worrell when he was first establishing it, and while this aspect perhaps did not eventuate the way he had hoped, with original research being conducted on site by park staff, he was generous in making his animals and facilities available to researchers. As an additional component of the research activities that Worrell tried to foster at the park, a short-lived journal called *Australian Reptile Park Records* began in 1963. However, only two issues were ever published, both of these being on snake taxonomy. These two publications were Worrell's final published works on taxonomy, although he maintained an interest in the classification of reptiles throughout his life. Herpetologist Richard Wells maintains that Worrell had a sophisticated understanding of the taxonomy of Australian reptiles but he only published a fraction of what he knew. Others in the herpetological realm were not as convinced by Worrell's taxonomic propositions and his *Reptiles of Australia* was given a poor review in the journal *Copeia* by American herpetologist Arnold Kluge. In contrast, other museum-based herpetologists with whom we spoke praised the book and Worrell's efforts to make sense of the taxonomy of venomous snakes in particular.

Worrell's niece Lynda Vereyt had worked casually at the park in the late 1960s. She had maintained a strong interest in animals and had made it known to her grandmother Rita Worrell that she was keen to work at the reptile park if there was ever an opening. When his mother approached Worrell about finding work at the park for his niece, he created a position for her as a mammal and bird keeper, even though putting on an additional staff member strained the already fragile business. Worrell was loyal to his family and was generous towards them, as he had been in building a house for his parents in the middle of the park. He had also assisted his sister financially and given her a home at the park.

One of the financial challenges facing the reptile park was the official opening of Old Sydney Town on Australia Day in 1975, a second major attraction on the Central Coast, in competition with the park. Old Sydney Town was a private concern but was supported by investment from the federal government, the National Australia Bank and, at a later stage, the NSW government. For many families wanting to have a day out, it did not really matter too much if they visited the reptile park or some other attraction such as Old Sydney Town, provided that it was accessible, would entertain the kids and provided a barbecue or picnic facility. In a difficult economic climate, any negative impact on gate takings was significant but it was not in the nature of Worrell, the reptile park and its staff to allow the competition to leave them behind. They started a relationship with Old Sydney Town by donating some grey kangaroos to add atmosphere to the site. Later, the two attractions undertook joint promotions and ultimately, the park would relocate to a site adjacent to Old Sydney Town, encouraging visits to both on the one-day trip. Thirty-five years later, it is the reptile park that continues to thrive, while Old Sydney Town closed as an attraction in 2003.

Worrell remained in the public eye for most of the 1970s. His encouragement of amateur herpetology was recognised with an invitation to be patron of the Australian Herpetological Society in 1971, an honour he later shared with Hal Cogger. (Worrell actively supported this society in a number of ways such as providing free admission to members and making the park available for annual general meetings.) He remained the star of the reptile park and was firmly established in the popular imagination as the Australian Snake Man. The media turned to him when they wanted comment or expert opinion regarding venomous snakes and spiders. He was the guest of honour of an episode of *This is Your Life* hosted by television journalist and presenter Mike Willesee in 1976. He regularly appeared on iconic television programs such as the *Mike Walsh Show* and the *Don Lane Show*. He even promoted Skippy Cornflakes in a television commercial for Sanitarium. In 1977, he was featured in an article in *People* magazine as Australia's 'foremost snake expert'. To reptile fanciers, Worrell was a superstar.

Given the relatively limited media outlets operating at the time, Worrell did extremely well in building his national profile as a naturalist of significance. It was a very different world to the media-conscious and image-saturated world of today where someone with similar talents – although with a very different performance style and personality – such as the late Steve Irwin could be elevated to international celebrity status.

Yet locally at least, the effects of Worrell's alcoholism had started to be noticed in the public sphere. Twice he was found driving under the influence of alcohol. He was unable to admit he had a problem but was intermittently hospitalised because of it during the 1970s, and would continue to be in the 1980s.

Robyn Worrell's level of responsibility increased substantially throughout the 1970s, culminating in her being made co-director. While Eric continued to operate at the big-picture level, it was Robyn – along with other key staff such as Jack Green and Lyn Abra, who by now had a high media profile as a spider expert and was known locally as the 'Spider Woman' – who made the vision a reality. Robyn was known for her close attention to detail and her methodical, common-sense approach to the operations of the park. She managed the accounts and payroll and was a productive milker of funnel-web spiders. She also took on some of the milking of the venomous snakes, played foster mum to innumerable orphaned baby marsupials and dealt with much of the growing volume of paperwork that was now part and parcel of operating the business. By the end of the decade, the park's licence to hold and exhibit native fauna was in Robyn's name, not Eric's. While Eric was the person associated with the park in the public imagination, it was Robyn who quietly went about running the business.

# 13

# **Decline and despair**

At the conclusion of the 1979–80 financial year, Robyn and Eric Worrell met with their business manager, William Ansell, and listened while he told them the bad news. Costs were soaring but visitation was declining and they were staring at an anticipated trading loss of $10 000. The park had just added a popular new drawcard, a purpose-built animal hospital for injured or sick native animals. The facility, built with financial assistance from Rita and Percy Worrell and Gosford North Rotary, would not be enough to improve the park's fortunes, but for Worrell it was the realisation of a long-sought goal. It allowed the park to continue doing the good work of saving native animals and it appealed to visitors, who enjoyed seeing vulnerable animals being looked after on the other side of a large glass window and the more able-bodied ones grazing in an enclosed yard. It also provided a supply of animals that could be added to park exhibits. Among those who attended the official opening of the animal hospital in October 1981 was a young man who, unbeknown to anyone at the time, would prove to be one of the park's saviours. He was John Weigel and he had just been employed as a keeper.

Born and raised in America, Weigel had developed a deep fascination with reptiles from an early age. As a young student, he would hide reptile books under his desk or inside a textbook while the teacher taught the class maths or some other cornerstone subject. He attended high school in Denver, Colorado, and recalls standing in fascination as a 15-year-old in front of the realistic Australian life zone dioramas at the Denver Museum. It was a pivotal moment in his life, igniting his interest in the nature of Australia. A few years later, he discovered Worrell's classic *Reptiles of Australia*, significantly deepening his fascination with Australian wildlife. He attended the University of Colorado on an athletics scholarship – he was an accomplished pole vaulter whose Colorado state record stood for nearly 20 years – studying first business and then biology. Midway through his degree, he entered an exchange program with the University of Veracruz in Mexico, where he studied Mesoamerican anthropology, in Spanish, while devoting all his spare time to the search for reptiles.

Upon returning to the University of Colorado, he contributed to three scientific papers on aspects of reptile biology and wrote to Worrell seeking employment as a reptile keeper. He expressed his keen interest in Australian herpetofauna and his eagerness to become involved with breeding programs and behavioural research. This letter came to nothing, so Weigel broadened his search for a job with reptiles, writing to many institutions around the world. The California Alligator Farm at Buena Park, Los Angeles, offered him a position. So strong was the pull to work with reptiles that he left university before completing his degree and worked at the alligator farm as a keeper for two years from 1979. In addition to his reptile-keeping responsibilities, he did two daily reptile demonstrations and two alligator and crocodile feeding shows, gaining confidence and showmanship skills. His commitment to a hands-on career with reptiles grew stronger – much to the concern of his parents.

Having saved enough money, in 1981 Weigel travelled to Australia, determined to work at the reptile park. Although there was interest in his skills and experience, there were no positions going at that time and Weigel went to Queensland for a working holiday, contacting the park from time to time to see if any openings had come up. All of this persistence paid off and he was offered the position of assistant to the noctarium keeper later that year. Weigel willingly took on this lowly position and soon worked his way up to keeper of mammals. His fresh approach and new ideas led to considerable immediate improvements to husbandry practices and the way exhibits were presented, but he looked forward to resuming his work with reptiles.

Unfortunately, the reptile park that Weigel had been so keen to join had by now been dramatically affected by Worrell's personal decline. By this stage, his drinking problem had worsened to the point where the usually gentle and quietly spoken man was often intoxicated, unpredictable and fractious.

After years of trying to ease the way for her husband, Robyn Worrell was feeling considerable strain. Her marriage had already deteriorated to the point of being unsalvageable and midway through 1982 she moved out of the family home. She and John Weigel had by this stage commenced a relationship. Robyn continued as an employee of the park and carried the weight of its day-to-day management; but Worrell's anger meant that Weigel's employment at the park was brought to an abrupt end.

Robyn and Eric Worrell's personal upheaval was matched by the substantial ongoing decline in the fortunes of the park. As predicted by Ansell, the park was sinking deeply into debt. In August 1982, the overdraft was more than $100000 and creditors were becoming uneasy. By 1984, things had reached a point where the Australian Reptile Park as a business had collapsed. A new business manager, local accountant Terry Dibben, produced a thorough financial report in January 1985. This showed problems at almost every level. There was markedly declining visitation and fewer sales in the gift shop and kiosk. Income from snake venom sales had slowed. The commercial relationship with John McLoughlin had been dissolved, on friendly terms, temporarily ending the park's long role as a supplier of snake venom to the CSL. The cost of animal feed was a significant burden, as the great range of mammals and, to a lesser extent, birds held by the park, in addition to the reptiles, required a broad variety of foods, including hay, grains, fruit, mice, mealworms, crickets, locusts, earthworms, silkworms and day-old chicks, the total cost of which rose from $21 391 in 1980–81 to $35 459 in 1983–84. Because bills were sometimes not paid, food supplies were erratic. Some staff secured spoiled fruit and stale bread from local grocers and brought in treats such as honey for the animals in their care.

The largest costs were for wages and for the charges and interest on loans, which had risen tenfold between 1981 and 1985. The 1980s saw the deregulation of the banks, allowing them to set their own rates of interest, and with high demands for funds in the economy, interest rates climbed. Overdraft rates were the highest of them all. They had been no greater than 10 per cent during the 1970s but climbed to 20 per cent in the mid-1980s, when the park was carrying a large amount of debt. The ability to borrow money against the security of the land and structures built on it helped to keep the park open, but the park's income was not sufficient to meet these interest repayments, let alone repay the capital. One expenditure that had seriously declined over the period was repairs and maintenance. The grounds and buildings were not receiving the attention they once did, leaving the park looking run-down and unkempt. Dibben's conclusion was stark and prophetic: 'The manner in which the Park is currently being run must change if you wish to keep control of its destiny.'[1]

Dibben recommended that visitation should be encouraged through more systematic marketing and promotion of the park and that visitors should be more actively engaged, rather than left to explore on their own. He suggested that staff members should give guided tours and talks on specific animals throughout the day. Money could be raised through encouraging the public to sponsor individual animals. Dibben also regularised some aspects of employment, including setting out a policy on sick leave and recommending that a full-time manager and management committee be established.

Instead of adopting the systematic approach advocated by Dibben, in desperation Worrell tried to take an unconnected series of measures intended to reduce costs or access outside funding that would allow the park to continue running as it was. He raised admission charges and lobbied to have his land tax waived, arguing that the park's contributions to venom research and the care of sick and injured native wildlife should give it tax-free status. His request was denied. He sought council approval to build a row of shops on a parcel of land owned by the park, which could be leased by businesses such as a tourist agency, plant nursery and pharmacy. Gosford City Council rejected the proposal because they did not see these businesses as being in keeping with the original zoning of the park's land as a research station. Worrell met with frustration at every turn.

There were problems on other fronts as well. Until now, Worrell's practice of collecting snakes from the wild to use in his venom program had been acceptable. With the rise in concern for native wildlife – partly fuelled by his own writings and the park's public education efforts – that approach to collecting snakes for milking had become less acceptable. When the reptile protection regulations were introduced in 1974, Worrell's special status as Australia's Snake Man had afforded him a level of immunity from close scrutiny. However, from the early 1980s onwards, the NSW National Parks and

Wildlife Service (NPWS) became increasingly reluctant to allow the Australian Reptile Park to annually collect large numbers of wild snakes to replace those that had died while in the venom extraction program. Nature conservation agencies in other states, such as Victoria and South Australia, were less willing to authorise the park to collect venomous snakes and lizards for display than they once were, which impacted on the capacity of the park to exhibit and keep these reptiles, limiting income from visitors and venom sales.

Several levels of government became involved in scrutinising the activities, policies and record keeping of animal parks and zoos. At the federal level, the Commonwealth Department of Health had to give its approval for the importation of antivenom used to protect staff in the case of bites by exotic snakes. The Australian Quarantine Service had to be notified when Australian snakes were transferred to the park from overseas or interstate and when exotic animals were being transferred within New South Wales. The NSW Division of the Animal Quarantine Service inspected the park's records and took Worrell to task over anomalies in the recording of imports, transfers, deaths and breeding of 13 exotic species. While Robyn Worrell and Lyn Abra did their best to ensure that proper records were kept, Worrell himself hated paperwork and didn't satisfactorily comply with the increasing bureaucracy. Worrell's greatest years of success were during a time when there were few restrictions or bureaucratic impediments; he did not adapt at all well to the changing circumstances in which he found himself.

Inspections of the park by the NSW NPWS revealed lapses in adherence to wildlife protection regulations, especially with regard to the operation of the animal hospital. The park's exhibitor's licence clearly stated that approval had to be secured from the NPWS for permission to keep any injured or orphaned native animal and that the animal remained the property of the Crown and could not be exhibited. However, when the general public brought injured or orphaned animals to the park or asked park employees to come and remove native animals from their properties, the park frequently retained those animals in its collection without seeking approval. The NPWS audit revealed almost 200 mammals, birds and reptiles apparently acquired in this way that weren't on the official lists of animals held by the park. In 1984, Worrell had to write to seek permission to retain them. The NPWS restated its preference that the park stop accepting injured animals and removing unwanted fauna from properties, although they permitted the park to keep the animals then in the collection.

As a long-term keeper and exhibitor of captive animals, and an expert who had been involved in the development of the wildlife protection regulations and the training of NPWS officers in reptile handling, Eric Worrell felt that he should not be bound by the increasingly restrictive measures imposed by the authorities. In 1981, he was frustrated and angered that the park's special licence to import exotic snakes for public education and venom production, granted in 1960, would no longer be honoured. He would now have to go through the normal process of applying for permission to import snakes and other reptiles. He wrote to the local member of the House of Representatives, Barry Cohen, arguing that the park's special licence was warranted due to his record of service to the community.

Worrell considered the lengthy process of obtaining permits to collect reptiles a personal affront and a threat to his livelihood, even writing to the prime minister, Bob Hawke, expressing his profound dismay at the way he was being treated. Low stocks of venomous snakes threatened his vision of the park as a research station, prevented the raising of income through venom sales and reduced gate takings because people were no longer being drawn by the spectacle of pits full of venomous snakes being milked. In a letter to the NSW NPWS asking why they were making him wait long periods for approval to collect snakes, Worrell noted that his collectors had been doing this for the park for between 20 and 30 years, implying that there was no need to change from this long precedent. His pleadings made little difference to the NPWS or other government authorities.

Worrell was outraged at the end of 1983 when a former employee wrote an article in an Australian zookeeping magazine critical of the keeping of snakes for venom extraction. In what became his oft-repeated justification, Worrell defended the deaths of snakes used for milking and the need to replace them annually. Dividing the number of snakes that died at the park by the number of people to whom life-saving antivenom was administered per year, he argued that 0.61 of a snake was sacrificed per human life saved. 'Is not .61 of a snake worth one human life?' he asked in an unpublished letter to the editor of the journal. Rejecting Worrell's position that the death of snakes used in the milking program was inevitable, the NPWS argued that the park had a high level of mortality compared to other similar operations. They attributed the discrepancy to 'inadequate and inappropriate husbandry practices' at the park.[2] Struan Sutherland had provided a reference for Worrell to the NPWS, calling Worrell and the park the 'backbone which allowed the development of new antivenoms, and improvement in existing preparations'. He ended his letter by saying, '[I] personally wholeheartedly support the continued concept of the Australian Reptile Park as a source of venoms, a place to care for and research our fauna and a superb education centre.'[3]

Nevertheless, the NSW NPWS sent the Australian Museum's Hal Cogger and Greg Mengden to inspect the park, then issued Worrell with 15 directives. Ten of these related to occupational health and safety matters and five to husbandry, including the employment of a 'trained and competent reptile keeper in addition to Mr Worrell' and a maximum number of ten snakes per snake pit.[4] The NPWS noted that the park did not have the fauna dealer's licence that was required by those trading in animals or their products. Worrell claimed established practice as justification for the park's ways of operating, although he undertook to make the changes required by the NPWS.

In September of 1984 the NSW NPWS gave the park approval to import or take from the wild 15 brown snakes, 15 tiger snakes, five death adders, four taipans, ten lowland copperheads and five black snakes, with the warning that this was not an endorsement of the park and its practices but was done to maintain venom supplies for the CSL. Even with the permission finally secured, getting the snake numbers back up again was slow, as each collector also needed to get approval from the NPWS and snake catching was seasonal. By the end of January 1985, of the snakes needed for antivenom production, there were only one taipan, two tiger snakes, 16 eastern brown snakes, two copperheads and two death adders in the collection.

The personal and emotional strains felt by all involved in the park at this time cannot be overstated. It was a time of great crisis. All the staff members were affected by the loss of direction at the top and the park's financial situation. They were frustrated to see the park running down while being able to do little about it. Jack Green's position as curator became increasingly difficult and there were a number of periods during the late 1970s and early 1980s when the park had no dedicated reptile keeper on staff. It was then left to Green to maintain the bulk of the reptile collection in addition to his usual responsibilities.

Employees' jobs were made more difficult by the lack of funds available for even basic necessities. Staff did what was necessary to keep the park functioning, patching up exhibits and pooling their limited skills in areas such as electrical work to maintain the lighting systems, which needed frequent repairs. Some chose to move on to other zoos or returned to previous careers. Many stayed at the park working for low wages because of their personal commitment to working with animals; some were aware that they would have been better off on the dole. Others took second jobs so they could afford to remain at the park, although Worrell frowned upon this. As the financial pressures mounted, staff had their hours cut, making it difficult for them not only to earn a living but to complete all of their daily tasks.

With declining park infrastructure, venom production virtually at a standstill, increased scrutiny from the authorities and falling patronage, Worrell faced the threat of foreclosure by his bank at the end of June 1985. The business, asset rich but cash poor, had been trading at a loss and was unable to meet the crippling interest payments. Worrell began to seek a new owner for what he believed was still a viable business. He initiated discussions with Gosford City Council, suggesting that they buy the park for $1 million, giving him $500 000 he needed to clear his debts upfront and the balance plus interest over five years. Taronga Zoo was also involved in these discussions and was considering the idea of making the Australian Reptile Park its specialist reptile facility. Director of Taronga Zoo Jack Throp suggested, somewhat hopefully, that the park could be declared an historic site based on its important contribution to venom research, attracting tax concessions and grants to assist with maintenance. In any of these scenarios, Worrell was still optimistic that he could continue as director and even prepared a document for staff detailing improvements he had long wanted to make, including a two-storey aquarium and a walk-through aviary for the larger birds.

Worrell considered proposals from a building company, an Asian amusement park firm and a local Aboriginal group who wanted to set up a display of Indigenous culture. A friend with a longstanding interest in reptiles, John Edwards, tried to pull together a consortium to take over the park and a Queensland woman who had met Worrell more than 30 years before expressed interest in providing him with financial backing. None of these possible solutions eventuated.

The difficulties the reptile park was facing and the proposed solutions were well covered in the local media and residents began to mobilise to support this local attraction. The park, the dinosaur and Worrell himself were very special to the local community, which had boomed in size during the lifetime of the park. The park's involvement in venom production gave it a national profile and made it a point of pride for residents, who liked to take visitors there. Almost every child in the area had visited the park with his or her parents or on a school excursion. Even for those locals who did not visit the park regularly, the dinosaur provided one of the few built landmarks on the Central Coast; it was a touchstone of home for residents using the busy Pacific Highway. A grassroots support group, Save the Reptile Park, was established, led by a couple of local women, Michelle Capewell and Wilma Rond. They received many letters of support and encouraged schoolchildren to make posters showing why they loved the park. They circulated a petition around the Central Coast area and presented 150 000 signatures to NSW Premier Neville Wran in the hope that the state government would provide funding to either the Gosford Council or Taronga Zoo to keep the park open. Their appeal came to nothing.

A second group, the Australian Reptile Park Steering Committee, headed up by the then chair of Central Coast Tourism Promotions, Bob Bourne, also came into being. It proposed that the park be taken over by a raft of federal, state and local government departments and institutions.

Perhaps sensing that these efforts were getting nowhere, Worrell acted in desperation, and without warning. He arranged to offer the park for sale by auction on 22 August 1985. He called it a step 'which I abhor [taking] after a lifetime of work placed into the Park'.[5] In order to raise the funds for the auctioneer's fee, he organised an auction of his own valuable collection of natural history books.

The announcement of the auction came as a devastating shock to Robyn Worrell, who had her own significant personal and financial investments in the park. The ground had shifted dramatically and she needed a new strategy. With few options available to her, she acted quickly by initiating divorce proceedings and then taking out a caveat preventing the sale of the park, on the grounds that it was part of the assets of the marriage, which had not yet been divided.

The NSW NPWS was also alarmed by the proposed auction, warning Worrell that he could not sell the park as an ongoing concern because any new owner would need to hold an exhibitor's licence. As Robyn held the exhibitor's licence, Worrell could not retain any of the animals if he sold the park. Nor could he continue to run the park without Robyn, as he was not considered suitable to hold the licence himself. The native animals themselves were property of the Crown and could not be sold with the park.

The caveat brought the financial crisis to a head, as the bank could extend no further credit while the asset was frozen. Worrell issued a terse retrenchment notice to most of the staff, in which he expressed his regret at having to terminate their employment and his concern over the hardship it would cause. Some of the staff continued to look after the animals, without pay. Worrell also invoked a notice of dismissal he had given Robyn the previous April, blaming her for his inability to continue to run the park.

Although Worrell alleged that Robyn was guilty of 'misconduct and insubordination', she had been the steadying force throughout this period of dramatic change for the park. Her decades of experience and increasing responsibility gave her the perspective and commitment to ensure that the park continued as a viable business, centre of research and tourist attraction. It had been a tumultuous and chaotic time for her as she dealt with the pressures of continuing to work in relatively close contact with Worrell after the breakdown of their marriage while trying to see a way forward for the park. In 1983, she had thought she'd had all she could take and notified the NSW NPWS that she would be leaving the park and that the exhibitor's licence would have to be transferred. She and John Weigel had even begun applying for work at other zoos in Australia. Instead, she stayed, a constant and invaluable member of the park team. Robyn had reached maturity in and with the park and her commitment to the place deepened with the years. Building the park into the best wildlife centre in Australia was her life's ambition, but her immediate aim was to stop it from being sold out from under her.

In a matter of just a few weeks, she engaged legal representation, commissioned reports by firms of accountants into the financial viability of the business, developed a business model that would return the reptile park to solvency and convinced her bank that its continued support was justified. This was a huge task, jointly carried out with her partner, John Weigel. Between the two of them, they managed to put together a business proposal showing that the park was viable provided various conditions were met. These included the sale of several parcels of land that weren't making a major contribution to the business; a sizeable injection of capital from Robyn; and a two-thirds reduction in the number of mammals and birds in order to return the focus of the park to reptiles. Reducing the number of other animals and expanding the reptile inventory would have the dual effects of lessening maintenance costs and enhancing the park's visitor appeal.

In the middle of all this chaos, on 8 September 1985, Worrell suffered a serious bite from a monocled cobra (*Naja kaouthia*). He had been performing a midafternoon snake handling display, probably something he shouldn't have been doing by this stage of his life. 'I was demonstrating the docility of a cobra in the main milking pit,' he later wrote.

I was explaining to visitors that despite their placid temperament there was a high death rate in their country of origin. Then I heard someone say at the back of the crowd 'Its fangs are out'. So I picked out a two-metre cobra, held it loosely so it could rise and spread its hood, then grabbed it by the neck. This of course irritated the snake and it was now prepared to bite. I chose a

beaker from the venom case but had misgivings when I saw that
all the beakers that had been prepared for me were too small
at the mouth. I introduced the beaker to the snake's open jaws.
The right fang clamped straight on to the rubber diaphragm and
spurted venom. The left fang overshot the beaker and sunk into
my right thumb. There was a lot of blood which I hoped had been
caused by the back teeth. I didn't want to alarm the visitors as last
time I was bitten by a tiger snake, two men fainted, so I casually
finished the demonstration and climbed out of the pit. Blood
was still flowing freely and I could overhear puzzled comments
because I completely ignored it. I walked over to my house across
the creek still feeling no effects.[6]

Within an hour, he suffered paralysis, pain at the site of the bite, a
severe rash and chest constriction accompanied by severe vomiting.
He had also gone blind, having what he termed a 'white-out', where
everything appeared to be white, with patches of pink where the light
was brightest. He collapsed on to his couch and decided to 'ride it out',
as he had done with bites from a variety of Australian species, but by
5.40 pm he knew his condition had begun to worsen. Robyn had been
told of the bite earlier in the afternoon and before she left work for
the day, she telephoned Eric to see if he was all right. According to
his account, he managed to crawl across to the telephone and flip the
receiver off the hook but was unable to speak. Alarmed, Robyn called
an ambulance and then drove to Gosford Hospital herself with the
park's supply of four ampoules of the appropriate antivenom. The life-
saving serum was administered and he was placed on a respirator,
lapsing in and out of consciousness over the following twelve hours.
Both Robyn and Lyn Abra spent much of the night at the hospital and
it was touch and go whether Eric would survive this bite. Medical
staff made contact with Worrell's friend and colleague Struan Sutherland of the CSL, who was able to provide the attending doctors with
his considerable expertise and advice.

By midday the following day, Worrell was out of danger. However, he was a diabetic and his blood sugar levels were erratic for a few days. He remained in hospital for another two days, was released, but then had to return for treatment of severe hives brought on by a reaction to the antivenom. He was left with severe aching in his joints for a number of weeks. The incident was widely reported in newspapers and magazines, and for some months after the bite the park enjoyed higher than normal gate takings. In a thank-you note to his friend Greg Churchill, who had sent him a card while he was recovering, Worrell wrote that he was 'nearly recovered from the bite and antidote, and have made friends with the Cobra again'. He still retained glimmers of his sense of humour.

Eric Worrell was a naturalist, reptile expert, a writer and a photographer and while he had entrepreneurial flair, he was not a natural businessman. The reptile park as an enterprise had grown in size and complexity and by the early 1980s, the park employed 13 full-time staff and the total number of animals held there was over 1000. Undoubtedly, Worrell had made a number of strategically significant decisions, such as assembling and nurturing a small team of loyal, dedicated staff; building the dinosaur; and handing over responsibility for the maintenance of the majority of snakes used in the venom program to John McLoughlin. However, he lacked the skills in financial management, strategic planning, industrial relations and marketing which were needed to ensure a successful business in an increasingly competitive marketplace. His commitment to his life's work in the supply of venoms and his fascination with venomous creatures remained, but unfortunately, he was burdened with an illness – an addiction to alcohol – that increasingly compromised his considerable abilities.


Robyn Worrell and John Weigel were successful in averting the auction and keeping the park open. An agreement was reached whereby

Robyn would take over the park and its liabilities as part of the divorce settlement. Until the divorce was finalised, Worrell would remain officially the sole proprietor of the Australian Reptile Park, and though she was owed a substantial amount of past wages and held the park's licences, Robyn would continue to be regarded as an employee. Had the auction gone ahead, with the difficulties of transferring the exhibitor's licence and custody of the native animal collection, the park would almost certainly have been dismantled. The resolution by Robyn Worrell, supported by Weigel, not to allow that to happen had saved the park. It was about to receive a new lease on life.

14

# Rescue and renewal

Under the 1986 divorce property settlement between Eric and Robyn, they became jointly registered proprietors of the Australian Reptile Park. Worrell was given the right to live in their house and have exclusive use of its contents for the rest of his life, and his mother and sister were also granted the right to remain as lifetime residents of the park. Worrell had to pay Robyn the wages she was owed.

Local businessman Ed Manners, a chartered accountant who had previously worked with companies under receivership, felt that he could help turn around the Australian Reptile Park's fortunes. He was the owner of the 55-hectare Askania Park, or Forest of Tranquillity, a nature tourism attraction located at nearby Ourimbah, and could see the advantages of co-promoting the two attractions. Manners, along with his partner in Askania Park, popular entertainer Bobby Limb, and Robyn and Eric formed a new company, the Australian Reptile Park Pty Ltd. In exchange for one of the parcels of land on which the park was situated, and the accompanying buildings, improvements, non-native animals and goodwill, Robyn and Eric received shares in this new company. Manners was appointed managing director for five years, with absolute authority for the first six months. The bank holding the loans made it clear that unless Manners was able to turn around the park's finances in that six-month period they intended to foreclose.

Limb, a member of the board of the Australian Tourism Commission, was given the title of first alternate director, though he was not to have an active role in management. Instead, he lent his celebrity status to the park, hosting special events and appearing in joint Askania Park and Australian Reptile Park television commercials. As Limb was an old acquaintance of Worrell's, his involvement helped to persuade Eric to accept the new arrangement, which left him as a non-executive director with no involvement in the running of the park. He still conducted occasional snake-handling and milking demonstrations, but he tended to spend more and more time in his house on the other side of the creek.

Robyn was officially made director of operations and continued in her role as the chief administrator of the park. With ongoing tension between himself and Eric, Weigel initially sidelined himself, acting in an advisory capacity to Ed Manners and making discreet evening visits to oversee the reptiles.

As part of the restructuring, in mid-1986 the employment of several full-time staff was terminated, including Jack Green. After more than 20 years' involvement with Worrell, first at Ocean Beach and then at the reptile park, Green was shocked by his dismissal. Worrell could offer him little but sympathy, as he now had no input into management decisions. In a letter to the local paper, Worrell expressed the hope that Green, like himself, would be satisfied by seeing that the park at least continued, albeit without them. Just like Worrell, Green had established a high level of recognition in the community through his work at the park and was frequently featured in stories in the local media. He had published a book,

*Australian Alligator Man*, in 1977, about his work with alligators, snakes and platypus at the park. Following his termination, Green carved out a career in nature photography and writing articles on a nature theme.

The re-launch of the park took place in October 1986 and featured popular singer Julie Anthony and the St George Building Society's mascot, the Happy Dragon, as Manners had negotiated a sponsorship deal with the financial institution. The new slogan developed for the park by Manners was 'Just for Fun', and a tremendous amount of effort had gone into creating a more interactive and enjoyable experience for visitors. The general facilities had been brought up to a much better standard. Gas barbecues were installed near new covered picnic tables. A slide was added to the pool and a children's playground established. Fresh paint was applied to the buildings and even to the dinosaur, which went from a traditional grey-green to bright golden yellow. Although sometimes referred to as Dino in the 1970s and again in 1987 after local senior high school students painted it with spots as an end-of-school prank, the dinosaur was usually referred to as just that, 'the dinosaur' – until 1996, when it was given the name Ploddy. It now became a symbol of change thanks to the new look Manners had given it. Worrell approved, noting that there were a number of golden-coloured frogs and reptiles. This was one of the few changes initiated by Manners that Worrell would actually agree with. However, the dinosaur's colourful transformation evoked some local debate. One man wrote a letter to the editor of the Central Coast Express suggesting that the involvement of 'show biz types' had made the reptile park into a garish, overly commercial eyesore.[1]

If all of the ideas that were floated at this time – including a restaurant, holiday units, retail plant nursery and a childcare centre – had gone ahead, there would have been a greater justification for such an opinion. But these ideas were not adopted. Instead, the focus remained on the education and conservation ethos that had underpinned the park from its inception, while the experience of observing the animals was enhanced with the extension of keeper talks, animal feeding displays and reptile shows.

To get to the stage of re-launching the park, a great deal of remedial work had been necessary to bring the exhibits back to an acceptable standard of maintenance. Improved systems of monitoring the animal collection on a day-to-day basis were needed, and the numbers of common snakes and some of the other reptiles such as freshwater turtles and common lace monitors needed to be reduced. Some enclosures had been allowed to run down until they posed a risk for staff. The cases holding venomous snakes known as the north and south corridors were among the first structures built at the park in the late 1950s. Over the years, the timber framework had begun to rot and gaps had appeared between the walls of the cases and the cement landscaping work, allowing snakes to slide into these crevices. This meant that sometimes staff were not sure where a snake was in its case, making servicing difficult and dangerous. Visiting in the evenings with some assistants, John Weigel quietly worked his way through the park making improvements. The number of snakes held in each pit was reduced and a regime was introduced to ensure that each individual was adequately fed. Pits were also landscaped to resemble much more closely the habitats from which the snakes originated, which also improved their appearance to visitors. New exhibits were developed to better showcase animals such as koalas, a Galapagos tortoise and crocodiles; the platypus enclosure was renovated and the noctarium divided into a desert section and a rainforest section, complete with waterfall.

The park reduced the number of birds and mammals in its collection. For instance, the mob of tame kangaroos and wallaroos that shared the picnic area had increased to an unsustainable level, so some of these animals were sent away to other wildlife parks and zoos. Removing these animals was difficult for staff. As they struggled to catch them, Worrell sometimes stood on the sidelines directing them to put the animals back.

With mammal and bird holdings shrinking, food costs were reduced, the keepers were allowed more time to look after the remaining animals and the park could re-focus on reptiles. To do that, they needed to acquire more reptiles, which Worrell had had such trouble doing earlier in the decade. Manners took a more conciliatory approach than Worrell to securing permission to acquire animals. For instance, before a formal application was made to transfer a group of rhinoceros iguanas from Taronga Zoo to the reptile park and dispose of surplus exotic specimens, he approached the quarantine authorities and asked for a meeting to discuss the proper procedure.

The NSW NPWS made it clear in late 1986 that it would not give approval for the acquisition of more reptiles until it could be shown that the park was operating effectively. When NPWS inspectors visited the park a year later, they were impressed with the upgrades and in particular with Weigel's improvements to the management of the animals. Accordingly, they were willing to allow the park to begin to rebuild its stocks of snakes, approving the collecting of a small number of copperheads, death adders, king brown snakes and western brown snakes. With Weigel's work behind the scenes, and that of reptile staff such as Grant Husband, Kevin Smith and Robert Montgomery, all of whom joined the staff between 1986 and 1987, the park was able to breed 20 species of reptiles in captivity. Some species of reptiles had been bred at the park prior to this time – such as the American alligators, which bred most years – but successes had tended to be ad hoc. This lack of breeding success wasn't unique to the reptile park, because up until the early 1980s, the factors that stimulate breeding in reptiles under captive conditions were generally not well understood, so when breeding did take place it was often more a matter of luck than design. But by the 1980s, the factors were much better understood and the park staff now began to see the results of improved husbandry.

More platypuses were added to the collection and plans were made to introduce koalas. Before any koalas could be acquired, though, the NSW NPWS required that the park demonstrate that it had the right accommodation and skilled staff to look after them. One-time employee Bev Drake – who worked at the park in the early 1970s, after which she moved on to become the keeper in charge of native mammals at Melbourne Zoo – was invited back late in 1986 to advise on the development of a koala exhibit. The next year, she and some Melbourne Zoo staff generously volunteered to spend a number of weekends contributing their expertise and labour to help with the renewal of the park. On one occasion, Drake was sitting at Worrell's desk when the once-great snake man came in on one of his now infrequent forays into the park. Somewhat embarrassed to be sitting at his desk, she greeted Worrell and explained what she was doing there. Worrell quietly welcomed her back home.

John Weigel was still giving Worrell a wide berth to avoid inflaming the friction between them. This continued for some time until Manners reminded Worrell of Manners's authority to employ anyone he chose. After this, Weigel's public role was restored; he was initially given the title of head reptile keeper. The park began to see modest but steady increases in visitor numbers and a sense of optimism overtook the darkness that had come to settle over it.

While all this was going on, Weigel was also becoming active in the wider world of reptile keeping in Australia. Coming from the United States, he was incredulous when he discovered the difficulties people faced if they wanted to keep Australian reptiles as pets. After attempting and failing to secure a licence to keep reptiles, in 1984 Weigel established the Reptile Keepers Association with Central Coast reptile enthusiast John Montgomery, who decades earlier had regularly supplied his friend Eric Worrell with a range of local reptiles. While Worrell had been involved to some degree in the setting up of the reptile protection legislation in the early 1970s that now made it very difficult for hobbyists to keep reptiles, Weigel, Montgomery and others worked hard to loosen the restrictions. They tried to convince politicians and bureaucrats that the legislation and the way it was enforced was unfair and not in the best interests of amateur herpetology and herpetoculture. The association became a lobby group, writing many letters and other documents putting forward their arguments, producing a regular newsletter and holding several large public meetings in Sydney of around 300 interested people keen to see a change in legislation. As part of this process, Weigel wrote a book, *Care of Australian Reptiles in Captivity*, which was published in 1988. It remains the most popular guide on the subject, having sold an astonishing 35 000 copies after ten printings.

The lobbyists were eventually successful in securing an agreement to change the legislation to make it easier to obtain and keep captive-bred reptiles. However, it was not until 1996, after additional work by herpetologists Gerry Swan and Glenn Shea, that the legislation was altered. Since then, anyone over the age of ten has been able to gain a Class I licence to keep relatively easily maintained, harmless reptiles from a list of over 100 species. Weigel has been a pioneer in the reptile hobbyist arena in Australia, strongly believing in the benefits that arise from enabling ordinary people to keep native reptiles properly, such as the enhancement of their appreciation of nature in general and of reptiles in particular. Weigel has played a big part in the increase in the number of licensed reptile keepers in New South Wales, which has grown from fewer than 100 in 1980 to over 15 000 in 2010.


Despite poor health caused by his problem with alcohol and his diabetes, Worrell did not relinquish his attachment to the park and not surprisingly found it difficult to reconcile his loss of control. Worrell even hoped that after Ed Manners had had his five years as managing director, he would be able to resume his active role in the park, as reptile curator.

Robyn had asked Irwynne Symington, a longtime friend of hers and Eric's, to be Eric's housekeeper and carer. She had worked at various times in the park in roles associated with the food kiosk and reception area, and had helped Eric's mother with domestic tasks. She readily agreed. Throughout her time as Eric's housekeeper, Irwynne, who remained a deeply caring friend of his, attempted to minimise his drinking, which exacerbated his other ailments, but with limited success. Gary Pryor, the park's groundsman, visited and helped with some chores around the house, and a young reptile keeper, Brian Starkey, was also a frequent visitor, spending time talking with, and learning from, Worrell.

By this stage, Worrell's circle of close friends had grown smaller. Russell Drysdale had passed on and Worrell had lost regular contact with others such as Vincent Serventy. He was visited from time to time by John Cann and John Dwyer and also by his son, Mark. Apart from outings to the Grange Hotel across the road and occasional forays into the park, Worrell rarely went anywhere.

Returning from the Grange one evening in July 1987, Worrell collapsed just metres from the front door of his house. A friend who had been doing some maintenance work around the house discovered his body at 9.20 am the following day.

While Lyn Abra set about ringing Worrell's friends and associates, Robyn gently broke the news to Worrell's mother, Rita, at the Wyoming nursing home where she now lived.

The funeral for Eric Worrell, MBE, Australian Snake Man, was held at a packed Christ Church, Gosford, four days later, following a coronial inquest that found he had suffered a serious heart attack and died almost immediately. He was 62 years old. Staff of the park attended in uniform, some acting as pallbearers. The eulogy was delivered by John Cann, whose father, George, the performing Snake Man of La Perouse, had taught and inspired young Eric in the 1930s. Cann also read a tribute to Worrell written by Struan Sutherland. Hal Cogger of the Australian Museum attended, as did Worrell's many friends and supporters from disparate groups including Rotary, the RSL and the RSPCA, as well as from the herpetology and zoo communities. A wake was later held back at Worrell's house. That night, his mother, who had for so long supported her son in his dreams of bringing reptiles to the people, passed away in her nursing home at the age of 86.

Worrell's life and contributions were celebrated in a number of obituaries in newspapers throughout Australia and Struan Sutherland published an obituary of his friend in the *Medical Journal of Australia*. His substantial legacy included valuable contributions to Australian herpetology, especially in relation to the taxonomy of venomous snakes, Australian natural history, wildlife tourism and, of course, to snake and funnel-web venom supply. Through the Australian Reptile Park, many hundreds of thousands of people had been encouraged to see reptiles as crucial elements in the nation's ecology. The park and Worrell's writings had stimulated interest in reptiles among many who today are professional wildlife scientists or hobbyists. Indeed, his *Reptiles of Australia*, written without the institutional support of a university or museum, and entirely self-funded, had remained the most comprehensive treatment of Australia's reptile fauna until the publication of *Reptiles and Amphibians of Australia* in 1975 by a man who, years earlier as a young teenager, had enjoyed visiting Ocean Beach Aquarium, Hal Cogger. While the new management structure of the park had removed him from any decision-making role, it did mean that his boyhood dream of founding his own reptile park could continue after his passing, and that his legacy would be an enduring one.

Balancing education and entertainment had been one of Worrell's key goals and the new park management brought a renewed effort to do that. A key to finding this balance was to offer visitors the opportunity for active involvement with the animals: getting close to them, touching them and watching them exhibit natural behaviours. Personal interactions helped to break down people's negative assumptions about reptiles and encourage empathy towards them. Educational programs at the park were expanded and invigorated. Some of these initiatives were led by former schoolteacher Bob Turner, who visited schools and ran programs for school groups visiting the park. Students from preschool to senior high school were given hands-on experiences with all sorts of strange and interesting animals in the belief that close interaction plays a critical role in 'arousing concern for wildlife appreciation which … translates into conservation'.[2] The enthusiastic and personal approach employed in the school programs proved popular with students and teachers. By the end of the 1980s, 20 000 schoolchildren a year were attending one of the park's programs. Turner reached further people by taking reptiles such as blue-tongue lizards, bearded dragons and harmless pythons out to shopping centres, convention centres and local events, and through media appearances. These programs not only increased people's understanding and appreciation of reptiles, they also increased awareness of and interest in the revitalised park.

Staff conducted a range of shows throughout the day for general visitors. Lynda Vereyt provided a daily feeding show in the noctarium, during which visitors learned about the habits of rarely seen animals such as magnificent greater gliders, spotted quolls and desert-dwelling hopping mice, while these animals were given food ranging from eucalypt blossoms through to dead day-old chicks. Reptile-keeping staff performed twice-daily reptile demonstrations with a range of species, from harmless local lizards through to venomous snakes such as tiger snakes and cobras, and giant reticulated pythons. These demonstrations kept alive the tradition established by Worrell and followed a similar format. At a designated time, the keeper would take centre stage in the demonstration pit, with an assortment of sacks and boxes containing various reptiles, which would be displayed one by one to visitors. Harmless lizards and pythons would be taken around the perimeter of the pit to enable people to gently touch these animals, helping to dispel any fears they may have had of them. The demonstration usually culminated in the milking of a venomous snake. There was always a focus on informing visitors of the correct way to apply first aid in the unlikely event that they should be bitten by a snake, something that Worrell was always very keen to do.

The signature tiger snake milking was a highlight for Kevin Smith, who had developed considerable flair in his demonstrations. On one occasion, though, he was bitten and rapidly became critically ill. He would have died relatively quickly if he hadn't received the right treatment and antivenom at Gosford District Hospital. The doctors there were invariably guided over the telephone by Eric Worrell's long-time supporter and friend Struan Sutherland, from the CSL. Between 1984 and 2010, six staff were envenomated while handling snakes at the park. Four of those bites occurred during public milking demonstrations. An exception was when Robert Montgomery was bitten by an Asian cobra during routine maintenance. Montgomery was rushed to the hospital for treatment and survived the cobra bite, though he lost his right index finger due to necrosis. Sadly, his life was cut short by cancer some years later.

Smith and Weigel took it in turns to present the alligator feedings, standing on the island in their pond. Edward and Edwina, who had been the stars when they performed with Jack Green, had both died in the early 1980s and another pair of large alligators now resided in their enclosure. The 3.4-metre male and the very cranky and dangerous female provided a riveting performance as they snapped at their food and the feeders. During all of these presentations, visitors could hear about, see and touch non-venomous snakes, lizards and baby alligators, and gain a more nuanced understanding of reptiles and their place in the ecology.

For the Bicentennial celebrations in 1988, the park joined in a co-promotion with Manners's other property, Askania Park, and Old Sydney Town, capitalising on the rise in interest in Australian history by marketing the three attractions as 'heritage tourism attractions'. Elements of the promotion included posters and billboards along Central Coast roads, the development of joint brochures distributed through the branches of the park's sponsor, St George, and a publicity campaign. An arrangement was made with Flick Pest and Weed Control in which they sponsored a spider display in return for naming rights.

Weigel was an accomplished photographer and he lent his skills to the production of a new guidebook for the revitalised park. The *Australian Reptile Park Story* looked at the history of the park and Eric Worrell, and detailed its attractions and activities. In keeping with the aim of re-focusing on the strength of the reptile collection, almost half of the book was devoted to information on the reptiles in the collection, accompanied by beautiful photographs of the animals. The mission of the park, according to the book's preface, was to 'chisel away at the myth that reptiles are loathsome creatures deserving of the shovels and heavy sticks with which they are often met'.[3]

Portions of the park's land were sold and the money put towards the bank loans and further developments at the park. A one-hectare block was used for medium-density housing units.

On a much more personal level, in December 1987, John Weigel and Robyn Worrell married, surprising the guests, including Weigel's parents, who thought that they had been invited to a house-warming party. Four years later, they would adopt three children from Columbia, Blanca, Anhel (Arnie) and Johanna.

Government restrictions on the keeping and exchange of animals, both from within and outside Australia, multiplied during the late 1980s and early 1990s. New bodies, including the federal government's Vertebrate Pest Committee, set restrictions on managing captive animals. While the Weigels and Manners were as cooperative as possible, they believed that the regulations often affected them more harshly as a private wildlife park than they did publicly funded zoos.

A major change during this period was the *Exhibited Animals Protection Act*, passed in 1986, which took responsibility for licensing animal-display facilities away from the NSW NPWS and gave it to the Zoological Parks Board of NSW. When a draft of the regulations through which the Act would be enforced was circulated, the Weigels led a critique of it, challenging aspects and suggesting alternative approaches, and their opinions were supported by others in the zoo community. One of their underlying objections was that instead of setting up an independent committee, the Act made the Zoological Parks Board, which was responsible for its own zoos, Taronga and Western Plains, ultimately responsible for the implementation of the Act, including advising the government on standards of animal care and display for the state. This was in keeping with the Zoological Parks Board's official role now as the authority representing zoological, fauna, marine parks and aquariums in New South Wales, but set up a potential conflict of interest with its role at Taronga Park Zoo. It appeared that private zoos would have no role in determining issues such as the minimum amount of space to be allocated to particular species, whether animals required air-conditioned enclosures, the provision of perches or hides for snakes and the composition of enclosures for crocodilians.

The legislation came into effect on 1 July 1989. Further concerns arose about the new process of applying for a licence to display animals. The application form included questions about the financial history of the applicants, including bankruptcies, receiverships and previous breaches of animal regulations. Manners, the Weigels and other members of the NSW Association of Fauna and Marine Parks found these questions commercially intrusive and unnecessary and refused as a group to complete them. The reptile park was threatened by the Zoological Parks Board with prosecution for not having a current licence, so the park eventually backed down and supplied the required financial information. After an inspection by the Zoological Parks Board in 1989, the licence was issued. Along with the licence came an Animal Record Book in which to record details of exhibited animals, acquisitions and disposals, replacing the NSW NPWS Fauna Record Book used up to that time. The board's prior approval was required for transactions of fauna between zoos and animal parks, and even for any modifications such as new exhibits or living quarters.

John and Robyn Weigel were determined to re-establish the park's role in the supply of snake venoms to the CSL. When venom supply from the park dried up in the early 1980s, other producers had come to dominate the market, including Charles Tanner, Peter Mirtschin in South Australia and John McLoughlin.

In 1987, Weigel had employed a cabinet-maker to build a state-of-the-art snake-keeping facility to his design, in the garage of the home he was renting in suburban Gosford. It comprised 145 cages of varying sizes and temperature and light combinations to suit a wide variety of snakes. They were arrayed in banks that stretched from the floor to the ceiling, which meant considerable savings in building materials and heating costs in comparison with less integrated cages. To ensure that the snakes could be cared for appropriately, each cage was to contain only one snake, except when the snakes were to be bred. A systematic, streamlined record-keeping system helped ensure that the entire collection of venom-producing snakes could be adequately cared for by just one person, further minimising expenses.

After Worrell's death, Weigel shifted the cages across to the reptile park, initially housing his own snakes in them. Once this new approach to the management of the venom-producing snakes had been established at the park, CSL officers were invited to inspect the facility some months afterwards and they were impressed by what they saw.

The procedure for exporting snake venom to overseas research scientists became more complex under the Commonwealth *Wildlife Protection (Regulation of Exports and Imports) Act* of 1982. No longer were blanket approvals given – each transaction had to be considered by the then Australian NPWS (ANPWS) separately. First, the receiving institution had to prove to the ANPWS that the venom was for use in a research program. Then the park had to apply to the ANPWS for permission to export. If the venom was to be used commercially, including for the production of antivenom, it would also need to be shown that the venom was supplied by captive-bred snakes or snakes taken from the wild under an approved management program. Most scientific institutions needed the venom relatively quickly, but it frequently took two to three months for the ANPWS to process an application and issue export approval.

Orders from the CSL declined sharply in the late 1980s, which was of concern to the park, especially after they had made a considerable investment in the new venomous snake facility. The annual orders for 1988 and 1989 were 60–70 per cent lower than those of the preceding decade. The CSL was undergoing its own transition during this period, as the federal government was gradually forcing it to pay its own way in the lead-up to the organisation's ultimate privatisation in 1994. Venom research and antivenom production, never a central concern of the CSL despite its high public profile, had been on a decline for some time.

Towards the end of 1989, Robyn and John Weigel wrote to the CSL in the hope of regaining the park's previous status as primary supplier of snake venoms. When this letter remained unanswered,

Robyn wrote again in February 1990, pointing out the very long association between herself and the CSL. She offered discounts on a range of snake venoms if CSL agreed to place a minimum of two-thirds of all of their future snake venom orders with the park. Most of this venom would originate at the park, but all of the brown snake venom and a quantity of several other venoms would be supplied by John McLoughlin in north Queensland. The CSL accepted the proposal and the park soon re-established itself as the CSL's primary supplier of snake venoms.

By the end of the decade, the management team of Director of Operations and Company Secretary Robyn Worrell and Managing Director Ed Manners, supported by John Weigel and staff, could look back with satisfaction at their achievements. Visitation, one of the main barometers of the success of their venture, had risen steadily, from 72 000 in 1985–86, to 93 000 in 1986–87, 101 000 in 1987–88 and 121 000 in 1988–89.[4] The Australian Reptile Park was back from the doldrums of the early 1980s. The late Eric Worrell's legacy was secure – and even greater changes were still to come.

# PART THREE
# SOMERSBY

# 15

# Pushing the boundaries

etting the park to an acceptable standard took hard work, innovative ideas and imaginative thinking. John and Robyn Weigel generated dozens of ideas and pitched them to Manners; those he agreed to guided the park's new direction. Cutting spending while increasing revenue was Manners' primary concern. He outlined ten actions for Robyn with the aim of doing just that. The final one was blunt and to the point: 'If in doubt about any expenditure say NO!'[1] A financially conservative approach was going to be essential if the park was to get back into the black and stay there.

At the same time that costs had to be reined in, the park had to compete with other similar attractions. John Weigel sought advice from Rick Shine, professor of ecology at the University of Sydney, regarding ways of enhancing the educational outcomes and research opportunities of the park, and many of his ideas were adopted.

Weigel and other staff had articles published in the professional zookeeping literature to restore and enhance the reputation of the park. An article in the prestigious *International Zoo Yearbook* focused on the park's educational and conservation work. Weigel and Abra had short articles about the changes to the park published in *Thylacinus*, the journal of the Australasian Region Zoological Parks and Aquaria Association. Weigel also published a new field guide titled *Australian Reptile Park's Guide to Snakes of Southeast Australia*, renewing the park's contribution to public education, which Worrell had established in the 1960s. Weigel realised that another of Worrell's ambitions, to display a massive saltwater crocodile, also had merit. Not since the short-lived display of Casanova in 1964 had the park displayed such an animal and Weigel was confident that it would draw the crowds.

Riding on the popularity of the 1986 film *Crocodile Dundee* and extensive media coverage of crocodile attacks in the Northern Territory and north Queensland, Weigel believed that the scale, power and novelty of an adult estuarine crocodile would be a strong draw-card for the park. Manners was less than convinced about the value of such an animal and was concerned about the costs of properly displaying it. However, Weigel's persuasiveness won the day and negotiations began in 1989 to transfer a suitable animal from the Northern Territory. The association with Bobby Limb opened doors, helping secure the approval of the NT Conservation Commission for the transfer of a crocodile and free passage on Ansett Airlines from Darwin to Sydney.

The Darwin Crocodile Farm, 40 kilometres south of Darwin, agreed to supply a crocodile in exchange for some of the park's alligators. Weigel's preference was for a crocodile around 5 metres in length. The farm supplied a problem crocodile that had been captured after lurking close to the Borroloola Aboriginal Community along the Macarthur River. The big fellow, about 4.7 metres long and estimated to weigh in at 500 kilograms, had chewed the heads off two smaller female crocodiles at the farm and had lost a foot in a fight with another male, so no doubt the farm's staff was not too unhappy to see him go. This crocodile was the largest taken to southern Australia by air.

On arrival at the park, in June 1989, the crocodile was lifted by crane into a refurbished enclosure opposite the pond where Edward and Edwina the large American alligators had lived for many years. As Weigel had hoped, the opportunity to see an impressively big, dangerous crocodile from Australia's north drew large crowds to the park, queuing outside the gates waiting to see him, just as they had for Casanova a quarter of a century earlier. This new crocodile had to do more than just lie placidly in its pool, however. Regular displays of crocodile feeding were introduced, inspired in part by those Weigel had been part of at the California Alligator Farm.

The park considered the idea of calling him Robbo, after media personality and park supporter Clive Robertson, but they decided in the end that the animal should be called Eric, as a fitting tribute to Eric Worrell. Eric was soon earning his keep not only from his performances but as the new symbol of a re-energised Australian Reptile Park. His toothy smile would appear on thousands of promotional bumper stickers and in advertisements.

Not content to simply bring the reptile park up to a new standard, Weigel began taking reptiles out to the people of Sydney. He ran a successful stall at the Beyond 2000 World Expo at Darling Harbour, staffed by park employees who carried reptiles such as pythons and bearded dragons among the showgoers.

Later, when he saw an 8-metre animatronic model of a crocodile made for a movie filmed in part at Old Sydney Town, Weigel convinced the owner to lend it to him for a few months to take a mould from it. Weigel had reptile park keeper Jan Nedved, a refugee from communist Czechoslovakia and a talented sculptor, create a magnificent fibreglass replica. Weigel's idea was to use the model crocodile as a focal point for an innovative reptile exhibit. He negotiated with the management of Centrepoint Tower, in Sydney, and the display 'Reptiles Alive!' was born. Weigel anticipated that the tower's central location and the large amount of passing foot traffic would allow him to reach a huge potential audience of school students, locals and tourists in the greater Sydney region. Initially, he intended the exhibit to run for six months before he took it to the other capital cities.

Weigel was to fund the project himself, use reptile park animals as exhibits, and put 10 per cent of the profits back into the park. With backing from his bank, he set about designing and constructing the exhibit. It was made up of modular components that were relatively straightforward to assemble so that the display could be relocated. Considerable innovation went into the moulding and casting of large expanses of fibreglass sandstone outcroppings, an impressive waterfall and replica trees up to a metre in diameter. It took Weigel and a team of workers four months to complete it.

Weigel was determined that 'Reptiles Alive!' would open with a welter of media coverage. He decided the most dramatic use of the model crocodile would be to haul it up the outside of Centrepoint Tower, using the window-washing hoist. The building's engineer said the model was too heavy to do that safely, so Nedved had to supply a second, lighter, version at short notice. Although the engineer was still concerned about the risks, the promotion went off without a hitch. Photographs of Nick the crocodile – named after the NSW premier, Nick Greiner – being hoisted up the side of Centrepoint Tower were published in newspapers throughout the country and the story featured in Sydney nightly television news reports. Bobby Limb did the honours of opening the exhibit in May of 1990.

The exhibit was not an immediate success. People were reluctant to pay to enter when they did not know what to expect inside the mouth of a 30-metre long fibreglass crocodile which Weigel had commissioned film-set designer and artist Dennis White to create. Weigel responded by having staff bring animals out of the exhibit area so that visitors to the tower could meet them and be persuaded to enter the exhibit, even going back to the old-time snake handler approach by inviting George Cann Jnr to conduct his show in and around the exhibit on weekends. Once the ice was broken, the visitors flowed in.

The exhibit was multisensory in that it incorporated bush sounds with lifelike replicas of sandstone cliff, buttressed tree and red sandy desert for context around the reptiles' enclosures. A variety of state-of-the art audiovisuals set the mood and created a dramatic and entertaining experience quite unlike any other to be had in the centre of the city. Staff conducted hourly talks and visitors were able to touch some of the animals. A dedicated education officer was on hand to tailor tours of the exhibit to meet the needs of specific groups of students. The exhibit was extremely popular, attracting almost half a million visitors over its five-year residency at Centrepoint Tower. It finally closed in 1995.

The exhibit's popularity with schoolchildren made Weigel think about taking a reduced version on the road. He began collaborating with Yanina Prager, a public relations officer for both the reptile park and Askania Park. A former teacher, she had been closely involved with 'Reptiles Alive!' as one of the education officers. Together, they developed a mobile education program called 'We're *All* Little Aussies' in 1992. It was one of the first such programs tied directly to the NSW primary and secondary school curricula in areas ranging from natural history to multiculturalism. Using reptiles as a means to explain concepts such as the importance of diversity and the rights of humans and reptiles to co-exist within Australian ecosystems, the program also made it clear that all students, regardless of ethnicity or gender, had an equal role in protecting Australia's fauna. The concept of stereotyping was explored by examining prejudices against reptiles and then this was used as a point from which to explore cultural- or ethnic-based prejudice. Four versions of the program were designed for different age groups and schools from all states and territories booked the program.

The service was run out of Rita and Percy Worrell's old house at the reptile park. Some 3500 schools were approached; one mail-out was a message tucked in the mouth of a plastic crocodile. Three sets of staff, working in pairs, conducted reptile shows in schools around the country, using music, humour and hands-on experiences. Students could see and touch a range of snakes, tortoises and lizards. The program proved a success. During the eight-year life of the program, well over a million Australian students were introduced to reptiles in a fresh context. A successor of the program continued for another eight years.

'Reptiles Alive!' and 'We're *All* Little Aussies' were groundbreaking for the time. They demonstrated the ability of the park's owners and staff to stretch the limits of what a wildlife park could be, taking reptiles into the heart of Australia's largest city and into classrooms in every state in the nation.

Although many significant improvements had been made to the park, the Wyoming site was burdened by decaying structures, some built in the late 1950s, and was hemmed in by increasingly dense suburban development. Tourist traffic had been almost entirely rerouted away from the famous Gosford dinosaur by further improvements to the F3 Freeway. Old buildings had been refitted for new purposes – including the gift shop, which had been made over as an education centre – with less than ideal results. Tastes in animal exhibition were changing and so, too, were visitor expectations. There was little space in which to experiment with new species or techniques for displaying them, or to conduct research.

There was also financial pressure, despite increased gate takings. After the death of his father, Mark Worrell brought a lawsuit contesting the establishment of the Australian Reptile Park Pty Ltd. Although the new company had been set up some eight months before Eric's death, his son was able to argue that it was an unfair diminution of Eric's estate. Mark was awarded a financial settlement, to be paid from the business over a six-month period. John concluded that selling the valuable Wyoming site for residential redevelopment and setting up a more streamlined park in a more suitable location was their best option.

Financially conservative Manners initially had strong reservations about moving the park. So, too, did Robyn Weigel, because of her long association with the original site. But John Weigel's excitement eventually won over Robyn and other key staff to his vision of a reinvigorated park in a new location. Manners agreed that the move was the best option but recognised that it would be a major undertaking. There would be many challenges along the way that he did not want to become involved with, so he sold his shares to the Weigels and his involvement with the park ended on good terms in 1995. Gosford City Council, eager to have the Wyoming site released for residential development and keen to have the reptile park remain as a successful tourist attraction within the municipality, raised no objections to rezoning the land.

The Weigels looked at possible locations for the park in the Gosford region. Just as they were preparing to sign a 50-year lease on land near the Mingara Recreation Club, at Terrigal, the park's accountant, Max Perry, suggested that a place the Weigels had already considered but disregarded, a disused turf farm on land adjacent to Old Sydney Town at Somersby, was a better location. The contract signing was delayed while John Weigel revisited the turf farm site the following day. It was a large tract of cleared land held by Old Sydney Town. He surreptitiously climbed a high-security fence to explore adjacent uncleared lands that had been earmarked by Old Sydney Town for future development as part of a golf course. Weigel was stunned by the natural beauty of the property. He selected a 6-hectare portion, envisioning it as the basis for a new type of reptile park. It was bushy and included a large dam and a beautiful creek running over a sandstone bed.

The owners and managers of Old Sydney Town were resistant to the idea of excising this block of land but eventually gave in to Weigel's enthusiastic pleading and cajoling. Just as in the 1950s Worrell had located his park near the main north-south traffic artery, 40 years later Weigel had chosen a site that was easily accessed by intercity travellers on the F3 Freeway. Locating the reptile park next to Old Sydney Town was strategic as well: tourists visiting one attraction might more easily be attracted to the other, especially if a collaborative approach was made to ticketing. In 1995, the Weigels negotiated an 80-year lease on the land at a fixed rental.

Such a huge undertaking as relocating the park could not be done without additional support. Rob Porter stepped into the role of park manager in 1995 with a considerable brief: to oversee the winding down of the old park and manage the staff's workload so that they could spend time at the new site without neglecting animals or visitors at the old. This would free up Weigel to focus on the design of the new park and the logistics of the move. Originally from the United Kingdom, Porter had emigrated to New Zealand with his family at the age of 12 and studied there up to masters level in zoology. He was always interested in reptiles and kept some of the small array New Zealand had to offer. Porter moved to Queensland and came to know John Weigel through a number of professional organisations and committees. Porter was also being sought by the Melbourne Zoo but was attracted by the opportunity to be involved with the development of the park in its new location.

He brought the reptile park into line with a more contemporary management style than the personal and informal styles that had been used by previous managers and curators. Communication with staff had generally taken place on a one-on-one basis, which meant that sometimes information did not circulate as it needed to. Working closely with Keith Rodrigues, head mammal keeper, Porter brought in weekly staff meetings and upgraded record keeping. Acting on Weigel's idea, Porter started up a Herp Readers Club, partly to open up the reptile park's extensive library to support staff development. It was like a weekly after-hours book club at which staff discussed reptile books and periodicals they had been reading.

Porter also overhauled the park's volunteer programs, making them more structured and effective. Volunteers had long been a valuable workforce for the park, willingly exchanging their labour for the opportunity to work with the animals they loved. Since they were not on the payroll, their comings and goings were not regulated and the park was not always able to depend upon their contributions. Porter re-envisioned volunteering at the park as professional development for people who wanted a career in zoo-keeping. He set up training programs, work routines and rosters as required by TAFE animal attendant educational programs, especially TAFE zookeeping programs. Aligning the volunteer program with TAFE programs led the way for many prospective zookeepers in the region to realise their ambitions. Today, most paid keepers at the park and a surprising proportion of keepers in zoos throughout Australia started out with a one-day-a-week volunteer position at the Australian Reptile Park.

Several evenings a week, Porter brainstormed ideas for the new park with the Weigels, Yanina Prager, who by now was the park's education officer, and marketing officer Mary Rayner. Unlike many of the staff, Rayner had not come to the park with a passionate interest in reptiles. As a local resident, she had been impressed on her occasional visits to the park but only became deeply interested in the early 1990s, when reptile park staff came to give a presentation at an environmental awareness program she had organised at her children's school. Looking to expand her involvement in environmental work, she approached the park and in 1994 began part-time work assisting them to secure sponsorships. She negotiated sponsorship of the spider exhibition but in general found that small-scale money-making projects such as children's birthday parties were more feasible. Although she had no formal training or background in marketing, Rayner rapidly grew into this role, which had not been part of the park's structure in the past. She became a highly effective promoter of the park through paid advertising and working with the media. Eventually, she was employed full time and given the title of marketing officer.

Up until the late 1980s, the park hadn't taken a strategic approach to promotion but had benefited from mostly feel-good stories in the Central Coast, Sydney and Newcastle media, which regularly covered the birth or hatching of appealing or unusual animals, the arrival of a new exhibit or the dangers of snake or funnel-web spider bites during summer. Following the arrival of Manners and Rayner, the park took a much more strategic and targeted approach to promotion. There was a deliberate emphasis on family fun and entertainment, in keeping with John Weigel's strongly held belief that the park had to provide an entertaining day out for the whole family before it could achieve its broader aims of conservation and education. Rayner worked very hard to create productive and resilient relationships with media and tourism organisations and the outcomes of all this effort were realised in increased visitor numbers prior to the move to the new site at Somersby. That was crucial during this period because of the costs associated with commencing work on the new park.

From 1989, a slightly changed name was used when the park was promoted to international tourists: the Australian Reptile Park and Wildlife Sanctuary. Even though Manners and the Weigels had refocused the park back to reptiles, Manners felt that for international tour groups, scaly animals had to be de-emphasised in favour of the broader range of animals that could be seen at the park. Images of kangaroos, koalas and wombats were used in advertisements as well as images of reptiles. Weigel did not agree with this initiative, arguing that it diminished the strong brand that they were building as a reptile park.

After protracted negotiations, the sales of land at the old reptile park site was settled in February 1996, allowing the park to clear its debts and establish a $1.2 million budget for the development at Somersby.

The site sloped gently down to a delightful creek that meandered across a bed of ochre-hued Hawkesbury sandstone and through a superb open forest of eucalypts growing over a dense understory of native plants typical of the area. A large dam on the northern boundary of the site would make a perfect lake for the park's large alligator population. Many of the site's mature eucalypts and other trees would be retained, to create a distinctive bushland feel. Regularly perching in the top of a well-positioned tree to gain a bird's eye view of the site, Weigel plotted the general layout of the park to make the most of its natural features. Numerous design layouts succeeded one another until a basic layout was produced.

Through using innovative design and construction methods, having park staff do as much of the work as possible and using recycled materials wherever appropriate, the Weigels were able to keep expenses to a fraction of what they otherwise might have been. Weigel's vision for the new park was one where the buildings would be in harmony with the setting, and where visitors were offered a high-quality experience that was interactive and educational but above all enjoyable. This vision was shared by Robyn Weigel, who also kept a critical eye on the finances and acted to keep John's grand ideas grounded in reality. She maintained her operational roles at the Wyoming park while also contributing to the ideas for the new site.

The main building housing the entry, offices, kiosk and the large reptile exhibit was constructed and smaller animal enclosures were built. Before the alligators were released into the lake, much of the nine million litres of water were drained and chlorine added to the remaining water to kill the mosquitofish that were flourishing there. By ridding the lake of these feral fish, it was hoped that species of endangered frogs such as the giant burrowing frog and the green and golden bell frog might have a better chance of establishing sustainable populations. The mosquitofish were eradicated and within months, five species of native fish had established sizeable populations. Unfortunately, on the eve of the official opening of the new park, the region experienced a once-in-a-hundred-years flooding event and the floodwaters brought with them more mosquitofish. Within a couple of years, the native fish and invertebrate species that had started to build their numbers had vanished.

After nine months of working sometimes literally around the clock, an opening date was set and the transfer of animals from the Wyoming site to the new park began. The logistics of moving so many animals had to be carefully thought through and plans made for any contingency, such as accidental injury to animals or keepers as they were caught and secured, sudden illness, poor weather and vehicle breakdowns. A wide array of containers was needed to transport the 600 or so animals. Many reptiles and birds were carried in cloth bags, including pillowcases, which both secured the animals and provided them with a calming, soft and darkened environment. Aquatic animals were placed in plastic tubs. The wombats and the sole Tasmanian devil the park owned at the time had to be lured with food into wooden crates. The four koalas were placed in cat carriers. Many of the animals were unproblematic and were easily caught and placed into their carry bags or crates. But large kangaroos, cassowaries and emus had to be wrestled to the ground and subdued before being loaded into suitable cages and crates. There was always the risk of someone being kicked or scratched by these potentially dangerous animals. Once in their crates, the animals were carried or wheelbarrowed to vehicles and delivered to their new enclosures as quickly as possible to minimise stress. The new park was only a 10-15–minute drive south of the Wyoming site, so the transportation between sites caused minimal stress.

Moving the park's population of forty-five alligators was a major operation. An alligator is a much less dangerous animal than a similar-sized crocodile, but it still takes several people to subdue one by throwing themselves on the alligator's back and holding it down until it can be restrained with rope. Up to seven people were required to secure the very largest of the alligators. Fortunately, because the move was being made in early spring, they were lethargic and easier to handle. Nevertheless, the job of shifting them was a challenging one and the park cleverly incorporated it into its advertising. One advertisement encouraged readers to go and 'see park keepers wrestle and restrain the first of the American alligators for their move to Somersby'.[2]

It was not only living animals that had to be carefully moved. There was also the monumental move of Ploddy. As the dinosaur had stood on the side of the Pacific Highway for three decades while the fortunes of the park waxed and waned, to relocate the Australian Reptile Park and leave Ploddy behind was unthinkable. Engineers were consulted about how to move all 34 tonnes of it. To free it from the steel reinforcing rods that connected it to the ground, the only option was to cut off its legs just above its feet and then carefully lift it onto the back of a truck using a crane. The relocation of such a beloved icon could not be done without appropriate ceremony. Mary Rayner organised the Walking the Dinosaur Parade, a tickertape procession down the main street of Gosford for locals to say farewell to the gentle giant, the park's staff and some of its live animals, such as dingoes, pythons and lizards. Shopkeepers along the main street joined in the fun by decorating their front windows and a local newspaper, the *Central Coast Express Advocate*, carried a five-page supplement a few days before the parade in which many local businesses wished the park and Ploddy well in the move. Two days before the event, dinosaur footprints were painted on the route of the parade in oil paint, ensuring that they lasted for a long time as a reminder of the occasion.

The one thing that could not be managed was the weather. On the day before the event, Gosford was inundated by torrential rain, complicating the protracted process of cutting Ploddy free and lifting it by crane onto the truck that would take it through the streets of Gosford to Somersby. The rain was so heavy that there were questions over whether the parade would even take place.

But the rains held off on the Saturday of the parade, which began early with a farewell sausage sizzle in the old reptile park. Then, as Ploddy slowly made its way from the place that it had called home for the past 35 years, a crowd of 10 000 people lined the streets of Gosford. The parade included colourful and creative floats produced by local clubs and businesses, from the Girl Guides to the local radio station. Olympic athletes, bands, local government representatives and formations of Harley-Davidson motorcycles all participated, with the park staff marching in newly designed uniforms near the rear. The finale was Ploddy, with an honour guard of army cadets. Gosford's dinosaur was slowly driven through the cheering crowds of well-wishers before heading up the Kariong hill to its new home, a highly visible site overlooking the F3 Freeway near the Gosford exit. The grand spectacle moved some to tears.

The new park would not be opened until the following week, after Eric the crocodile had been moved. His transportation was approached with great deliberation and took place several days after the hubbub and crowds of the parade had died away, because Eric posed a potential danger to staff moving him between sites and the sedation and move could be harmful to him.

Eric's pool was drained of water and longtime veterinarian to the park Peter Nosworthy bravely entered his enclosure, carrying a special stabbing spear gun arrangement with a syringe and needle at the end. When the needle was jabbed into the crocodile's hide, a spring mechanism would compress the syringe to dispense the drug. With John Weigel providing a distraction to the crocodile by moving within the animal's field of vision, Nosworthy stabbed the needle into Eric's neck. One shot of the sedative Hypnovel rendered the crocodile unconscious, allowing the vet to then approach him and inject the muscle relaxant Flexedril. This enabled the staff to approach him in safety. His snout was secured shut with cloth and duct tape, then 12 men lifted his half-tonne bulk onto a board stretcher. He was bound to it with ropes and lifted onto the back of a truck by a crane. Covered with shade cloth, he began his journey to the new park.

At Somersby, he was lifted by crane and deposited on the grass of his new enclosure, to the delight of the small crowd of staff and volunteers that had gathered to witness the moment. To bring him gradually back to consciousness, the vet administered a shot of Atrosine and Neostigmine. A tense hour of careful monitoring followed, as the vet closely watched the crocodile's breathing and periodically checked his eye response with a small torch. For Weigel, the whole experience became highly stressful when he couldn't detect Eric breathing for a period following the injection of the drugs to bring him around. He was greatly relieved when Eric's nostrils could again be seen flaring as he rhythmically breathed in and out a couple of times per minute. One of the most nerve-racking aspects of the move had been accomplished without harm to staff or to Eric.

The next major challenge was to organise an official opening that would herald the start of a new era for the Australian Reptile Park.

16

# The new Australian Reptile Park

The official opening of the new park took place in September 1996. John Weigel described it as a rather tragic affair. Unbelievably heavy rain – 200 millimetres over a four-hour period – the previous night had washed away much of the landscaping. The reptile section was incomplete and some of the enclosures were devoid of animals. It was a nightmare way to begin.

However, with its bushland setting, innovative displays and emphasis on fun and enjoyment, the new park struck a chord with the public. Soon, visitation soared and the park was winning tourism awards, including the highly prestigious national award for Best Regional Attraction in both 1998 and 1999, as well as a special NSW Tourism award for Best Regional Attraction of the Decade for the 1990s.

Visitors' expectations of the new wildlife park were rather

different from when Eric Worrell opened the first park at the end of the 1950s. The leisure market was now highly competitive and the park found itself vying with an array of sophisticated entertainment options. It would have to constantly find ways of engaging and re-engaging people. There was also a greater degree of scepticism generally about the relevance of captive animal exhibits, which also impacted on attendance figures. Visitors now were more sophisticated and discerning and many had a greater awareness of animal welfare. Government regulations affected all aspects of the park's operations, something Worrell would not have welcomed. The expectation that the park would contribute to public education and wildlife conservation was also very high. All of these factors shaped the new Australian Reptile Park.

Understanding the importance of building up visitors' expectations as they approach an attraction, Weigel's design for the entry to the new park was dramatic. Visitors enter the park through a portal embossed with low relief images of reptiles. This is topped with a highly realistic giant model of a frilled lizard, sculptured by natural history exhibition specialist David Joffe. They stroll down sandstone steps and along a path, beside which flows a small, realistic-looking stream. Gardens of native plants add to the natural look. Visitors then enter a spacious lobby in the main building, where they can buy tickets. The administration offices are also located here. A large window provides an excellent view across the alligator lake, giving visitors a taste of what is to come when they have paid their entry fee. The earlier and heavier version of the saltwater crocodile that was to have been hoisted up Centrepoint Tower lies positioned in an inside pool and the gift shop adds more colour to the space.

The majority of the reptile collection is housed within this main building, with visitors entering the exhibit through the mouth of a giant crocodile, just as they had at the 'Reptiles Alive!' exhibit at Centrepoint. Much of the knowledge and skills acquired in creating

'Reptiles Alive!' was used in the design and construction of the new permanent exhibits.

Outside, visitors first see a series of pits housing a variety of local species of snakes and lizards. Unlike the formal cement-rendered pits of the old reptile park, these are designed to look like natural sandstone formations. Carefully designed hides have viewing windows in them so that the snakes and lizards can be looked at while at the same time feeling secure. While the venomous snakes are under the care of a separate keeper, the rest of the reptile collection is managed by Liz Vella, who has made an important contribution to the park over many years.

Another lesson that was driven home by 'Reptiles Alive!' was the importance of lively interpretation. While talks and demonstrations had been on offer in the old park, at Somersby they were increased in number and sophistication. Michael Tate, in the character of Ranger Mick, has proven popular with visitors who crowd around the demonstration pit to watch him handle various reptiles and talk about their habits, infusing humour and showmanship into his talks. He fulfils the role of a ringmaster, directing the attention of visitors to the spectacles the park has to offer and heightening the enjoyment of their visit. When hiring new staff to provide interpretive talks, Weigel made people skills a top priority along with skills in working with animals.

Eric the crocodile became a larger-than-life character at the new park, living in a specially designed enclosure next to the Hard Croc Café. Eric was honoured with a friendly and fun-loving cartoon character in his likeness, his own bumper stickers and a section of the park website named Eric's Swamp. He even had an annual birthday party, at which John Weigel would attempt to give him a birthday kiss, showbags were distributed and birthday cake was shared with all of the visitors. Eric's fan club grew to almost 10 000 members and television personality Rove McManus was one of his sponsors for several years, cementing the croc's status as a celebrity. Eric became one of the key marketing icons for the new park, a stylised graphic of him appearing on signs and brochures. Extra seating had to be built around his enclosure to cope with the enthusiastic crowds of onlookers as he performed for his lunch by obediently lunging out of the water to take hold of a fish or chicken from the hand of a keeper.

When we look at the reptile park today with an understanding of its history, we can see evidence of change, but also constancy. One example is the commitment to carrying out the extraction of venom from snakes. It was always an important part of the park's mission, but the systems in place to manage it have become greatly more sophisticated. In contrast to Worrell's procedures, which grew out of the kerosene tin and masking tape days of the 1950s, since the 1990s the highest priority in the snake-milking program has shifted to the welfare of the snakes and the staff who deal with them. The snakes used for milking are no longer displayed to the public in pits or cases. Instead, they are held in a series of off-exhibit snake rooms where each snake is individually housed in one of a bank of cages. Just as John McLoughlin, Worrell's north Queensland venom supplier, had discovered in the 1960s when he began keeping his snakes in individual enclosures, isolation is better for their health. It greatly reduces stress, curbs the spread of disease and parasites, and enables the keeper to more closely monitor the snakes and their food intake. The venom-producing snakes in the new Australian Reptile Park are healthier and more robust than in the past, with a life span that typically exceeds a decade, rather than a year or so, which was typical in Worrell's day.

The handling of the snakes for milking has also improved. No longer are they milked in an outdoor pit. Handlers have also left behind the practice of using a jigger – a Y-shaped metal frame with a narrow leather strap across it – to apply pressure to the back of the snake's neck before picking it up and forcing it to bite a membrane over the mouth of a beaker. Using a jigger was the standard way of securing dangerous snakes in Worrell's time, but it is now known that being handled in this way produces a high level of agitation and stress in the animals. As John Weigel has written, 'possibly the most frightening thing any reptile can experience is a sudden "whack" to the back of its neck or head by a narrow object, since in nature that usually signals impending death.'[1]

Building upon the experience of early venom producers John McLoughlin and Charles Tanner, Weigel developed a new technique so that instead of a narrow band of rubber pinning the snake's neck, a broad flat surface applied more equal pressure across the snake's head and neck. Eventually, after some experimentation, Weigel found that the best implement to use was a perspex disc with a diameter of about 15 centimetres attached to the end of a metre-long shaft. The clear perspex greatly reduced the visual threat that the approaching pinning device represented to the snake, as well as evenly spreading the pressure across its head and neck. The result is a far less visible defensive response from the snake, and a safer process for the keeper, particularly in executing the next step of the operation: the disk is slid forward until the only part of the snake being compressed is the head, allowing the handler to secure a grip behind the head. Once the snake is securely held, its head is brought close to the beaker and the snake is then encouraged to bite through the membrane stretched across the top of the beaker.

Several other measures are used to reduce the risk of the handler being bitten and reduce the stress on the snake. Before a snake is handled, the temperature inside its cage is reduced to the low 20s to slow down its responses. The snake is placed on a soft pinning pad with a non-slip surface. The pad is covered by a plastic sheet, on which a dry adhesive is sprayed, creating a tacky surface. This tacky surface helps to keep the snake where the handler wants it and the snake is less able to make rapid movements. As a result of all these improvements in procedure, the snake is more relaxed and doesn't attempt to bite and thus waste some of the precious venom. Since these changes were introduced, there has not been a single envenomation accident in the venom production program. This is a remarkable outcome, considering that over 60000 milkings have been undertaken at the park in the past 20 years, about half by Weigel. He claims to have extracted venom from more snakes in Australia than anyone except John McLoughlin and possibly Tim Nias, the park's venomous snake keeper between 2005 and his death from leukaemia in 2007. Nias had worked at Venom Supplies Pty Ltd at Tanunda, in South Australia, prior to taking up his appointment at the park; he was widely regarded as an outstanding manager of venomous snakes and someone who was generous with his time, knowledge and skills. The three men also shared the rare status of never having been envenomated during their entire venom extraction careers.

If the Australian Reptile Park had a life span like one of its animals, this final chapter might be expected to tell the tale of a comfortable retirement, with the park slowing down, content to rest on its laurels. But the story is very different. The park is more like the phoenix, the mythical bird that once every 500 years burns in its own nest so that a young bird may arise from the ashes, to live its allotted span and be consumed by flames in the same way. If injured during its lifetime, the phoenix is able to regenerate, making it invulnerable to its enemies. Over its own lifetime, the park has been resurrected three times. The first rebirth was the return from economic collapse in 1986; the second was the move from the Wyoming site to Somersby in 1996; the third, like the phoenix, was after a devastating fire in 2000.

Shortly after midnight on Sunday, 16 July 2000, fire broke out in the main building. It took hold before motion detectors on the security system triggered an alarm and it spread rapidly, devouring the building, including the entire reptile display and holding facility, where reptiles that were not on display, such as the snakes used in the venom program, were maintained. Mary Rayner received a call from John Weigel in the early hours of the morning to be told that he was watching the park going up in flames. She vividly remembers the pall of smoke suspended in the early morning sky as she drove up to the park soon after.

Apart from a pig-nosed turtle, a small American corn snake and a snapping turtle that was found crawling through the embers – and which was renamed Terminator – none of the reptiles, amphibians or spiders survived the flames and toxic smoke. Hundreds of snakes and lizards, hundreds of funnel-webs, a platypus and frogs had perished. Animals housed away from the main building, including the koalas, birds and alligators whose enclosures were positioned along a bush pathway through the park, were unaffected.

For the staff who had cared for the animals for so many years, the fire was a terrible blow. It shattered John and Robyn Weigel. The utter distress and anguish they felt was clear. Yet both were determined that they would not let the fire close the park down permanently. Within 12 hours, together with their loyal staff and Glen Balneaves of Raybal, the firm that had constructed the building just four years earlier, they had come up with a basic plan to get the park operational again.

The fire moved members of the general public. Visitors sent cards of condolence and a few days after the fire, about a hundred people – staff, volunteers and friends of the park – attended a memorial service at Somersby Falls, just down the road from the reptile park. On that cold, windy day, John Weigel and Mary Rayner addressed the gathering, reassuring them that the park would be rebuilt and inviting everyone to play a role in getting the park back on track. There were opportunities to reflect on, to remember and pay tribute to the lives of the animals that had been taken by the fire. While the staff tried to be positive about the future, John Weigel noted that he, at least, felt empty and hollow for some time afterward. Considerable support came from the Central Coast community, from individuals as well as from the various levels of government.

One of the Weigels' many concerns was about the disruption to the snake and spider venom supply to the CSL, as a shortage could impede the production of antivenom and ultimately cost human lives. Peter Mirtschin, of South Australia-based Venom Supplies, assisted by supplying venom in the short term. A senior CSL representative toured the park's ruined facility and agreed to provide $100 000 to restart the venom supply program and to pay higher prices for each unit of venom.

The Gosford coroner investigated the fire, finding that it was due to an electrical fault associated with one of the reptile exhibits and could not have been foreseen. He suggested that monitored smoke alarms could have drawn attention to the fire at an earlier stage. The park was not obliged to have such alarms, but they decided they would install them during the rebuilding process.


The Sydney Olympic Games were to open on 15 September 2000, just two months after the fire. To have missed out on the extra visitors the games would bring would have been a disaster on top of a tragedy, so the Weigels, their staff and their builders set a time frame to reopen the park that most businesses would have discounted as being completely impossible.

Amazingly, the park did reopen in time, using a large temporary building on loan from the Department of Education to house the few surviving reptiles, others that they had been able to obtain on short notice, the gift shop and smaller site offices for the administrative staff. Visitors showed their support and returned. A new platypus display had been built but not yet opened before the fire. The new building had escaped the flames but lost one of its intended inhabitants, Barney, in the blaze, so its opening was post-poned. After a pair of young platypuses was captured in the Macquarie River, near Bathurst, the platypusary was opened on Boxing Day 2000. Designed to encourage breeding, the display includes a large tank viewable by visitors in a darkened building as well as connecting tunnels and three open-air ponds where the platypuses can live away from the public eye.

Even as the park was being salvaged from the embers, a new vision was taking shape. When moving the park from Wyoming, Weigel had been unable to achieve some of his grander plans because of time pressures and limited funds. Now, instead of simply replacing what had been lost, he used the rebuilding process as an opportunity to build the main reptile exhibit as he had originally imagined it. Weigel wanted to move away from the traditional viewing of specimens in a neutral display space. Instead, he imagined the reptiles displayed in fantastical settings that evoked other times and places. It was to be a dramatic exhibit combining elements of theme park and museum.

The 'Lost World of Reptiles' was to become a major component of a visit to the park. While zoos have embraced the idea of themed exhibits, they have tended to base them on ecosystems such as rainforests or savanna. The 'Lost World of Reptiles' instead creates a narrative of adventure, fantasy and escapism. The exhibit houses most of the park's display reptiles and includes 40 individual enclosures. Reptiles are housed in naturalistic exhibits but these are set into exhibition spaces that incorporate relics of ancient Greek, Roman and Egyptian civilisations as well as the Australian outback. Spectacle, playfulness and irony reinforce the focus on entertainment at the park.

Mary Rayner, now the park's general manager, wrote, 'Australia's only themed reptile exhibit, 'Lost World of Reptiles', reminiscent of an Indiana Jones movie scene, takes visitors on a journey to ancient civilizations.' She describes it as 'dark and mysterious, displaying reptiles that are deadly, weird looking and giant, accompanied by a 5-metre-high icon of Sobek the Egyptian crocodile god … and mummies that actually talk with glowing eyes or are positioned in crazy poses.'[2]

To create something so innovative was not easy. As the construction progressed, seemingly unsolvable design problems came to the fore. Help was found in the form of creative designer Graham Wade, who had experience in everything from creating Bible-story comic books to futuristic movie set designs. Graham and John would successfully work together on a range of projects, including an innovative interpretive panel in the shape of a crocodile that depicted the life history of Eric the crocodile.

Raybal returned to the job with the same single-minded resolve they had displayed when building the park in 1996. Artisans Erika Joffe and Murray Scott, who had also worked on the 1996 exhibition spaces, transformed ideas into reality, decorating the interior with expansive areas of realistic-looking artificial sandstone and creating striking murals. David Joffe produced large and inspirational sculptures of extinct giant pythons and monitors from Australia's prehistory, which bring an additional sense of drama to the exhibit. Dedicated park staff contributed to the rebuilding, swapping their park uniforms for work clothes to help out in the evenings pouring concrete or painting until well after midnight, because they were determined to open the new exhibit just after Christmas. It was a physically and emotionally exhausting time for all, but particularly for the Weigels who, in addition to the physical and intellectual work of designing and building the new reptile exhibit, were involved in complicated negotiations over insurance and finance arrangements as well as rebuilding the animal collection.

The good standing in which the park was held meant that following the fire, many zoos and wildlife parks around the country gave them surplus reptiles, some of which were the progeny of animals that had lived at the old park. A number of exotic species were imported from overseas institutions, a process that involved protracted negotiations. To facilitate the transfer of these animals, Weigel established a quarantine facility at the park. Included among the exotic species were four king cobras, which hadn't been exhibited in Australia since the park had last held them, in the early 1980s. A large female, 3.6 metres in length, residing in a spacious exhibit replete with a flowing stream and bamboo grove, is one of the highlights of the 'Lost World of Reptiles'. The other three snakes are kept singly in an off-exhibit area. Under the particular care and management of the venomous snake keeper, Tim Nias, the park successfully bred this species in 2007.

Another key addition was a large American alligator snapping turtle which, unbelievably, was found lumbering down a stormwater drain in an inner-Sydney suburb late in 2000. While it is illegal for any exotic reptile to be kept as a pet in Australia, it is likely that this turtle had once been someone's pet but had either escaped or been released into a suburban creek or pond when it became too big to be managed safely. It is also just possible that this turtle was one of a number of hatchlings that had been bred at the park in the 1970s and which were subsequently stolen. When sufficiently annoyed, snapping turtles can bite with such force that they can easily sever a human finger.

A large display case was given over to housing several rough-scaled pythons, a snake that has a special place in John Weigel's own history. Weigel had organised several expeditions to the Kimberley region of Western Australia in the early 1990s in search of this python, which until then had been known only through some preserved specimens held in the Western Australian Museum. Although the python proved elusive, Weigel and other staff from the reptile park managed to find several during trips made over a number of years. Weigel then applied for and received a permit in 1998 to take five snakes so that a study could be made of their habits and attempts made to breed them in captivity. While there were some initial problems in getting them to feed, the snakes thrived and were successfully bred at the beginning of 2001. Some of the progeny were given to other zoos in Australia and overseas, while others entered the hobbyist market. It proved to be a relatively straightforward species to keep and breed in captivity and today, just a little over a decade since he caught the founder snakes, they are a readily available species.

From 1984 to 2010, Weigel conducted over 20 extended explorations of the northern Kimberley region and found a number of reptile and frog species that were new to science. He was rewarded, and surprised, when two of those species were named after him: a frog known as Weigel's toad, *Notaden weigeli*, and the pygmy king brown snake, *Pseudechis weigeli*.

A companion display to the 'Lost World of Reptiles' was built at the same time as the 'Lost World' exhibit – 'Spider World', which draws upon the long association of the park with the extraction of funnel-web venom but takes it much further. It was funded in part by a national tourism grant of $90 000, which was matched with funds from the park.

As it is for the introduction of other animals, the importation of exotic spiders is carefully controlled to avoid their establishment in the wild and to prevent the introduction of pests or diseases that might affect Australian animals. It took a year and a half to gain approval from the Australian Quarantine and Inspection Service to import exotic spiders for the exhibit. The park had to prove that it had secure facilities, including display cases with shatter-proof glass, three independently sealed doors and restricted access. Eventually, specimens of the goliath birdeater, Antilles pinktoe, king baboon, sun tiger and Brazilian red spiders were put on display. In total, twenty-four tarantulas and four scorpions were imported from the Memphis Zoo in exchange for a shipment of funnel-webs. Only female spiders could be imported, to avoid the possibility of breeding. The park received specimens on loan from other institutions, including

Mexican red-knee and Peruvian pink-toe tarantulas from the Australian Museum. In the 'Spider World' display, visitors can safely view the care and milking of the funnel-web spiders through reassuringly thick glass. After the retirement of Lyn Abra in June 1999, the process of milking funnel-webs has been continued by a number of enthusiastic spider keepers.

'Spider World' explores the uneasy relationship between people and spiders through nursery rhymes, songs and oversized versions of familiar backyard settings. There is even a model of a funnel-web spider – 3 metres in diameter. Leavening the serious messages with humour, 'Spider World' encourages people to learn about a much-feared animal. Many visitors initially avoided the exhibit, believing that they would feel uncomfortable looking at spiders. An animatronic rap-dancing and singing red-back spider was added near the entrance to overcome the hesitation on the threshold.

The entertaining exhibits, animal shows, alligator and crocodile feeding, keeper talks and opportunities for visitors to touch reptiles and other animals have helped the park break down the boundaries between visitors and the animals on exhibit. The success of the new approach saw an increase in visitation numbers – rising some 20 per cent in the year after the reopening. There is also evidence suggesting that the interactive approach taken at the Australian Reptile Park is likely to be more effective in shifting visitors' attitudes to snakes and other reptiles than when visitors are able only to look at animals.[3]

Al Mucci succeeded Rob Porter as general manager, from 2001 to 2005. Al provided a vibrant environment for staff, with high-energy meetings, coupled with a grounded and personable management style. He built on Porter's progress, advancing the reptile park in a wide range of areas, from staff training and relations to the park's participation and role in regional zoo programs. Al's wife, Peggy, was head mammal and bird keeper during this same period. Another husband and wife team have been of tremendous importance to the park, Craig and Jacky Adams-Maher, who both worked as keepers and managers from 1996 to 2009, following several years as travelling rangers in the 'We're *All* Little Aussies' mobile education program. Craig and Jacky advanced the reptile park's educational and entertainment content, and the design and presentation of the live animal displays and interactive displays. They often provided a public face for a young and vibrant 'new' Australian Reptile Park after the move, appearing on numerous television shows, including on Kerri-Anne Kennerley's program and *The Footy Show*.

During the Queen's Birthday long weekend in June 2007, the Central Coast of New South Wales was battered by a ferocious storm which caused widespread flooding and damage. A couple of kilometres from the reptile park, a young family died when the Pacific Highway collapsed underneath their car. It was a horrific weekend.

The storm led to long disruptions to the power supply, which shut down the heating for the pool in which Eric the salt water crocodile lived. The cooling of his pond, coupled with the hours of chilly rain that teemed down on him overnight, took a toll on the animal's body and it appears that an infection took hold. He lost condition over the next couple of weeks but there was little that could be done for him.

On 30 June, he hauled his sick body out of his pool and lay with flaccid legs on the ground, while worried staff and volunteers took it in turns to pour bucketfuls of warm water over him. A gas heater provided extra warmth, but such tender treatment was just not enough to overcome the infection that had rendered the magnificent and powerful predator listless and vulnerable. He died later that afternoon.

His death was a hard blow to staff and volunteers, and to people for whom a visit to the park was not complete until they had seen Eric have a feed. Newspapers and television news programs across the country carried the story of his death, in a way that was some-what reminiscent of the concern shown over the death of the park's other crocodile celebrity, Casanova, in the 1960s. One of many trib-utes to Eric was an anonymous poem written shortly after his death, recalling the phrase the audience was encouraged to chant when Eric was being fed. It concluded:

'One … Two … Three … ERIC!!' echoed with excitement,
rain or shine;
To rousing applause he slammed his prize – he never did decline.
Amid the birthday cakes and the silliness the three-legged
croc never lost his cred;
He was a reptile – cold-blooded, after all – you could see that
when he fed.
Yet it was from the heart and warm-blooded when 'one, two,
three' we chanted;
And his passing left a hard-love lesson: embrace our time,
and take nothing at all for granted.
'One … Two … Three … ERIC!!'

A stone garden in the shape of Eric has been created at the park as a memorial to this great animal that had become so much a part of the life of the park and it is here that he is buried.

The park has continued to innovate, introducing evening functions in 2003. Visitors can enjoy croctails and engage with keepers who walk among guests with animals to view and touch. These events, which range from private birthday parties through to corporate net-working functions, form an additional source of revenue during the hours the park is not usually open to the public. Many similar insti-tutions now do the same, capitalising on the novelty of being at a wildlife park or zoo when visitors have gone home. Other programs focus on team building, in which groups of employees from a range of industries get to enjoy backstage tours and then playfully cooperate in teams to test their courage as tarantulas walk across their hands, or to correctly apply first aid for a mock snakebite.

For those with anxieties about spiders, there is the Bust-a-Phobia program. This program is facilitated by a trained psychologist who uses neurolinguistic programming techniques to address serious snake and spider phobias that can have significant impacts on people's day-to-day lives.

A large part of John Weigel's efforts from 2004 until 2008 involved the establishment of Snake Ranch, a large-scale commercial reptile-breeding facility on a rural property on the mid-north coast of New South Wales. The property is owned by Kevin Smith, an ex-Reptile Park staff member, while the business itself is owned by Weigel and research herpetologist Gavin Bedford. The ranch produces over 1000 hatchling pythons per year, in a range of species and colour morphs, to supply the fast-growing Australian hobby of reptile keeping.

An important new project managed by Weigel and head curator Tim Faulkner is the keeping and breeding of Tasmanian devils. Devils were first exhibited at the reptile park in the early 1960s. There has been concern about the species in the 2000s, as devil facial tumour disease (DFTD), a form of cancer that can be passed on when the animals bite one another, has decimated many wild populations in Tasmania. It is estimated that about 60 per cent of the devils have been killed by the disease and some predict that they will soon be extinct in the wild. The loss of the devils would not only be a tragedy in itself but would leave their ecological niche – as carnivores – open for introduced predators such as dogs, foxes and cats.

A national captive breeding program was initiated to build up a population of the species as insurance against extinction. Since 2006, the reptile park has played a major role in the program, overseen by the Tasmanian Government and the Australasian Regional

Association of Zoological Parks and Aquaria (now known as the Zoo and Aquarium Association). The park is one of 17 mainland institutions that hold and breed devils sourced from various populations that have been determined to be free of DFTD. Some funding was provided by the state government but the costs of building pens and maintaining the devils, some $500 000 by 2010, have been absorbed by the park. It is a considerable financial outlay, but just as Eric Worrell did with funnel-web venom production, the park believes that it should make the contribution to the Tasmanian devil breeding program for ethical reasons. The reptile park's initiative has been very successful, producing more than 40 devil joeys in the first three years of the program – more than any other institution involved in the program. The park's population of 50 Tasmanian devils is the largest captive population of the species anywhere.

The next ambitious stage of this project is Devil Ark. The reptile park has leased several hundred hectares in the Barrington Tops, along the rim of the Hunter Valley, in New South Wales. The plan is to maintain a population of up to 1000 Tasmanian devils there in relatively free-range enclosures, of approximately 2 to 5 hectares each. It has become clear that an insurance population of sufficient genetic diversity will need to be maintained over an extended period – perhaps as long as 30 years – while DFTD runs its course in Tasmania. The Save the Tasmanian Devil Program run by the Tasmanian government recommends the devils be kept in free-range enclosures such as those planned at Devil Ark. At the site, researchers will be able to observe the devils and experiment to find the optimal sex ratios and paddock size for keeping and breeding the animals. The ultimate aim is to create a surplus population of disease-free devils that can be placed back in the wild in Tasmania once DFTD is no longer a threat.

The Tasmanian devil project and Snake Ranch have absorbed much of the creative energy of the park in the past few years. Weigel's passion for reptile keeping means that he has maintained a close involvement in the operations of Snake Ranch and he derives a great sense of achievement from breeding the pythons that are kept there. He looks forward to a time when the efforts to save the Tasmanian devil have picked up their own momentum. Then the program can be taken over by a trust or other responsible body, allowing him to focus again on his core project, the reptile park. Not that it is in any way faltering. The summer of 2010 exceeded all others in terms of visitor numbers and income. The ideas and innovations keep flowing and Weigel and other staff have plans for ongoing improvement and innovation in the coming decade. Another large crocodile, named Elvis, has proven to be a suitable replacement for Eric, and thrills visitors with his impressive bulk and eagerness to accept food from the hands of staff.

In the first chapter of this book we evoked an image of a green python shedding its outer layer to reveal the fresh, vivid skin beneath. The reptile park has had an uncanny ability to do the same – a great capacity to repair, renew and regenerate. But of course it is not the park itself that has driven its repeated renewals, it is the people involved in the park, only some of whom we've been able to mention in this book. They have shown tremendous resilience and courage in the face of at times incredible adversity. The history of the park is the history of people who have, through their ideas and actions, helped shape it and bring it to life.

The Australian Reptile Park began as a big idea of a young lad growing up in the depression years in inner-city Sydney. He had an ambition and he resolutely followed through on it from adolescence into adulthood, with limited formal education, no stores of family capital and a lack of any real models to follow. He didn't reach his goal alone but with the support of family, staff and friends. His professional output in the 1950s and 1960s was astonishing and he involved himself in a range of disparate and demanding activities:

writing, field expeditions, work on reptile taxonomy, filmmaking, managing the park and establishing and maintaining the snake and funnel-web spider milking programs. Eric Worrell was the great populariser of reptiles who also made contributions to the specialist science of taxonomy. His interest in venoms lasted throughout his life and he took pride in his, and the park's, considerable and sustained involvement in the production of life-saving antivenoms.

Today, the site of the original Australian Reptile Park is occupied by a supermarket and a residential development named as a tribute to the land's former owner: Worrell Park. Echoes of the former park are seen in an artist's whimsical renderings of a gecko and a python in the open space area encircled by Worrell Circuit and of a crocodile sunning its stony back along the banks of Wingello Creek. There are living reptiles along the creek as well. Eastern water dragons scuttle among the undergrowth and shiny fat land mullets slide under the protection of mossy logs, ready to inspire new generations with the interest in reptiles that motivated Eric Worrell throughout his life.

While Eric Worrell left the park's story in 1987, his legacy has not only continued but has flourished, as the park remains significant as a national tourist attraction and for wildlife education and conservation. John Weigel is another man with the capacity for dreaming big, realising those dreams and gathering around him a team of supportive and talented people. He was named a Member of the Order of Australia in 2008 for his outstanding contributions to wildlife display and conservation, to Australian herpetology and herpetoculture, and to the snake and spider venom program.

Throughout it all, Robyn Weigel has continuously and unobtrusively made sure that the park is going in the right direction. Nobody but Robyn has had such a sustained and intimate involvement in the park, which has been part of her life for almost five decades. She came to the park not with the grand visions or ambitions of Eric Worrell or John Weigel but rather with the intention of supporting others to achieve their dreams. But by the 1970s, she had become an integral part of the park's fabric and it in turn became part of her own. It is this relationship that has given her such a deep belief in the value of the park and the resilience to overcome the challenges and tragedies the park has faced. Her achievements have been acknowledged and celebrated with awards, too. As well as receiving the National Australia Bank's Humanitarian Award for her and Eric Worrell's work on the funnel-web venom project, she was named the Central Coast Woman of the Year in 2000. And just as Eric Worrell was, John and Robyn have been supported by family, staff, friends and associates who have all helped them achieve the ambitious goals that they have set for themselves and the park.

After half a century of existence, the Australian Reptile Park continues to strive towards the ambitious goals that Eric Worrell articulated back in 1959: conservation, education and research. While these goals are now widely shared by contemporary zoos and similar organisations, Worrell was ahead of his time in advancing them, particularly at a privately owned wildlife park. The fun and excitement of close encounters with animals at the park, the entertaining patter of Ranger Mick and the element of surprise when a crocodile god speaks coexist happily with the more serious face of the park. Through the activities of the Weigels, the management team, staff and volunteers, the Australian Reptile Park is at the forefront of the husbandry and exhibition of captive reptiles; and in its production of venom for the making of life-saving antidotes, it is unique among wildlife parks and zoos.

# Publications
# by Eric Worrell

The following is a list of articles and books by Eric Worrell. In some instances, Worrell used pseudonyms such as Sapengro, Eric Sapengro, Karliboodi, Belvedere, Craig Warrell, or was simply called 'Our Northern Correspondent'. There are also many typescripts by Worrell held in the Australian Reptile Park archives that may or not have been published and we are certain that the list we present here is incomplete. However, it has been included to assist those readers who are interested in reading Worrell's work and to give an indication of his productivity and the range of topics on which he wrote.


1945    'Trapping diving lizards', *Walkabout*, 1 January: 16.

1945    'The crocodile hunters', *Walkabout*, 1 December (with AJ Mathews).

1945    'Freshwater crocodile in captivity', *Proceedings of the Royal Zoological Society of NSW*, 1944–45: 30–32.

1945    'The orange-naped whipsnake', *PRZSNSW*, 1944–45: 32–33.

1946    'Snake-catching in the Northern Territory', *Walkabout*, 1 April: 37–38.

1946    'A strange tortoise', *PRZSNSW*, 1945–46: 29–31.

1946    'The northern river snake', *PRZSNSW*, 1945–46: 32.

1947    'On tiger snakes', *Walkabout*, 1 December: 12–13.

1947    'Emotion in reptiles', *PRZSNSW*, 1946–47: 31–32.

1947    'A new sanctuary', *PRZSNSW*, 1946–47: 33–36.

1948    'Python', *Outdoors and Fishing*, April, 104, 147.

1948    'Snake bite need not be fatal', *Outdoors and Fishing*, May, 186–87.

1948    'Talking crocodiles'. *Outdoors and Fishing*, June, 283–85.

1948    'Buffalo hunters must be tough', *Outdoors and Fishing*, July, 348–49, 373 (author Karliboodi).

1948    'Murder by magic', *Outdoors and Fishing*, August, 415, 440 (author Karliboodi).

1948    'Snake catching a lively occupation', *Outdoors and Fishing*, August, 422–24.

1948    'Learn from the blackfellow', *Outdoors and Fishing*, September, 40–41 (author Karliboodi).

1948    'Art and the Aborigine', *Outdoors and Fishing*, October, 98–99.

1948    'Song of the snake', *Outdoors and Fishing*, December, 236–37.

1949    'Know them by their tracks', *Outdoors and Fishing*, February, 361–63.

1949    'Be your own taxidermist', *Outdoors and Fishing*, February, 386 (author Belvedere).

1949    'Eagles are pests – sometimes', *Outdoors and Fishing*, May, 172–74 (author Belvedere).

1949    'Have your fish and eat it!' *Outdoors and Fishing*, May, 174–76 (author Karliboodi).

1950    'Who'd be a dingo, the story behind the Australian wild dog', *Outdoors and Fishing*, January, 294–95.

1951    'Classification of the Australian Boidae', *PRZSNSW*, 1949–50: 20–25.

1952    *Dangerous Snakes of Australia*, Angus & Robertson, Sydney.

1952    'Hold that tiger!' *Outdoors and Fishing*, March, 334–36, 366.

1952    'On Randall's ranch', *Outdoors and Fishing*, April, 402– 405 (author Our Northern Correspondent)

1952    'The case for the croc', *Outdoors and Fishing*, April, 423.

1952    'Don't you believe it', *Outdoors and Fishing*, May, 30–33.

1952    'Snakes of Australia', *Outdoors and Fishing*, August, 236–39.

1952    'Is the Northern Territory a tourist's paradise?', *Outdoors and Fishing*, September, 304–307 (author Our Northern Correspondent).

1952    'Distribution of venomous snakes in Australia', *Walkabout*, 1 September: 29–31.

1952    'Snakes of Australia – No. 2: Snakebite', *Outdoors and Fishing*, September, 320–23.

1952    'Our dangerous snakes, snakes of Australia, Part 3', *Outdoors and Fishing*, November, 38–40.

1952    'Snake-men play with death', *Man Junior*, December, 47–51 (author Craig Warrell).

1952    'A race to the Gulf in search of crocs', *Outdoors and Fishing*, December, 110–12, (author Karliboodi).

1952    'Do you know these snakes?', *Outdoors and Fishing*, December, 116–19.

1952    'The Australian crocodiles', PRZSNSW, 1951–52: 18–23.

1953    *Dangerous Snakes of Australia*, Angus & Robertson, Sydney, 2nd edition.

1953    'Killer of the Queensland cane fields', *Outdoors and Fishing*, August, 226–229.

1953    'Pythons in the gorge', *Outdoors and Fishing*, September, 300–302, 337.

1953    'Man-eater in the making', *Walkabout*, 1 November, 46.

1953    'Life with "LIFE" in Arnhem Land', *Outdoors and Fishing*, December, 516–19, 546, 548.

1954    'Australian pythons', *Walkabout*, 1 February 1954, 32–33.

1954    'I hunted with the blacks', *Outdoors and Fishing*, September, 18–22 (author Karliboodi).

1954    'Crocodile man', *Outdoors and Fishing*, September, 38–41.

1954    'Cry poacher!', *Outdoors and Fishing*, November, 26–29, 60, 69.

1955    'A new elapine snake from Queensland', PRZSNSW, 1953–54: 41–43.

1955    'A young giant in spiked shoes', *Sporting Life*, March, 22–23 (anonymous).

1956    'Snake safari', *Modern Motor*, January, 42–44, 70, 72.

1956    'Sorry … my mistake!', *Australasian Aqualife*, April–May.

1956    'Adventure on Mutton Bird Island', *Australian Country*, December.

1956    'A new snake from Queensland', *Australian Zoologist*, 12(3):202–205.

1956    'Notes on skull characters of some Australian snakes', *Australian Zoologist*, 12(3):205–210.

1957    *Dangerous Snakes of Australia*, Angus & Robertson, Sydney, 3rd edition.

1957    'It's a bad summer for snakes', *Australian Country*, January 1957, 28–30.

1957    'Stop … spare that goanna!', *Australian Country*, February 1957, 40.

1957    'Life behind the blare of side show alley', *People*, 7 April  (author Eric Sapengro).

1957    'Great snakes … how big do they grow', *Australian Country*, July 1957, 34, 75.

1958   *Song of the Snake*, Angus & Robertson, Sydney.

1958   'A new Papuan python', *PRZSNSW*, 1956–57: 26–27.

1960   'The snake-men (they're mostly dead now)', *People*, 27 April, 42–45.

1960   'The Sydney funnel-web spider', *Cycad* 1(2):11–12.

1960   'Are zoos really necessary?', *Cycad* 1(3–4):6–9.

1960   'A new generic name for a nominal species of *Denisonia*', *PRZSNSW*, 1958–59: 54–55.

1960   'A new insular brown snake', *PRZSNSW*, 1958–59: 56–58.

1961   *Dangerous Snakes of Australia and New Guinea*, Angus & Robertson, Sydney, 4th edition.

1961   Review of H. Cogger's *Frogs of NSW* (1960), *Cycad* 2(2):16–21.

1961   'The need to protect native lizards', *Cycad* 2(3):3–7.

1961   'Herpetological name changes', *Western Australian Naturalist* 8(1):18–27.

1963   *Dangerous Snakes of Australia and New Guinea*, Angus & Robertson, Sydney, 5th edition.

1963   *Reptiles of Australia*, Angus & Robertson, Sydney.

1963   'A new elapine generic name', *Australian Reptile Park Records* 1:2–7.

1963   'Some new Australian reptiles, Part I: Two new subspecies of the genus *Notechis* from Bass Strait', *Australian Reptile Park Records* 2:2–11.

1966   *Australian Wildlife*, Angus & Robertson, Sydney.

1966   *Australian Snakes, Crocodiles, Turtles, Tortoises and Lizards*, Angus & Robertson, Sydney.

1966   *The Great Barrier Reef*, Angus & Robertson, Sydney.

1966   'The unpopular ones', in Jock Marshall (ed), *The Great Extermination*, William Heinemann, 90–111.

1967   *Trees of the Australian Bush*, Angus & Robertson, Sydney (with Lois Sourry).

1968   *Making Friends with Animals*, Angus & Robertson, Sydney.

1969   *Dangerous Snakes of Australia and New Guinea*, Angus & Robertson, Sydney, 6th edition.

1970   *Reptiles of Australia*, Angus & Robertson, Sydney, revised, 2nd edition.

1970   *Eric Worrell's Australian Birds and Animals*, Angus & Robertson, Sydney.

1975   'Breeding the Australian cassowary at the Australian Reptile Park', *International Zoo Yearbook*, 15(1):94–97 (with B Drake and R Krauss).

1976   *Things That Sting*, Angus & Robertson, Sydney.

1977   *Eric Worrell's Australian Birds and Animals*, Angus & Robertson, Sydney, revised, 2nd edition.

# Notes

## Introduction

1   Watkin Tench (1996) 1788, *Comprising A narrative of the expedition to Botany Bay and A complete account of the settlement at Port Jackson*, Tim Flannery (ed.), Text Publishing, Melbourne, 242.
2   William Noah (1798) in Stephen Martin (1993) *A New Land, European Perceptions of Australia, 1788–1850*, Angus & Robertson, Sydney, 113.
3   Joseph Arnold in Martin, *A New Land*, 87.
4   Annie Baxter Dawbin (1998) *Journal of Annie Baxter Dawbin*, July 1858 – May 1868, Lucy Frost (ed.), UQP, Brisbane, 67, 78.
5   Alan Moorehead, 'Snake phobia', *Sun Herald*, 13 January 1968.

## Chapter 1

1   Eric Worrell (1966) 'The unpopular ones', in AJ Marshall (ed.) *The Great Extermination*, Heinemann, Melbourne.
2   *Canberra Times*, 6 February 1935 and 5 February 1948.
3   Eric Worrell (1958) *Song of the Snake*, Angus & Robertson, Sydney, 146. With permission from HarperCollins Australia as Publisher.

## Chapter 2

1   Eric Worrell, 'East Point episodes', unpublished undated typescript, Australian Reptile Park collection.
2   Eric Worrell, Addendum to *Song of the Snake*, unpublished undated typescript, ARP collection.

3  Worrell, 'East Point episodes'.
4  Eric Worrell to Roy Kinghorn, undated, probably 1944 or 1945, Australian Reptile Park collection.
5  Eric Worrell (1956) 'A new water-monitor from northern Australia', *Australian Zoologist*, Vol. XXII, Part 3, 201–202.
6  Eric Worrell (1945–46) 'A strange tortoise', *Proceedings of the Royal Zoological Society of New South Wales*, 29–31.
7  Eric Worrell, 'Is the war influencing our wildlife?', unpublished undated typescript, Australian Reptile Park collection.
8  Eric Worrell 'Trapping diving lizards', *Walkabout,* 1 January 1945, 16.
9  Eric Worrell, 'Trapping diving lizards', 16.
10  Eric Worrell, 'Snake catching in the Northern Territory', *Walkabout*, 1 April 1946, 38.
11  Eric Worrell and AJ Matthews, 'The crocodile hunters', *Walkabout*, 1 December 1945, 15.
12  Eric Worrell (1944–45) 'Freshwater crocodiles in captivity', *PRZSNSW*, 30. Worrell spelled the scientific name of this species *Crocodylus johnsoni*, however, the accepted spelling is *C. johnstoni*.
13  A place now signposted as Stevie's Hole is the only billabong within several kilometres of the homestead and is probably the site of their camp. The large figs are missing but these could have been lost during severe floods.
14  This species is now known as *Craterocephalus stercusmuscarum* (Gunther 1867) and its common name is flyspecked hardyhead. See <http://www.environment.gov.au/cgi-bin/abrs/fauna/details.pl?pstrVol=PISCES3;pstrTaxa=5796;pstrChecklistMode=2>.
15  Eric Worrell (1958) *Song of the Snake*, Angus & Robertson, Sydney, 21. With permission from HarperCollins Australia as Publisher.
16  Ibid.
17  Roland Robinson (1973) *The Drift of Things*, MacMillan, Melbourne, 341.
18  Robinson, *The Drift of Things*, 341.
19  Robinson, *The Drift of Things*, 342.
20  Eric Worrell (1946–47) 'A new sanctuary', *PRZSNSW*, 33–36.
21  Eric Worrell (1958) *Song of the Snake*, Angus & Robertson, Sydney, 35.
22  Eric Worrell (1952) *Dangerous Snakes of Australia, A Handbook for Bushmen, Bushwalkers, Mission Workers, Servicemen, Boy Scouts, New Australians and Naturalists on the identification and venoms of Australian snakes, with directions for first aid treatment of snake bite*, Angus & Robertson, Sydney and London.
23  'Lilyfield's got cold shivers', *The Sun*, 26 November 1947, 2.
24  'Croc catcher's off on the job again', *The Sun*, 4 February 1948, 3.

## Chapter 3

1  'Snake woman finds a mate', *Sunday Sun*, 25 July 1948, 23.
2  'The Snake Park now open to public, interesting display of reptiles', *The Register*, Adelaide, 19 March 1927, 13.

3    'The Snake Park', *The Register*, Adelaide, 23 March 1927, 4.

4    'Anti-humbug' in 'Notes and Queries', *The Register*, Adelaide, 24 January 1927, 12.

5    The building was still standing on the corner of West and Augusta Streets, Umina, at the time of writing this book. It now houses a financial advisory service rather than crocodiles and taipans.

## Chapter 4

1    Eric Worrell (1958) *Song of the Snake*, Angus & Robertson, Sydney, 87. With permission from HarperCollins Australia as Publisher.

2    AC Cheng and K Winkel (2001) 'Snakebite and antivenoms in the Asia-Pacific: wokabaut wantaim raka hebau', *The Medical Journal of Australia*, 175:648–651.

3    CH Campbell (1969) 'A clinical study of venomous snake bite in Papua', a thesis submitted for the degree of Doctor of Medicine, University of Sydney, 6.

4    Alan Moorehead, 'Snake phobia', *Sun Herald*, 13 January 1968.

5    Eric Worrell history, <http://www.reptilepark.com.au/park_history_eric.asp> (accessed 17 August 2007).

6    CSL snake venom stock register held at the Australian Venom Research Unit, University of Melbourne.

7    As Worrell would tell this story, the bite came from the very first snake that he milked for the CSL. However, this seems unlikely given that the first test batch that he sent to the CSL was from four 'small tiger snakes', according to the CSL's records. The bite presumably occurred following the test run.

8    'Mr. Worrell, Woy Woy', undated memo probably written between 18 January 1952 and 15 February 1952, contained in the CSL snake venom stock register held at the Australian Venom Research Unit, University of Melbourne.

9    Dr Forbes, quoted in letter from Director, School of Public Health and Tropical Medicine, to Dr FG Morgan, Director, CSL, 14 March 1952, contained in the CSL snake venom stock register held at the Australian Venom Research Unit, University of Melbourne.

10   Mare Carter, (2007) *Landmarks in a Travelling Life*, Glenrock Books, Foxground, NSW.

11   Jeff Carter, 'Milking a taipan, *People*, 2 May 1956, 23.

12   'She plays with snakes', *Sun-Herald*, 19 December 1954, 25.

13   Jeff Carter, 'Milking a taipan', *People*, 2 May 1956, 23.

14   Jack Green, email to authors, May 2007.

## Chapter 5

1    'He wants to catch venomous killers', *Daily Telegraph*, 31 July 1950, 9.

2    *Cairns Post*, 4 August 1950 and 11 June 1956.

3    *North Australian Monthly*, January 1957, 16.

4    Eric Worrell to Roy Kinghorn, 28 January 1944, Eric Worrell correspondence file, Australian Museum 101/44.
5    Worrell to Kinghorn, 11 May 1944, Australian Museum.
6    D Thomson, 'A new snake from North Queensland', *Proceedings of the Royal Society of London*, 1934:529–531.
7    '£50 for Taipan', *The Tablelander*, 18 July 1952, 1.
8    Eric Worrell, 'Killer of the Queensland cane fields', *Outdoors and Fishing*, August 1953, 226–229.
9    For a full account, see John Cann (2001) *Snakes Alive! Snake Experts and Antidote Sellers of Australia*, John Cann, Sydney. Worrell also wrote a number of articles on snake handlers and the show circuit, including 'The snake-men', *People*, 27 April 1960, 42–45; and 'Snake-men play with death', *Man Junior*, December 1952, 47–51.
10   From our investigations, it appears that this snake was accessioned into the Australian Museum collection on 23 May 1955 and has the accession number R.14420.
11   Eric Worrell (1952) *Dangerous Snakes of Australia*, Angus & Robertson, Sydney and London, 26.
12   *Cairns Post*, 1 August 1955 , 3.
13   Eric Worrell, 'Snake safari, Part II', *Modern Motor*, February 1956, 57.
14   Eric Worrell (1958) *Song of the Snake*, Angus & Robertson, Sydney, 130. With permission from HarperCollins Australia as Publisher.
15   Ibid.

## Chapter 6

1    Head Teacher, Flinders Island Area School, to George Cann, 10 March 1954, reproduced in *Monitor* (1994) 5 (3): 95.
2    Eric Worrell, 'Tiger Snake Island', *Outdoors and Fishing*, February 1956, 40.
3    Eric Worrell, 'Some new Australian reptiles, Part I: Two new subspecies of the genus *Notechis* from Bass Strait', *Australian Reptile Park Records* (1963) 2: 2–11.
4    The Australian Film and Sound Archives holds 38 films Worrell made from 1954 to 1956. Some are finished documentaries, while others are raw footage of Australian animals and bush life.
5    Eric Worrell (1958) *Song of the Snake*, Angus & Robertson, Sydney, 200.
6    Mary Gillham (2000) *Island Hopping in Tasmania's Roaring Forties*, Arthur Stockwell Ltd, Elms Court, UK, 205.
7    Gillham, *Island Hopping*, 336.
8    Simon Fearn, 'Six weeks among the "giant slugs" of Chappell Island', *Monitor* (1994) 5 (3): 121–130.
9    Mary Gillham (1960), IV, 'Subsequent report to Tasmanian Fauna Board after visit paid to Chappell Island on 23.3.60', unpublished report, Australian Reptile Park collection.
10   Eric Worrell to secretary, Tasmanian Fauna Board, 14 May 1960, Australian Reptile Park collection.

## Chapter 7

1    'Australian Reptile Park Guide to the Exhibits and Souvenir of Your Visit', 1st edition, 1962.
2    <www.nevilleburns.net/overview.html>, accessed 17 February 2009.
3    '14 deadly New Guinea taipans for reptile park', *The Gosford Times*, 6 October 1959, 1.
4    Ken Slater to Eric and Carol Worrell, 10 September 1957, Australian Reptile Park collection.
5    Ken Slater to John Graydon, 22 April 1960, Struan Sutherland Archive, Melbourne.
6    Eric Worrell (1968) *Making Friends with Animals*, Angus & Robertson, Sydney, 6.
7    Eric Worrell (1963) *Australian Reptile Park Guide to the Exhibits and Souvenir of Your Visit*, 3rd edition.

## Chapter 8

1    'Monster for Wyoming', *Central Coast Express*, 4 January 1963, 21.
2    'Dinosaur is landmark', *Central Coast Express*, 24 June 1963, 10.

## Chapter 9

1    '30 crocodiles snared for reptile park', *Central Coast Express*, 2 September 1964, 33.
2    Eric Worrell (1966) 'The unpopular ones', in AJ Marshall (ed.) *The Great Extermination*, Angus & Robertson, Sydney, 82.
3    '"Casanova" shot – may die soon', *Central Coast Express*, 21 September 1964, 1.
4    '"Casanova shot – may die soon"', *Central Coast Express*, 1.

## Chapter 10

1    'Teenager of the week', *Gosford Star*, 15 July 1964, 10.
2    Eric Worrell (1968) *Making Friends with Animals*, Angus & Robertson, Sydney, 20.
3    Eric Worrell (1968) *Making Friends with Animals*, Angus & Robertson, Sydney, 20–23.
4    Taronga Zoo memo, 26 October 1962, Taronga Zoo archives.
5    Jock Marshall to RM Watts, Chief Quarantine Officer (Animals), Department of Agriculture Sydney, 30 August 1962, Jock Marshall Papers, National Library of Australia, Canberra.
6    Struan Sutherland to Eric Worrell, 3 November 1967, Australian Reptile Park collection.
7    Struan Sutherland to Eric Worrell, 8 February 1968, Australian Reptile Park collection.

8   Struan Sutherland to Eric Worrell, 8 February 1968.
9   Eric Worrell to Struan Sutherland, 24 December 1968, Australian Reptile Park collection.

## Chapter 11

1   Eric Worrrell, 'The snake's strike', addendum to *Song of the Snake*, unpublished, undated, Australian Reptile Park collection.
2   Ibid.
3   Jack Green (1977) *Australian Alligator Man*, J Green, Woy Woy, 47–48.
4   Ibid., 46–47.

## Chapter 12

1   Eric Worrell to JW Gibson, chief purchasing officer, CSL, 11 October 1966, Australian Reptile Park collection.
2   Eric Worrell to EH Breen, CSL, 21 July 1967, Australian Reptile Park collection.
3   EH Breen, CSL, to Eric Worrell, 2 August 1967, Australian Reptile Park collection.
4   Charles Tanner to Eric Worrell, dated 23 January 1977 in error, should be 1974, Australian Reptile Park collection.
5   Struan Sutherland (1998) *A Venomous Life*, Hyland House, 215.
6   GM Nicholson, A Graudins, H Wilson, M Little and K Broady, 'Arachnid toxinology in Australia: From clinical toxicology to potential applications' *Toxicon* (2006) 48: 882; Geoffrey K Isbister, et al, 'Funnel-web spider bite: a systematic review of recorded clinical cases', *Medical Journal of Australia* (2005) 182 (8): 407–11.
7   Hal Cogger to Eric Worrell, 6 January 1972, Australian Reptile Park collection.
8   Park inventory, January 1979, reptiles file, Australian Reptile Park collection.

## Chapter 13

1   Terry M Dibben and Associates, Public Accountants, 'ARP Financial Report, 1 July 1984 to 31 January 1985', Australian Reptile Park collection.
2   DA Johnson, Director, NPWS, to Eric and Robyn Worrell, 17 July 1984, Tourism management file, Australian Reptile Park collection.
3   Struan Sutherland to P Brinsley, Wildlife Division, NPWS, 19 October 1983, Australian Reptile Park collection.
4   P Brinsley, Executive Officer, Wildlife, National Parks and Wildlife Service, to Robyn Worrell, 20 September 1984, Tourism management file, Australian Reptile Park collection.

5   Eric Worrell to Jack Throp, 26 May 1985, Taronga file, Australian Reptile Park collection.
6   Eric Worrell, addendum to *Song of the Snake*, unpublished, undated, Australian Reptile Park collection.

## Chapter 14

1   CJ Duggan, Letter to Editor, *Central Coast Express*, 17 December 1986.
2   J Weigel and B Turner (undated), 'The education program at the Australian Reptile Park', typescript, Education service file, Australian Reptile Park collection.
3   John Weigel (1988) *The Australian Reptile Park Story*, Weigel photoscript, Gosford.
4   EP Manners to W Meikle, Exhibited Animals Advisory Panel, 2 October 1989, Licence application file, Australian Reptile Park collection.

## Chapter 15

1   Ed Manners to Robyn Worrell-Weigel, memo, 19 May 1990, Australian Reptile Park collection.
2   *Central Coast Express*, 12 July 1996, 63.

## Chapter 16

1   John Weigel, 'Venom, Part 3: snake venom program at Australian Reptile Park', *Reptiles Australia* (2009) 5 (5):37-45.
2   Mary Rayner, 'More bite', *Australian Leisure Management* (2004) 47: 16-18.
3   JM Morgan and JH Granmann, 'Predicting effectiveness of wildlife education programs: A study of students' attitudes and knowledge towards snakes', *Wildlife Society Bulletin* (1989) 17 (2): 501–509.

# Index

Printed and bound by CPI Group (UK) Ltd, Croydon, CR0 4YY

07/07/2026

14916230-0003